T0325608

Multi–Agent Applications with Evolutionary Computation and Biologically Inspired Technologies:
Intelligent Techniques for Ubiquity and Optimization

Shu–Heng Chen
National Chengchi University, Taiwan

Yasushi Kambayashi
Nippon Institute of Technology, Japan

Hiroshi Sato
National Defense Academy, Japan

MEDICAL INFORMATION SCIENCE REFERENCE

Hershey · New York

Director of Editorial Content:	Kristin Klinger
Director of Book Publications:	Julia Mosemann
Acquisitions Editor:	Lindsay Johnston
Development Editor:	Julia Mosemann
Publishing Assistant:	Casey Conapitski, Travis Gundrum
Typesetter:	Casey Conapitski, Travis Gundrum, Michael Brehm
Production Editor:	Jamie Snavely
Cover Design:	Lisa Tosheff

Published in the United States of America by
Medical Information Science Reference (an imprint of IGI Global)
701 E. Chocolate Avenue
Hershey PA 17033
Tel: 717-533-8845
Fax: 717-533-8661
E-mail: cust@igi-global.com
Web site: http://www.igi-global.com

Library of Congress Cataloging-in-Publication Data

Multi-agent applications with evolutionary computation and biologically inspired technologies : intelligent techniques for ubiquity and optimization / Yasushi Kambayashi, editor.
      p. cm.
   Includes bibliographical references and index.   Summary: "This book compiles numerous ongoing projects and research efforts in the design of agents in light of recent development in neurocognitive science and quantum physics, providing readers with interdisciplinary applications of multi-agents systems, ranging from economics to engineering"-- Provided by publisher.
ISBN 978-1-60566-898-7 (hardcover) -- ISBN 978-1-60566-899-4 (ebook) 1. Multiagent systems. 2. Evolutionary computation. I. Kambayashi, Yasushi, 1958- QA76.76.I58M78 2010
006.3'2--dc22
                                    2010011642

British Cataloguing in Publication Data
A Cataloguing in Publication record for this book is available from the British Library.

All work contributed to this book is new, previously-unpublished material. The views expressed in this book are those of the authors, but not necessarily of the publisher.

Table of Contents

Section 1
Multi-Agent Financial Decision Systems

Section 2
Neuro-Inspired Agents

Section 3
Bio-Inspired Agent-Based Artificial Markets

Section 4
Multi-Agent Robotics

Section 5
Multi-Agent Games and Simulations

Section 6
Multi-Agent Learning

Detailed Table of Contents

Section 1
Multi-Agent Financial Decision Systems

Chapter 1

Mak Kaboudan, University of Redlands, USA

Successful decision-making by home-owners, lending institutions, and real estate developers among others is dependent on obtaining reasonable forecasts of residential home prices. For decades, home-price forecasts were produced by agents utilizing academically well-established statistical models. In this chapter, several modeling agents will compete and cooperate to produce a single forecast. A cooperative multi-agent system (MAS) is developed and used to obtain monthly forecasts (April 2008 through March 2010) of the S&P/Case-Shiller home price index for Los Angeles, CA (LXXR). Monthly housing market demand and supply variables including conventional 30-year fixed real mortgage rate, real personal income, cash out loans, homes for sale, change in housing inventory, and construction material price index are used to find different independent models that explain percentage change in LXXR. An agent then combines the forecasts obtained from the different models to obtain a final prediction.

Chapter 2

Yukiko Orito, Hiroshima University, Japan
Yasushi Kambayashi, Nippon Institute of Technology, Japan
Yasuhiro Tsujimura, Nippon Institute of Technology, Japan
Hisashi Yamamoto, Tokyo Metropolitan University, Japan

Portfolio optimization is the determination of the weights of assets to be included in a portfolio in order to achieve the investment objective. It can be viewed as a tight combinatorial optimization problem that has many solutions near the optimal solution in a narrow solution space. In order to solve such a tight problem, this chapter introduces an Agent-based Model. The authors employ the Information Ratio, a well-known measure of the performance of actively managed portfolios, as an objective function. This agent has one portfolio, the Information Ratio and its character as a set of properties. The evolution of agent properties splits the search space into a lot of small spaces. In a population of one small space, there is one leader agent and several follower agents. As the processing of the populations progresses, the agent properties change by the interaction between the leader and the follower, and when the iteration is over, the authors obtain one leader who has the highest Information Ratio.

Section 2
Neuro-Inspired Agents

Chapter 3

Shu-Heng Chen, National Chengchi University, Taiwan
Shu G. Wang, National Chengchi University, Taiwan

Recently, the relation between neuroeconomics and agent-based computational economics (ACE) has become an issue concerning the agent-based economics community. Neuroeconomics can interest agent-based economists when they are inquiring for the foundation or the principle of the software-agent design, normally known as agent engineering. It has been shown in many studies that the design of software agents is non-trivial and can determine what will emerge from the bottom. Therefore, it has been quested for rather a period regarding whether we can sensibly design these software agents, including both the choice of software agent models, such as reinforcement learning, and the parameter setting associated with the chosen model, such as risk attitude. This chapter starts a formal inquiry by focusing on examining the models and parameters used to build software agents.

Chapter 4

Germano Resconi, Catholic University Brescia, Italy
Boris Kovalerchuk, Central Washington University, USA

This chapter models quantum and neural uncertainty using a concept of the Agent–based Uncertainty Theory (AUT). The AUT is based on complex fusion of crisp (non-fuzzy) conflicting judgments of agents. It provides a uniform representation and an operational empirical interpretation for several uncertainty theories such as rough set theory, fuzzy sets theory, evidence theory, and probability theory. The AUT models conflicting evaluations that are fused in the same evaluation context. This agent approach gives also a novel definition of the quantum uncertainty and quantum computations for quantum gates that are realized by unitary transformations of the state. In the AUT approach, unitary matrices are interpreted as logic operations in logic computations. The authors show that by using permutation operators any type of complex classical logic expression can be generated. With the quantum gate, the authors introduce classical logic into the quantum domain. This chapter connects the intrinsic irrationality of the quantum system and the non-classical quantum logic with the agents. The authors argue that AUT can help to find meaning for quantum superposition of non-consistent states. Next, this chapter shows that the neural fusion at the synapse can be modeled by the AUT in the same fashion. The neuron is modeled as an operator that transforms classical logic expressions into many-valued logic expressions. The motivation for such neural network is to provide high flexibility and logic adaptation of the brain model.

Section 3
Bio-Inspired Agent-Based Artificial Markets

This chapter investigates the dynamics of trader behaviors using an agent-based genetic programming system to simulate double-auction markets. The objective of this study is two-fold. First, the authors seek to evaluate how, if any, the difference in trader rationality/intelligence influences trading behavior. Second, besides rationality, they also analyze how, if any, the co-evolution between two learnable traders impacts their trading behaviors. The authors have found that traders with different degrees of rationality may exhibit different behavior depending on the type of market they are in. When the market has a profit zone to explore, the more intelligent trader demonstrates more intelligent behaviors. Also, when the market has two learnable buyers, their co-evolution produced more profitable transactions than when there was only one learnable buyer in the market. The authors have analyzed the trading strategies and found the learning behaviors are very similar to humans in decision-making. They plan to conduct human subject experiments to validate these results in the near future.

This chapter presents agent-based simulations as well as human experiments in double auction markets. The authors' idea is to investigate the learning capabilities of human traders by studying learning agents constructed by Genetic Programming (GP), and the latter can further serve as a design platform in conducting human experiments. By manipulating the population size of GP traders, the authors attempt to characterize the innate heterogeneity in human being's intellectual abilities. They find that GP traders are efficient in the sense that they can beat other trading strategies even with very limited learning capacity. A series of human experiments and multi-agent simulations are conducted and compared for an examination at the end of this chapter.

Chapter 7

Hiroshi Sato, National Defense Academy, Japan
Masao Kubo, National Defense Academy, Japan
Akira Namatame, National Defense Academy, Japan

This chapter conducts a comparative study of various traders following different trading strategies. The authors design an agent-based artificial stock market consisting of two opposing types of traders: "rational traders" (or "fundamentalists") and "imitators" (or "chartists"). Rational traders trade by trying to optimize their short-term income. On the other hand, imitators trade by copying the majority behavior of rational traders. The authors obtain the wealth distribution for different fractions of rational traders and imitators. When rational traders are in the minority, they can come to dominate imitators in terms of accumulated wealth. On the other hand, when rational traders are in the majority and imitators are in the minority, imitators can come to dominate rational traders in terms of accumulated wealth. The authors show that survival in a finance market is a kind of minority game in behavioral types, rational traders and imitators. The coexistence of rational traders and imitators in different combinations may explain the market's complex behavior as well as the success or failure of various trading strategies.

Chapter 8

Hiroshi Takahashi, Keio University, Japan
Takao Terano, Tokyo Institute of Technology, Japan

This chapter describes advances of agent-based models to financial market analyses based on the authors' recent research. The authors have developed several agent-based models to analyze microscopic and macroscopic links between investor behaviors and price fluctuations in a financial market. The models are characterized by the methodology that analyzes the relations among micro-level decision making rules of the agents and macro-level social behaviors via computer simulations. In this chapter, the authors report the outline of recent results of their analysis. From the extensive analyses, they have found that (1) investors' overconfidence behaviors plays various roles in a financial market, (2) overconfident investors emerge in a bottom-up fashion in the market, (3) they contribute to the efficient trades in the market, which adequately reflects fundamental values, (4) the passive investment strategy is valid in a realistic efficient market, however, it could have bad influences such as instability of market and inadequate asset pricing deviations, and (5) under certain assumptions, the passive investment strategy and active investment strategy could coexist in a financial market.

Section 4
Multi-Agent Robotics

Chapter 9
Masanori Goka, Hyogo Prefectural Institute of Technology, Japan
Kazuhiro Ohkura, Hiroshima University, Japan

Artificial evolution has been considered as a promising approach for coordinating the controller of an autonomous mobile robot. However, it is not yet established whether artificial evolution is also effective in generating collective behaviour in a multi-robot system (MRS). In this study, two types of evolving artificial neural networks are utilized in an MRS. The first is the evolving continuous time recurrent neural network, which is used in the most conventional method, and the second is the topology and weight evolving artificial neural networks, which is used in the noble method. Several computer simulations are conducted in order to examine how the artificial evolution can be used to coordinate the collective behaviour in an MRS.

Chapter 10
Yasushi Kambayashi, Nippon Institute of Technology, Japan
Yasuhiro Tsujimura, Nippon Institute of Technology, Japan
Hidemi Yamachi, Nippon Institute of Technology, Japan
Munehiro Takimoto, Tokyo University of Science, Japan

This chapter presents a framework using novel methods for controlling mobile multiple robots directed by mobile agents on a communication networks. Instead of physical movement of multiple robots, mobile software agents migrate from one robot to another so that the robots more efficiently complete their task. In some applications, it is desirable that multiple robots draw themselves together automatically. In order to avoid excessive energy consumption, the authors employ mobile software agents to locate robots scattered in a field, and cause them to autonomously determine their moving behaviors by using a clustering algorithm based on the Ant Colony Optimization (ACO) method. ACO is the swarm-intelligence-based method that exploits artificial stigmergy for the solution of combinatorial optimization problems. Preliminary experiments have provided a favorable result. Even though there is much room to improve the collaboration of multiple agents and ACO, the current results suggest a promising direction for the design of control mechanisms for multi-robot systems. This chapter focuses on the implementation of the controlling mechanism of the multi-robot system using mobile agents.

Section 5
Multi-Agent Games and Simulations

Chapter 11

Robert G. Reynolds, Wayne State University, USA
John O'Shea, University of Michigan-Ann Arbor, USA
Xiangdong Che, Wayne State University, USA
Yousof Gawasmeh, Wayne State University, USA
Guy Meadows, University of Michigan-Ann Arbor, USA
Farshad Fotouhi, Wayne State University, USA

This chapter investigates the use of agile program design techniques within an online game development laboratory setting. The proposed game concerns the prediction of early Paleo-Indian hunting sites in ancient North America along a now submerged land bridge that extended between Canada and the United States across what is now Lake Huron. While the survey of the submerged land bridge was being conducted, the online class was developing a computer game that would allow scientists to predict where sites might be located on the landscape. Crucial to this was the ability to add in gradually different levels of cognitive and decision-making capabilities for the agents. The authors argue that the online component of the courses was critical to supporting an agile approach here. The results of the study indeed provided a fusion of both survey and strategic information that suggest that movement of caribou was asymmetric over the landscape. Therefore, the actual positioning of human artifacts such as hunting blinds was designed to exploit caribou migration in the fall, as is observed today.

Chapter 12

Thillainathan Logenthiran, National University of Singapore, Singapore
Dipti Srinivasan, National University of Singapore, Singapore

The technology of intelligent Multi-Agent System (MAS) has radically altered the way in which complex, distributed, open systems are conceptualized. This chapter presents the application of multi-agent technology to design and deployment of a distributed, cross platform, secure multi-agent framework to model a restructured energy market, where multi players dynamically interact with each other to achieve mutually satisfying outcomes. Apart from the security implementations, some of the best practices in Artificial Intelligence (AI) techniques were employed in the agent oriented programming to deliver customized, powerful, intelligent, distributed application software which simulates the new restructured energy market. The AI algorithm implemented as a rule-based system yielded accurate market outcomes.

Section 6
Multi-Agent Learning

The multiagent reinforcement learnig approach is now widely applied to cause agents to behave rationally in a multiagent system. However, due to the complex interactions in a multiagent domain, it is difficult to decide the each agent's fair share of the reward for contributing to the goal achievement. This chapter reviews a reward shaping problem that defines when and what amount of reward should be given to agents. The author employs keepaway soccer as a typical multiagent continuing task that requires skilled collaboration between the agents. Shaping the reward structure for this domain is difficult for the following reasons: i) a continuing task such as keepaway soccer has no explicit goal, and so it is hard to determine when a reward should be given to the agents, ii) in such a multiagent cooperative task, it is difficult to fairly share the reward for each agent's contribution. Through experiments, this chapter finds that reward shaping has a major effect on an agent's behavior.

Chapter 14

In open multiagent systems, individual components act in an autonomous and uncertain manner, thus making it difficult for the participating agents to interact with one another in a reliable environment. Trust models have been devised that can create level of certainty for the interacting agents. However, trust requires reputation information that basically incorporates an agent's former behaviour. There are two aspects of a reputation model i.e. reputation creation and its distribution. Dissemination of this reputation information in highly dynamic environment is an issue and needs attention for a better approach. The authors have proposed a swarm intelligence based mechanism whose self-organizing behaviour not only provides an efficient way of reputation distribution but also involves various sources of information to compute the reputation value of the participating agents. They have evaluated their system with the help of a simulation showing utility gain of agents utilizing swarm based reputation system.

Exploitation-oriented Learning XoL is a new framework of reinforcement learning. XoL aims to learn a rational policy whose expected reward per an action is larger than zero, and does not require a sophisticated design of the value of a reward signal. In this chapter, as examples of learning systems that belongs in XoL, the authors introduce the rationality theorem of profit Sharing (PS), the rationality theorem of reward sharing in multi-agent PS, and PS-r*. XoL has several features. (1) Though traditional RL systems require appropriate reward and penalty values, XoL only requires an order of importance among them. (2) XoL can learn more quickly since it traces successful experiences very strongly. (3) XoL may be unsuitable for pursuing an optimal policy. The optimal policy can be acquired by the multi-start method that needs to reset all memories to get a better policy. (4) XoL is effective on the classes beyond MDPs, since it is a Bellman-free method that does not depend on DP. The authors show several numerical examples to confirm these features.

Section 7
Miscellaneous

Pheromones are the important chemical substances for social insects to realize cooperative collective behavior. The most famous example of pheromone-based behavior is foraging. Real ants use pheromone trail to inform each other where food source exists and they effectively reach and forage the food. This sophisticated but simple communication method is useful to design artificial multiagent systems. In this chapter, the evolutionary pheromone communication is proposed on a competitive ant environment model, and the authors show two patterns of pheromone communication emerged through co-evolutionary process by genetic algorithm. In addition, such communication patterns are investigated with Shannon's entropy.

Cellular Automata (CAs) have been investigated extensively as abstract models of the decentralized systems composed of autonomous entities characterized by local interactions. However, it is poorly understood how CAs can interact with their external environment, which would be useful for implementing decentralized pervasive systems that consist of billions of components (nodes, sensors, etc.) distributed in our everyday environments. This chapter focuses on the emergent properties of CAs induced by external perturbations toward controlling decentralized pervasive systems. The authors assumed a minimum task in which a CA has to change its global state drastically after every occurrence of a perturbation period. In the perturbation period, each cell state is modified by using an external rule with a small probability. By conducting evolutionary searches for rules of CAs, the uathors obtained interesting behaviors of CAs in which their global state cyclically transited among different stable states in either ascending or descending order. The self-organizing behaviors are due to the clusters of cell states that dynamically grow through occurrences of perturbation periods. These results imply that the global behaviors of decentralized systems can be dynamically controlled by states of randomly selected components only.

Preface

ABSTRACT

From a historical viewpoint, the development of multi-agent systems demonstrates how computer science has become more social, and how the social sciences have become more computational. With this development of cross-fertilization, our understanding of multi-agent systems may become partial if we only focus on computer science or only focus on the social sciences. This book with its 17 chapters intends to give a balanced sketch of the research frontiers of multi-agent systems. We trace the origins of the idea, a biologically-inspired approach to multi-agent systems, to John von Neumann, and then continue his legacy in this volume.

1. GENERAL BACKGROUND

Multi-agent system (MAS) is now an independent, but highly interdisciplinary, scientific subject. It offers scientists a new research paradigm to study the existing complex natural systems, to understand the underlying mechanisms by simulating them, and to gain the inspiration to design artificial systems that can solve highly complex (difficult) problems or can create commercial value. From a historical viewpoint, the development of multi-agent systems itself demonstrates how computer science has become more social, and, in the meantime, how the social sciences have become more computational. With this development of cross-fertilization, our understanding of multi-agent systems may become partial if we only focus on computer science or only focus on the social sciences. A balanced view is therefore desirable and becomes the main pursuit of this editing volume. In this volume, we attempt to give a balanced sketch of the research frontiers of multi-agent systems, ranging from computer science to the social sciences.

While there are many intellectual origins of the MAS, the book "Theory of Self-Reproducing Automata" by von Neumann (1903-1957) certainly contributes to a significant part of the later development of MAS (von Neumann, 1966). In particular, it contributes to a special class of MAS, called cellular automata, which motivates a number of pioneering applications of MAS to the social sciences in the early 1970s (Albin, 1975). In this book, von Neumann suggested that an appropriate principle for designing artificial automata can be productively inspired by the study of natural automata. Von Neumann himself spent a great deal of time on the comparative study of the nervous systems or the brain (the natural automata) and the digital computer (the artificial automata). In his book "The Computer and the

Brain", von Neumann demonstrates the effect of interaction between the study of natural automata and the design of artificial automata.

This biologically-inspired principle has been further extended by Arthur Burks, John Holland and many others. By following this legacy, this volume has this biologically-inspired approach to multi-agent systems as its focus. The difference is that we are now richly endowed with more natural observations for inspirations, from evolutionary biology, and neuroscience, to ethology and entomology. The main purpose of this book is to ground the design of multi-agent systems in biologically-inspired tools, such as evolutionary computation, artificial neural networks, reinforcement learning, swarm intelligence, stigmergic optimization, ant colony optimization, and ant colony clustering.

Given the two well-articulated goals above, this volume covers six subjects, which of course are not exhaustive but are sufficiently representative of the current important developments of MAS and, in the meantime, point to the directions for the future. The six subjects are multi-agent financial decision systems (Chapters 1-2), neuro-inspired agents (Chapters 3-4), bio-inspired agent-based financial markets (Chapters 5-8), multi-agent robots (Chapters 9-10), multi-agent games and simulation (Chapters 11-12), and multi-agent learning (Chapters 13-15). 15 contributions to this volume are grouped by these subjects into six section of the volume. In addition to these six sections, a "miscellaneous" sectiont is added to include two contributions, each of which addresses an important dimension of the development of MAS. In the following, we would like to give a brief introduction to each of these six subjects.

2. MULTI-AGENT FINANCIAL SYSTEMS

We start with the multi-agent financial system. The idea of using multi-agent systems to process information has a long tradition in economics, even though in early days the term MAS did not even exist. In this regard, Hayek (1945) is an influential work. Hayek considered the market and the associated price mechanism as a way of pooling or aggregating the market participants' limited knowledge of the economy. While the information owned by each market participant is imperfect, the pool of them can generate prices with any efficient allocation of resources. The assertion of this article was later on coined as the Hayek Hypothesis by Vernon Smith (Smith 1982) in his double auction market experiments. The intensive study of the Hayek hypothesis in experimental economics has further motivated or strengthened the idea of prediction markets. A prediction market essentially generates an artificial market environment such that forecasts of crowds can be pooled so as to generate better forecasts. Predicting election outcomes via what is known as political future markets becomes one of the most prominent applications.

On the other hand, econometricians tend to pool the forecasts made by different forecasting models so as to improve their forecasting performance. In one literature, this is known as the combined forecasts (Clement 1989). Like prediction markets, combined forecasts tend to enhance the forecast accuracy. The difference between prediction markets and combined forecasts is that agents in the former case are heterogeneous in both data (the information acquired) and models (the way to process information), whereas agents in the latter case are heterogeneous in models only. Hybrid systems in machine learning or artificial intelligence can be regarded as a further extension of the combined forecasts, for example, Kooths, Mitze, and Ringhut (2004). Their difference lies in the way they integrate the intelligence of the crowd. Integration in the case of a combined forecast is much simpler, most of the time, consisting of just the weighted combination of forecasts made by different agents. This type of integration can function well because the market price under certain circumstances is just this simple linear combination of a pool of forecasts. This latter property has been shown by the recent agent-based financial markets. Nevertheless, the hybrid system is more sophisticated in terms of its integration. It is not just the horizontal combination of the pool, but also involves the vertical integration of it. In this way, heterogeneous agents do not just behave independently, but work together as a team (Mumford and Jain, 2009).

Chapter 1 *"A Multi-Agent System Forecast of the S&P/Case-Shiller LA Home Price Index"* authored by Mak Kaboudan provides an illustration of the hybrid systems. He provides an agent-based forecasting system of real estate. The system is composed of three types of agents, namely, artificial neural networks, genetic programming and linear regression. The system "aggregates" the dispersed forecasts of these agents through a competition-cooperation cyclic phase. In the competition phase, best individual forecasting models are chosen from each type of agent. In the cooperation phase, hybrid systems (reconciliatory models) are constructed by combining artificial neural networks with genetic programming, or by combining artificial neural networks with regression models, based on the solutions of the first phase. Finally, there is a competition again for individual models and reconciliatory models.

Chapter 2 *"An Agent-based Model for Portfolio Optimization Using Search Space Splitting"* authored by Yukiko Orito, Yasushi Kambayashi, Yasuhiro Tsujimura and Hisashi Yamamoto proposes a novel version of genetic algorithms to solve the portfolio optimization problem. Genetic algorithms are population-based search algorithms; hence, they can naturally be considered to be an agent-based approach, if we treat each individual in the population as an agent. In Orito et al.'s case, each agent is an investor with a portfolio over a set of assets. However, the authors do not use the standard single-population genetic algorithm to drive the evolutionary dynamics of the portfolios. Instead, the whole society is divided into many sub-populations (clusters of investors), within each of which there is a leader. The interactions of agents are determined by their associated behavioral characteristics, such as leaders, obedient followers or disobedient followers. These clusters and behavioral characteristics can constantly change during the evolution: new leaders with new clusters may emerge to replace the exiting ones. Like the previous chapter, this chapter shows that the wisdom of crowds emerges from complex social dynamics rather than just a static weighted combination.

3. NEURO-INSPIRED AGENTS

Our brain itself is a multi-agent system; therefore, it is natural to study the brain as a multi-agent system (de Garis 2008). In this direction, MAS is applied to neuroscience. However, the other direction also exists. One recent development in multi-agent systems is to make software agents more human like. Various human factors, such as cognitive capacity, intelligence, personality attributes, emotion, and cultural differences, have become new working dimensions for software agents. Since these human factors have now been intensively studied in neuroscience with regard to their neural correlates, it is not surprising to see that the design of autonomous agents, under this influence, will be grounded deeper into neuroscience. Hence, the progress of neuroscience can impact the design of autonomous agents in MAS. The next two chapters are written to feature this future.

Chapter 3 *"Neuroeconomics: A Viewpoint from Agent-Based Computational Economics"* by Shu-Heng Chen and Shu G. Wang gives a review of how the recent progress in neuroeconomics may shed light on different components of autonomous agents, including their preference formation, alternatives valuation, choice making, risk perception, risk preferences, choice making under risk, and learning. The last part of their review covers the well-known dual system conjecture, which is now the centerpiece of neuroeconomic theory.

Chapter 4 "*Agents in Quantum and Neural Uncertainty*" authored by Germano Resconi and Boris Kovalerchuk raises a very fundamental issue: does our brain fuzzify the received signals, even when they are presented in a crispy way? They then further inquire into the nature of uncertainty and propose a notion of uncertainty which is neural theoretic. A two-layered neural network is proposed to be able to transform crisp signals into multi-valued outputs (fuzzy outputs). In this way, the source of fuzziness comes from the conflicting evaluations of the same inputs made by different neurons, to some extent, like Minsky's society of minds (Minsky, 1998). Using various brain image technologies, the current study of neuroscience has already explored various neural correlates when subjects are presented with vague, incomplete and inconsistent information. This mounting evidence may put the modal logic under a close examination and motivate us to think about some alternatives, like dynamic logic.

4 BIO-INSPIRED AGENT-BASED ARTIFICIAL MARKETS

The third subject of this volume is bio-inspired agent-based artificial markets. Market is another natural demonstration of multi-agent systems. In fact, over the last decade, the market mechanism has inspired the design of MAS, known as the market-based algorithm. To some extent, it has also revolutionized the research paradigm of artificial intelligence by motivating the distributed AI. However, in a reverse direction, MAS also provides economists with a powerful tool to explore and to test the market mechanism. This research helps them to learn when markets may fail and hence learn how to do market designs. Nevertheless, the function of markets is not just about the institutional design (the so-called structuralism); a significant number of studies of artificial markets have found that institutional design is not behavior-free or culture-free. This behavioral awareness and cultural awareness has now also become a research direction in experimental economics and agent-based computational economics.

The four chapters contributing to this section all adopt a behavioral approach to the study of artificial markets. Chapter 5 "*Bounded Rationality and Market Micro-Behaviors: Case Studies Based on Agent-Based Double Auction Markets*" authored by Shu-Heng Chen, Ren-Jie Zeng, Tina Yu and Shu G Wang can be read as an example of the recent attempt to model agents with different cognitive capacities or intelligence. It is clear that human agents are heterogeneous in their cognitive capacity (intelligence), and the effect of this heterogeneity on their economic and social status has been found in many recent studies ranging from psychology and sociology to economics; nevertheless, conventional agent-based models paid little attention to this development, and in most cases agents were explicitly or implicitly assumed to be equally smart. By using genetic programming parameterized with different population sizes, this chapter provides a pioneering study to examine the effect of cognitive capacity on the discovery of trading strategies. It is found that larger cognitive capacity can contribute to the discovery of more complex but more profitable strategies. It is also found that different cognitive capacity may coordinate different matches of strategies of players in a co-evolutionary fashion, while they are not necessarily the Nash equilibria.

Chapter 6 "*Social Simulation with both Human Agents and Software Agents: An Investigation into the Impact of Cognitive Capacity on Their Learning Behavior*" authored by Shu-Heng Chen, Chung-Ching Tai, Tzai-Der Wang and Shu G Wang. This chapter can be considered to be a continuation of the cognitive agent-based models. What differs from the previous one is that this chapter considers not only software agents with different cognitive capacity which is manipulated in the same way as in the previous chapter, but also considers human agents with different working memory capacity. A test borrowed from psychology is employed to measure the working memory capacity of human subjects. By placing software agents and human agents separately in a similar environment (double auction markets, in this case) to play against the same group of opponents (Santa Fe program agents), they are able to examine whether the economic significance of intelligence observed from human agents can be comparable to that observed in the software agents, and hence to evaluate how well the artificial cognitive capacity has mimicked the human cognitive capacity.

Chapter 7 *"Evolution of Agents in a Simple Artificial Market"* authored by Hiroshi Sato, Masao Kubo and Akira Namatame is a work devoted to the piling-up literature on agent-based artificial stock markets. As Chen, Chang and Du (2010) have surveyed, from the viewpoint of agent engineering, there are two major classes of agent-based artificial stock markets. One comprises the H-type agent-based financial models, and the other, the Santa-Fe-like agent-based financial models. The former has the agents whose behavioral rules are known and, to some extent, are fixed and simple. The latter has the agents who are basically autonomous, and their behavior, in general, can be quite complex. This chapter belongs to the former, and considers two types of agents: rational investors and imitators. It uses the standard stochastic utility function as the basis for deriving the Gibbs-Boltzmann distribution as the learning mechanism of agents and shows the evolving microstructure (fraction) of these two types of agents and its connection to the complex dynamics of financial markets.

Chapter 8 *"Agent-Based Modeling Bridges Theory of Behavioral Finance and Financial Markets"* authored by Hiroshi Takahashi and Takao Terano is another contribution to agent-based artificial stock markets. It shares some similarities with the previous chapter; mainly, they both belong to the H-type agent-based financial markets, categorized in Chen, Chang and Du (2010). However, this chapter distinguishes itself by incorporating the ingredients of behavioral finance into agent-based financial models, a research trend perceived in Chen and Liao (2004). Specifically, this chapter considers passive and active investors, overconfident investors, and prospects-based investors (Kahneman-Tversky investors). Within this framework, the authors address two frequently-raised issues in the literature. The first one is the issue pertaining to survival analysis: among different types of agents, who can survive, and under what circumstances? The second issue pertains to the traceability of the fundamental prices by the market price: how far and for how long can the market price deviate from the fundamental price. Their results and many others in the literature seem to indicate that the inclusion of behavioral factors can quite strongly and persistently cause the market price to deviate from the fundamental price, and that persistent deviation can exist even after allowing agents to learn.

5 MUTLI-AGENT ROBOTICS

Section 4 comes to one of the most prominent applications of multi-agent systems, i.e., multi-agent robotics. RoboCup (robotic soccer games) which was initiated in the year 1997 provides one of the exemplary cases (Kitano, 1998). In this case, one has to build a team of agents that can play a soccer game against a team of robotic opponents. The motivation of RoboCup is that playing soccer successfully demands a range of different skills, such as real-time dynamic coordination using limited communication bandwidth. Obviously, a formidable task in this research area is how to coordinate these autonomous agents (robots) coherently so that a common goal can be achieved. This requires each autonomous robot to follow a set of behavioral rules, and when they are placed in a distributed interacting environment, the individual operation of these rules can collectively generate a desirable pattern. This issue is so basic that it already exists in the very beginning of MAS, such as pattern formation in cellular automata. The simple cellular automata are homogeneous in the sense that all automata follow the same set of rules, and there is a mapping from these sets of rules to the emergent patterns. Wolfram (2002) has worked this out in quite some detail.

Multi-robot systems can be considered to be an extension of the simple cellular automata. The issue pursued here is an inverse engineering problem. Instead of asking what pattern emerges given a set of rules, we are now asking what set of rules are required to generate certain kinds of patterns. This is the coordination problem for not only multi-agent robots but also other kinds of MAS. Given the complex structure of this problem, it is not surprising to see that evolutionary computation has been applied to tackle this issue. In this part, we shall see two such studies.

Chapter 9 *"Autonomous Specialization in a Multi-Robot System using Evolving Neural Networks"* authored by Masanori Goka and Kazuhiro Ohkura gives a concrete coordination problem for robots. Ten autonomous mobile robots have to push three packages to the goal line. Each of these autonomous robots is designed with a continuous-time recurrent artificial neural network. The coordination of them is solved using evolutionary strategies and genetic algorithms. In the former case, the network structure is fixed and only the connection weights evolve; in the latter case, the network structure is also evolved with the connection weights. It has been shown that in the latter case and in the later stage, the team of robots develops a kind of autonomous specialization, which divides the entire team into three sub-teams to take care of each of the three packages separately.

Chapter 10 *"A Multi-Robot System Using Mobile Agents with Ant Colony Clustering"* authored by Yasushi Kambayashi, Yasuhiro Tusjimura, Hidemi Yamachi, and Munehiro Takimoto presents another coordination problem of the multi-robot systems. In their case, the robots are the luggage carts used in the airports. These carts are picked up by travelers at designated points and left in arbitrary places. They are then collected by man one by one, which is very laborious. Therefore, an intelligent design is concerned with how these carts can draw themselves together at designated points, and how these gathering places are determined. The authors apply the idea of mobile agents in this study. Mobile agents are programs that can transmit themselves across an electronic network and recommence execution at a remote site (Cockayne and Zyda, 1998). In this chapter, mobile agents are employed as the medium before the host computer (a simulating agent) and all these scattered carts via the device of RFID (Radio Frequency Identification). The mobile software agent will first collect information with regard to the initial distribution of these luggage carts, and this information will be sent back to the host computer, which will then use ant colony clustering, an idea motivated by ant corps gathering and brood sorting behavior, to figure out the places to which these carts should return. The designated place for each cart is then transmitted to each cart again via the mobile software agent.

The two chapters in this part are in interesting sharp contrast. The former involves the physical movement of robots during the coordination process, whereas the latter does not involve physical move-ment until the coordination problem has been solved via the simulation. In addition, the former uses the bottom-up (decentralized) approach to cope with the coordination problem, whereas the latter uses the top-down (centralized) approach to cope with the coordination problem, even though the employed ant colony clustering itself is decentralized in nature. It has been argued that the distributed system can coordinate itself well, for example, in the well-known El Farol problem (Arthur, 1994). In the intelligent transportation system, it has also been proposed that a software driver be designed that can learn and can assist the human drives to avoid traffic routes, if these software drives can be properly coordinated first (Sasaki, Flann and Box, 2005). Certainly, this development may continue and, after reading these two chapters, readers may be motivated to explore more on their own.

6 MULTI-AGENT GAMES AND SIMULATION

The analysis of social group dynamics through gaming for sharing of understanding, problem solving and education can be closely tied to MAS. This idea has been freshly demonstrated in Arai, Deguchi and Matshi (2006). In this volume, we include two chapters contributing to gaming simulation.

Chapter 11 *"Agile Design of Reality Games Online"* authored by Robert Reynolds, John O'Shea, Farshad Fotouhi, James Fogarty, Kevin Vitale and Guy Meadows is a contribution to the design of online games. The authors introduce agile programming as an alternative to the conventional waterfall model. In the waterfall model, the software development goes through a sequential process, which demands that every phase of the project be completed before the next phase can begin. Yet, very little communication occurs during the hand-offs between the specialized groups responsible for each phase of development. Hence, when a waterfall project wraps, this heads-down style of programming may create product that is not actually what the customer want. The agile-programming is then proposed as an alternative to help software development teams react to the instability of building software through an incremental and iterative work cycle, which is detailed in this chapter. The chapter then shows how this incremental and iterative work cycle has been applied to develop an agent-based hunter-deer-wolf game. In these cases, agents are individual hunters, deer, and wolves. Each of these individuals can work on his own, but each of them also belongs to a group, a herd or a pack so that they may learn socially. Social intelligence (swarm intelligence) can, therefore, be placed into this game; for example, agents can learn via cultural algorithms (Reynolds, 1994, 1999). The results of these games are provided to a group of archaeologists as an inspiration for their search for human activity evidences in ancient times.

Chapter 12 *"Management of Distributed Energy Resources Using Intelligent Multi-Agent System"* authored by T Logenthiran and Dipti Srinivasan is a contribution to the young but rapidly-growing literature on the agent-based modeling of electric power markets (Weidlich, 2008). The emergence of this research area is closely related to the recent trend of deregulating electricity markets, which may introduce competition to each constituent of the originally vertically-integrated industry, from generation, transmission to distribution. Hence, not surprisingly, multi-agent systems have been applied to model the competitive ecology of this industry. Unlike those chapters in Part III, this chapter is not direct involved in the competitive behavior of buyers and sellers in the electricity markets. Instead, it provides an in-depth description of the development of the simulation software for electricity markets. It clearly specifies each agent, in addition to the power-generating companies and consumers, of electricity markets. Then they show how this knowledge can be functionally integrated into simulation software using a multi-agent platform, such as JADE (Java Agent DEvelopment Framework).

7 MULTI-AGENT LEARNING

The sixth subject of the book is about learning in the context of MAS. Since the publication of Bush and Mosteller (1955) and Luce (1959), reinforcement learning is no longer just a subject of psychology itself, but is proved to be important for many other disciplines, such as economics and games. Since the seminal work of Samuel (1959) on the checkers playing program, the application of reinforcement learning to games is already 50 years on. The recent influential work by Sutton and Barro (1998) has further pushed these ideas so that they are being widely used in artificial intelligence and control theory. The advancement of various brain-image technologies, such as fMRI and positron emission tomography, has enabled us to see how our brain has the built-in mechanism required for the implementation of reinforcement learning. The description of reinforcement learning systems actually matches the behavior of specific neural systems in the mammalian brain. One of the most important such systems is the dopamine system and the role that it plays in learning about rewards and directing our choices that lead us to rewards (Dow, 2003; Montague, 2007).

However, like other multi-disciplinary development, challenging issues also exist in reinforcement learning. A long-lasting fundamental issue is the design or the determination of the reward function, i.e., reward as a function of state and action. Chapter 13 *"Effects of Shaping a Reward on Multiagent Reinforcement Learning"* by Sachiyo Arai and Nobuyuki Tanaka addresses two situations which may make reward function exceedingly difficult to design. In the first case, the task is constantly going on and it is not clear when to reward. The second case involves the learning of team members as a whole instead of individually. To make the team achieve its common goal, it may not be desirable to distribute the rewards evenly among team members, but the situation can be worse if the rewards are not properly attributed to the few deserving individuals. Arai and Tanaka address these two issues in the context of keepaway soccer, in which a team tries to maintain ball possession by avoiding the opponent's interceptions.

Trust has constantly been a heated issue in multi-agent systems. This is so because in many situations agents have to decide with whom they want to interact and what strategies to use. By all means, they want to be able to manage the risk of interacting with malicious agents. Hence, evaluating the trustworthiness of "strangers" become crucial. People in daily life would be willing to invest to gain information to deal with this uncertainty. Various social systems, such as rating agencies, social networks, etc., have been constructed so as to facilitate the acquiring of the reputations of agents. Chapter 14 *"Swarm Intelligence Based Reputation Model for Open Multiagent Systems"* by Saba Mahmood, Assam Asar, Hiroki Suguri and Hafiz Ahmad deals with the dissemination of updated reputations of agents. After reviewing the existing reputation models (both centralized and decentralized ones), the authors propose their construction using ant colony optimization.

Chapter 15 *"Exploitation-oriented Learning XoL: A New Approach to Machine Learning Based on Trial-and-Error Searches"* by Kazuteru Miyazaki is also a contribution to reinforcement learning. As we have said earlier, a fundamental challenge for reinforcement learning is the design of the reward function. In this chapter, Miyazaki proposes a novel version of reinforcement learning based on many of his earlier works on the rationality theorem of profit sharing. This new version, called XoL, differs from the usual one in that reward signals only require an order of importance among the actions, which facilitates the reward design. In addition, XoL is a Bellman-free method since it can work on the classes beyond Markov decision processes. XoL can also learn fast because it traces successful experiences very strongly. While the resultant solution can be biased, a cure is available through the multi-start method proposed by the author.

8 MISCELLANEOUS

The last part of the book has "Miscellaneous" as its title. There are two chapters in the part. While these two chapters can be related to and re-classified into some of the previous parts, we prefer to make them "stand out" here to not blur their unique coverage. The first one in this part (Chapter 16) is related to the multi-agent robotics and also to multi-agent learning, but it is the only chapter devoted to the simulation of the behavior of insects, namely, the ant war. It is an application of MAS to entomology or computational entomology, and a biologically inspired approach that is put back to the study of biology. The last chapter of the book (Chapter 17) is devoted to cellular automata, an idea widely shared in many other chapters of the book, but it is the only chapter which exclusively deals with this subject with an in-depth review. As we have mentioned earlier, one of the origins of the multi-agent systems is von Neumann's cellular automata. It will indeed be aesthetic if the whole book has the most recent developments on this subject as its closing chapter.

Chapter 16 *"Pheromone-Style Communication for Swarm Intelligence"* authored by Hidenori Kawamura and Keiji Suzuki simulates two teams of ants competing for food. What concerns the authors is how ants effectively communicate with their teammates so that the food collected can be maximized. In a sense, this chapter is similar to the coordination problems observed in RoboCup. The difference is that insects like ants or termites are cognitively even more limited than robots in RoboCup. Their decisions and actions are rather random, which requires no memory, no prior knowledge, and does not involve learning in an explicit way. Individually speaking, they are comparable to what is known to economists as zero-intelligent agents (Gode and Sunder, 1993). Yet, entomologists have found that they can communicate well. The communication is however not necessarily direct, but more indirect, partially due to their poor visibility. Their reliance on indirect communication has been noticed by the French biologists Pierre-Paul Grasse (1895-1985), and he termed this style of communication or interaction stigmergy (Grosan and Abraham, 2006)

He defined stigmergy as: "Stimulation of workers by the performance they have achieved." Stigmergy is a method of communication in which the individuals communicate with each another via modifying their local environment. The price mechanism familiar to economists is an example of stigmergy. It does not require market participants to have direct interaction, but only indirect interaction via price signals. In this case the environment is characterized as the price, which is constantly changed by market participants and hence constantly invites others to take actions further.

In this chapter, Kawamura and Suzuki use genetic algorithms to simulate the co-evolution processes of the emergent stigmergic communication among ants. While this study is specifically placed in a context of an ant war, it should not be hard to see its potential in a more general context, such as the evolution of language, norms and culture.

In Chapter 17 *"Evolutionary Search for Cellular Automata with Self-Organizing Properties toward Controlling Decentralized Pervasive Systems"* authored by Yusuke Iwase, Reiji Suzuki and Takaya Arita bring us back to where we begin in this introductory chapter, namely, cellular automata. As we have noticed, from a design perspective, the fundamental issue is an inverse engineering problem, i.e., to find out rules of automata by which our desired patterns can emerge. This chapter basically deals with this kind of issue but in an environment different from the conventional cellular automata. The cellular automata are normally run in a closed system. In this chapter, the authors consider an interesting extension by exposing them to an open environment or a pervasive system. In this case, each automaton will receive external perturbations probabilistically. These perturbations will then change the operating rules of the interfered cells, which in turn may have global effects. Having anticipated these properties, the authors then use genetic algorithms to search for rules that may best work with these perturbations to achieve a given task.

The issues and simulations presented in this chapter can have applications to social dynamics. For example, citizens interact with each other in a relatively closed system, but each citizen may travel out once in a while. When they return, their behavioral rules may change due to cultural exchange; hence they will have an effect on their neighbors that may even have a global impact on the social dynamics. In this vein, the other city which hosts these guests may experience similar kinds of changes. In this way, the two systems (cities) are coupled together. People in cultural studies may be inspired by the simulation presented in this chapter.

7. CONCLUDING REMARKS

When computer science becomes more social and the social sciences become more computational, publications that can facilitate the talks between the two disciplines are demanded. This edited volume demonstrates our efforts to work this out. It is our hope that more books or edited volumes as joint efforts among computer scientists and social scientists will come, and, eventually, computer science will help social scientists to piece together their "fragmental" social sciences, and the social sciences will constantly provide computer scientists with fresh inspiration in defining and forming their new and creative research paradigm. The dialogue between artificial automata and natural automata will then continue and thrive.

Shu-Heng Chen
National Chengchi University, Taiwan

Yasushi Kambayashi
Nippon Institute of Technology, Japan

Hiroshi Sato
National Defense Academy, Japan

REFERENCES

Albin, P. (1975). The Analysis of Complex Socioeconomic Systems. Lexington, MA: Lexington Books.

Arai, K., Deguchi, H., & Matsui, H. (2006). Agent-Based Modeling Meets Gaming Simulation. Springer.

Arthur, B. (1994). Inductive reasoning and bounded rationality. American Economic Review, 84(2), 406–411.

Bush, R.R., & Mosteller, F. (1955). Stochastic Models for Learning. New York: John Wiley & Sons.

Hayek, F. (1945). The use of knowledge in society. American Economic Review, 35(4), 519-530.

Chen S.-H, & Liao C.-C. (2004). Behavior finance and agent-based computational finance: Toward an integrating framework. Journal of Management and Economics, 8, 2004.

Chen S.-H, Chang C.-L, & Du Y.-R (in press). Agent-based economic models and econometrics. Knowledge Engineering Review, forthcoming.

Clement, R. (1989). Combining forecasts: A review and annotated bibliography. International Journal of Forecasting, 5, 559-583.

Cockayne, W., Zyda, M. (1998). Mobile Agents. Prentice Hall.

De Garis, H. (2008). Artificial brains: An evolved neural net module approach. In J. Fulcher & L. Jain (Eds.), Computational Intelligence: A Compendium. Springer.

Dow, N. (2003). Reinforcement Learning Models of the Dopamine System and Their Behavior Implications. Doctoral Dissertation. Carnegie Mellon University.

Grosan, C., & Abraham, A. (2006) Stigmergic optimization: Inspiration, technologies and perspectives. In A. Abraham, C. Gorsan, & V. Ramos (Eds.), Stigmergic Optimization (pp. 1-24). Springer.

Gode, D., & Sunder, S. (1993). Allocative efficiency of markets with zero intelligence traders: Market as a partial substitute for individual rationality. Journal of Political Economy, 101,119-137.

Kitano, H. (Ed.) (1998) RoboCup-97: Robot Soccer World Cup I. Springer.

Kooths, S., Mitze, T., & Ringhut, E. (2004). Forecasting the EMU inflation rate: Linear econometric versus non-linear computational models using genetic neural fuzzy systems. Advances in Econometrics, 19, 145-173.

Luce, D. (1959). Individual Choice Behavior: A Theoretical Analysis. Wiley.

Minsky, M. (1988). Society of Minds. Simon & Schuster.

Montague, R. (2007). Your Brain Is (Almost) Perfect: How We Make Decisions. Plume.

Mumford, C., & Jain, L. (2009). Computational Intelligence: Collaboration, Fusion and Emergence. Springer.

Reynolds, R. (1994). An introduction to cultural algorithms. In Proceedings of the 3rd Annual Conference on Evolutionary Programming (pp. 131-139). World Scientific Publishing.

Reynolds, R. (1999). An overview of cultural algorithms. In D. Corne, F. Glover, M. Dorigo (Eds.), New Ideas in Optimization (pp. 367-378). McGraw Hill Press.

Samuel, A. (1959). Some studies in machine learning using the game of checkers. IBM Journal, 3(3), 210-229.

Sasaki, Y., Flann, N., Box, P. (2005). The multi-agent games by reinforcement learning applied to on-line optimization of traffic policy. In S.-H. Chen, L. Jain & C.-C. Tai (Eds.), Computational Economics: A Perspective from Computational Intelligence (pp. 161-176). Idea Group Publishing.

Smith, V. (1982). Markets as economizers of information: Experimental examination of the "Hayek Hypothesis". Economic Inquiry, 20(2), 165-179.

Sutton, R.S., & Barto, A.G. (1998). Reinforcement Learning: An Introduction. Cambridge, MA: MIT Press.

von Neumann, J. (1958). The Computer and the Brain. Yale University Press.

von Neumann, J. completed by Burks A (1966). Theory of Self Reproducing Automata. Univ of Illinois Press.

Weidlich, A. (2008). Engineering Interrelated Electricity Markets: An Agent-Based Computational Approach. Physica-Verlag.

Wolfram, S. (2002). A New Kind of Science. Wolfram Media.

Acknowledgment

The editors would like to acknowledge the assistance of all involved in the collection and review process of this book, without those support the project could not have been completed. We wish to thank all the authors for their great insights and excellent contributions to this book. Thanks to the publishing team at IGI Global, for their constant support throughout the whole process. In particular, special thanks to Julia Mosemann for her patience in taking this project to fruition.

Shu-Heng Chen
National Chengchi University, Taiwan

Yasushi Kambayashi
Nippon Institute of Technology, Japan

Hiroshi Sato
National Defense Academy, Japan

Section 1
Multi-Agent Financial Decision Systems

Chapter 1
A Multi–Agent System Forecast of the S&P/Case– Shiller LA Home Price Index

Mak Kaboudan
University of Redlands, USA

ABSTRACT

Successful decision-making by home-owners, lending institutions, and real estate developers among others is dependent on obtaining reasonable forecasts of residential home prices. For decades, home-price forecasts were produced by agents utilizing academically well-established statistical models. In this chapter, several modeling agents will compete and cooperate to produce a single forecast. A cooperative multi-agent system (MAS) is developed and used to obtain monthly forecasts (April 2008 through March 2010) of the S&P/Case-Shiller home price index for Los Angeles, CA (LXXR). Monthly housing market demand and supply variables including conventional 30-year fixed real mortgage rate, real personal income, cash out loans, homes for sale, change in housing inventory, and construction material price index are used to find different independent models that explain percentage change in LXXR. An agent then combines the forecasts obtained from the different models to obtain a final prediction.

INTRODUCTION

The economic impacts of temporal changes in residential home prices are well documented. Changes in home prices play a significant role in determining homeowners' abilities to borrow and spend and therefore general economic conditions. Case and Shiller (2003, p. 304) state: "There can be no doubt that the housing market and spend-ing related to housing sales have kept the U.S. economy growing and have prevented a double-dip recession since 2001." Earlier, Case *et al.* (2000) investigated the effects of the real estate prices on the U.S. economic cycles. The economic benefits of home building were measured in a study by the National Association of Home Builders (2005). They maintain that home building generates lo-cal income and jobs for residents and increases governments' revenues. The economic impacts of home-price changes are not unique to the U.S.

DOI: 10.4018/978-1-60566-898-7.ch001

housing market. For example, Ludwig and Torsten (2001) quantify the impact of changes in home prices on consumption in 16 OECD countries, and in Australia the government expected "moderating consumption growth as wealth effects from house price and share price movements stabilise" (Commonwealth of Australia, 2001). Changes in home prices in a relatively large economy (such as that of the U.S.) also affect economic conditions in others. For example, Haji (2007) discussed the impact of the U.S. subprime mortgage crisis on the global financial markets. Reports in the Chinese news (e.g., China Bystanders, 2008) reveal that banks in China are experiencing lower profits due to losses on trading mortgage-related securities. On August 30, 2007, a World Economy report published by the *Economist* stated that "subprime losses are popping up from Canada to China". On May 16, 2008, CNN Money.com (2008) published a summary report of the mid-year update of the U.N. World Economic Situation and Prospects 2008. The U.N. report stated that the world economy is expected to grow only 1.8% in 2008 and the downturn is expected to continue with only a slightly higher growth of 2.1% in 2009. The slow growth is blamed on further deterioration in the U.S. housing and financial sectors that is expected to "continue to be a major drag for the world economy extending into 2009." Given the economic impacts of changes in home prices, accurate predictions of future changes probably help project economic conditions better.

Since April 2006, forecasting home prices gained additional importance and therefore attention after the Chicago Mercantile Exchange (CME) began trading in futures and options on housing. Investors can trade the CME Housing futures contracts to profit in up or down housing markets or to protect themselves against market price fluctuations, CME Group (2007). Initially, prices of those contracts were determined according to indexes of median home prices in ten metropolitan statistical areas (MSAs): Boston, Chicago, Denver, Las Vegas, Los Angeles, Miami, New York, San Diego, San Francisco, and Washington, D.C. as well as a composite index of all 10 cities (CME, 2007). A second composite index was later introduced to include twenty metropolitan areas. Additionally, it is calculated for Atlanta, Charlotte, Cleveland, Dallas, Detroit, Minneapolis, Phoenix, Portland, Seattle, and Tampa as well as a composite index of all 20 MSAs. These are financial tools to trade U.S. real estate values and are based on the S&P/Case-Shiller Indexes (CSIs) for all 20 cities and the two composites. CSIs are recognized as "the most trustworthy and authoritative house price change measures available" (Iacono, 2008). Case and Shiller (1989 and 1990) presented early introduction of the index, its importance, and forecasts.

This chapter focuses on forecasting the S&P/Case-Shiller index for Los Angeles MSA (LXXR). Two reasons explain why LXXR is selected. First, modeling and predicting only one index is a challenge to be addressed first before tackling 20 indexes in different locations that are characteristically heterogeneous markets and before predicting either composite. Second, the Los Angeles area housing market is one of the hardest hit by the unprecedented subprime financial problems. The plan is to predict monthly percentage changes in LXXR for 24 months (April of 2008 through March of 2010). Monthly percentage change in LXXR = $\%D\_LXXR_t = 100*\{Ln(LXXR_t)-Ln(LXXR_{t-1})\}$, where Ln = natural logarithm, and $t = 1, \ldots,$ T months. (Hereon, $\%D\_X_t = 100*\{Ln(X_t)-Ln(X_{t-1})\}$.) Input data used is monthly and covers the period from January of 1992 through March of 2008. The forecast is developed in stages. In the first, variables (X_i where $i = 1, \ldots, n$) suspected of creating temporal variations in LXXR are logically and intuitively identified, data of those variables are collected, then variables that best explain variation in LXXR (X_j where $j = 1, \ldots, k$ and $k \subseteq n$) are determined using genetic programming (GP). Variables identified as best in the first stage are forecasted for 24 months in the second. A multi-agent system (MAS) is finally developed to model the monthly percent change in LXXR (%D_LXXR). In this third stage, MAS is a network of computational techniques employed first to obtain several independent "best" forecasts of %D_LXXR. By assuming that each of the techniques employed captures the variable's dynamics over history (1992-2008) at least partially, a single agent then independently takes the forecasts produced (by the techniques employed) as input to produce a single forecast as output.

Ideally, the best forecast should be evaluated relative to others published in the literature. However, an extensive literature search failed to find any monthly forecast of LXXR or %D_LXXR. (Most probably this is the first study to model and predict LXXR monthly.) Only independent annual changes expected in CA were found. For example, the California Association of Realtors (C.A.R.) publishes an annual forecast for the entire state. Their quarterly forecast (C.A.R., 2008) projects a modest price increase in the second half of 2008 and in 2009. Their annual forecast (Appleton-Young, 2008) projects a decline of about 5% for 2008. A different forecast is produced by Housing Predictor (2008) who predicts that home prices will decline by 12.8% in 2008. Only two forecasts of the S&P/Case-Shiller indexes were found: Moody's Economy.com (2008) and Stark (2008). Moody's forecast is of the 20-city composite index and is only presented graphically. It suggested that housing prices will continue to decline through the middle of 2009 when they are expected to start bottoming out. Stark (2008) predicted that the composite index will decline by 12% in 2008, decline by 0.3% in 2009, and increase by 3.8% in 2010. Only one city forecast of CSI was found. The BostonBubble.com (2007) published a forecast of the index for Boston MSA through 2011. They project that the Boston CSI will continue to decline until April of 2010.

The balance of this chapter has four sections. Before describing the methodology used to produce a forecast of %D_LXXR using multi-agent systems, the S&P/Case-Shiller Index is briefly introduced. Estimation results are presented next followed by the forecasts obtained. The final section has the conclusion.

THE S&P/CASE-SHILLER INDEX

The Case/Shiller indexes are designed to measure changes in the market value of residential real estate in each Metropolitan Statistical Area (MSA) by tracking the values of single-family housing within the United States. It measures changes in housing prices with homes sold held at a constant level of quality by utilizing data of matched sale pairs for pre-existing homes. In short, its calculation is based upon repeat sales of existing homes. For each MSA, a three-month moving average is calculated. The monthly moving average is of sales pairs found for that month and the preceding two months. A Standard & Poor's report (2008a) contains a full description of how the indexes are calculated. The indexes are published by Standard & Poor's (2008b). Without going into details here, LXXR is designed to measure changes in the total value of all existing single-family housing stock in the Los Angeles Metropolitan Statistical Area which includes Los Angeles, Long Beach, and Santa Ana.

The Los Angeles MSA index (LXXR) depicts the sharp declines in housing prices experienced 2007 and 2008. Changes in housing prices around the Los Angeles area were much stronger than most of the nation. Figure 1 shows a comparison between the monthly percentage changes in LXXR and the 10- and the 20-composite indexes (CSXR and SPCS20R) over the period 1992-2008. The more aggressive volatility in %D_LXXR (higher % increases and more pronounced % decreases) relative to the two composite indexes is evident.

Figure 1.

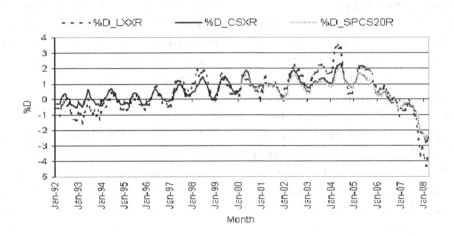

METHODOLOGY

Modeling the percentage change in LXXR (or %D_LXXR) is rather challenging given the dynamics portrayed in Figure 1. Seasonal patterns as well as irregular variations are evident. Given such nonlinear complexity of temporal variations in %D_LXXR, it is reasonable to assume that employing several types of modeling techniques simultaneously can deliver a more reliable forecast of the variable than one delivered by a single technique. Each modeling technique can be viewed as an autonomous agent, with one common goal for all: minimize each fitting model's mean square error (MSE). It is assumed here that the best models the agents construct produce forecasts that may be successful during given periods and not in others. If this is the case, a logical modeling strategy would be to use the outputs of those best models as inputs for others and re-estimate and forecast new ones. Thus, a multi-agent system is constructed to capture interactions between the obtained best models. The objective is to determine whether a multi-agent system can produce a single forecast that outperforms any single agent prediction.

In constructing models to forecast %D_LXXR, all available data (January 1992 through March 2008) is used. Typically, a small sample of the data is withheld to determine the forecasting efficacy of a model. Instead, it is assumed here that even if a model succeeds in producing an impressive *ex post* forecast (i.e., a forecast of outcomes already known) there is no guarantee that it will produce a reliable *ex ante* forecast (i.e., a forecast of unknown outcomes). Further, under conditions when market volatility is evident, it is more logical to use the most recent information (or data) to obtain models than to use such information in testing them. Given that the objective is to ultimately forecast LXXR for a period of two years (or 24 month), withholding any of the data to test the models may be at the cost of obtaining relatively larger forecast errors toward the end of the 24-month period.

In the multi-agent system utilized here, several modeling techniques or agents are employed. Bio-inspired computational techniques (genetic programming or GP and artificial neural networks or ANN) and a statistical technique (a regression model or RM using standard ordinary least squares or OLS) are candidate estimation and prediction agents. An independent model will first be produced by each technique. The combined fitted values are then the input variables into a GP, an

ANN, and an OLS to obtain future (predicted) values of %D_LXXR.

Use of multi-agent systems when developing models that explain dynamics of systems is not new. Chen and Yeh (1996) used GP learning of the cobweb model. Barto (1996) used ANN multi-agent reinforcement learning. Chen and Tokinaga (2006) used GP for pattern-learning. Vanstone and Finnie (2007) used ANN for developing stock market trading system. As mentioned in the introduction, modeling %D_LXXR to forecast LXXR is completed in three stages. Before describing each stage, brief introductions to GP and ANN follow.

Genetic Programming

GP is an optimization search technique. Koza (1992) provides foundations of GP. Examples of its use in forecasting are in Chen and Yeh (1995), Tsang *et al.* (1998), and Warren (1994). The GP software used in this study is TSGP (for Time Series Genetic Programming) written in C++ for Windows environment and runs on a standard PC (Kaboudan, 2004). TSGP is used because it is designed specifically to produce regression-type models and to compute standard regression statistics. Statistical properties of models TSGP produces were analyzed in Kaboudan (2001). Two types of input files are needed for executing TSGP: data files and a configuration file. Data input files contain values of the dependent and each of the independent variables. The configuration file contains execution information a user controls. TSGP assembles an initial population of individual equations (say 1000 of them) with random specifications, computes their fitness (MSE = mean squared error), and then breeds new equations as members of a new generation with the same population size. Each individual – member of a population – is a regression-like model represented by a parse tree. The key run parameters specified in the configuration file are: population size =1000, fitness measure = MSE,

maximum tree depth = 100, maximum number of generations = 100, mutation rate = 0.6, crossover rate = 0.3, self reproduction = 0.10, and operators = +, -, x, /, sin, & cos.

GP-evolved equations are in the form of a parse tree. Trees are randomly assembled such that if an operator is selected, the tree grows. Operators are thus its inner nodes. A tree continues to grow until end nodes (or terminal) contain variables or constant terms. Once a population of equation is assembled, a new generation is then bred using mutation, crossover, and self reproduction. Fitter equations in a population get a higher chance to participate in breeding. In mutation, a randomly assembled sub-tree replaces a randomly selected existing part of a tree. In crossover, randomly selected parts of two existing trees are swapped. In self reproduction, a top percentage of the fittest individuals in one population (usually top 10%) are passed on to the next generation. For all bred individuals, if the offspring are fitter than their parents, they survive; else the parents survive. The idea in GP is to continue generating new populations while preserving "good genes" in a Darwinian sense. After completing a specified number of generations (100 to 200), the program terminates and saves the fittest model to an output file. Actual, fitted, forecasted values, residuals, as well as evaluation statistics (R^2, MSE, and MAPE = mean absolute percent error) are written to another output file.

A GP algorithm has its characteristics. The program randomly selects the explanatory variables and the coefficients. The iterative process produces coincidental very strange specifications. It is based on heuristics and lacks theoretical justification. Further, during execution the computerized algorithm occasionally gets trapped at a local minimum MSE in the search space and never reaches a global one. This necessitates conducting a large number of searches (say 100) to find the 100 fittest equations. One or more of them should actually produce a superior fit (and forecast) that may not be otherwise obtainable.

Perhaps this explains why GP-evolved equations have strong predictive abilities.

GP and conventional statistical regressions have differences. Because equation coefficients are not computed (they are computer-generated random numbers between -128 and 127 for TSGP), problems of multicollinearity, autocorrelation, and heteroscedasticity are nonexistent. Further, there are also no degrees of freedom lost when more explanatory variables are added.

Neural Networks

Artificial neural networks (ANN) have been around for more than thirty years now. Principe *et al*. (2000) among many others describe how ANN is used in forecasting. The most commonly used neural network structures are multilayer perceptrons (MLP) and generalized feedforward networks (GFF). MLP is a layered feedforward network that learns nonlinear function mappings. It employs nonlinear activation functions. Networks are typically trained with static backpropagation and require differentiable, continuous nonlinear activation functions such as *hyperbolic tangent* or *sigmoid*. A network takes explanatory variables as input to learn how to produce the closest fitted values of a dependent variable. Although MLP trains slowly and requires a large number of observations, it seems to approximate well. GFF is a generalization of MLP with connections that jump over layers. GFF also trains with static backpropagation. MLP is used in this study to produce forecasts. The forecasts were produced here using NeuroSolutions™ software (2002).

Parameters used to complete ANN searches here are determined by trial and error. To obtain a forecast, first, several configurations of MLP are tested to identify the suitable network structure to use. Hidden layers use *hyperbolic tangent* transfer function, employ 0.70 learning momentum, and train a minimum of 1000 epochs. First, searches are completed using one hidden layer. Testing those starts with 1000 training epochs then the

number of epochs is increased by increments of 500 until the best network is identified. The search is repeated using networks with two hidden layers if no reasonably acceptable output is obtained. The configuration with the best estimation statistics is then used in forecasting. The fitness parameter is MSE to be consistent with GP.

Stage I: Determining the Explanatory Variables

Determining variables to employ in explaining the dynamics of changes in LXXR and help predict its future variations dependents on correctly identifying possible candidate variables and availability of data. Identifying a logical set of these variables is guided here by economic theory. Carefully evaluating factors (in this case, measurable variables) that impact the demand and supply sides of the housing market furnishes a good number of variables that logically impose possible upward or downward price pressures. Rather than constructing a system of market equilibrium that encompasses separate demand and supply functions, a system of market disequilibrium (Fair and Jaffee, 1972) that befits the housing market conditions is assumed. Under disequilibrium conditions, the market rarely ever reaches equilibrium. This means that prevailing quantities traded and transaction prices may be anywhere on a disequilibrium trading curve where the quantity traded is the lesser of the quantities demanded and supplied. (This accommodates the nature of housing markets where there is always more homes to sell than those bought. To account for what might seemingly be perpetual excess inventory in the housing market, an assumption is made that the market has an average inventory level; and excess supply or excess demand would be that number of homes for sale above or below that average inventory level.) Therefore, a single price trading equation that combines demand and supply determinants should be sufficient under disequilibrium conditions. Economic theory suggests including demand determinants such as income and mortgage rate and supply determinants such as construction cost and changes in inventory for example in the equation.

The identified complete set of explanatory variables are then X_i for $i = 1, ..., n$ possible demand and supply determinants as well as their lagged values. Lagged explanatory variables are logical since the decision making process when buying a home tends to be rational and conditional upon verification of income, inspection of homes, among other transaction-completing progression routine that can take anywhere from three months to a year. To determine X_j for $j = 1, ..., k$ (that subset of X_i) variables to select when modeling changes in LXXR, a single agent is employed. It is a GP regression-like model-assembler. That agent is given all possible variables identified as candidates to explain variation in %D_LXXR. Its task is to generate a number of promising models. Thus, given X_i, a GP agent will evolve a reasonably good number of equations (say 200) first. The fittest of those equations (say best 10%) are identified. Explanatory variables included in these equations are then harvested to be used in re-estimating models of %D_LXXR employing GP as well as other agents.

The idea of selecting the best explanatory variables using GP is new. The process starts with all the identified variables and their lagged values (with $\lambda = 3, 13, ..., 24$ monthly lags considered and included). The GP agent then generates models to explain historical variations in the dependent variable %D_LXXR. Specifications of each of the 20 best evolved models (10% of all GP models evolved) are obtained and whatever variables they contain are tabulated. Any variable in an equation gets one vote regardless of the number of times it appears in that same equation. This means that the number of votes a variable gets is the number of GP models that it appears in, and therefore, the maximum number of votes a variable can have is 20. The votes are tallied and those variables repeatedly appearing in the equations are selected to employ when searching for the final %D_LXXR

forecasting model. Only those variables occurring often in the generated GP equations are reported here (given that there is no obvious benefit from discussing others). The GP agent determined that variations in %D_LXXR are best explained by: COL = cash out loans or percent of amount borrowed above the amount used to finance a purchase (Freddie Mac, 2008a). FS = number of houses for sale in thousands (U.S. Census Bureau, 2008a). SOLD = number of units sold in thousands (U.S. Census Bureau, 2008a). ES = excess supply = $FS_{t-1} - SOLD_t$. CMI = construction material index (U.S. Census Bureau, 2008b). CHI = change in housing inventory = $FS_t - FS_t$. MR = 30-year real mortgage rate (Freddie Mac, 2008b). LOAN = indexed loan = LXXR * LPR (LPR = loan-to-price ratio). LAPI = Los Angeles real personal income (U.S. Bureau of Economic Analysis, 2008). ESDV = excess supply dummy variable where $ESDV_t = 1$ if $ES_t <$ average ES_t and = zero otherwise. These variables are taken at different lags $\lambda = 3, ..., 24$ months: $COL_{t-\lambda}$, $FS_{t-\lambda}$, $SOLD_{t-\lambda}$, $ES_{t-\lambda}$, $CMI_{t-\lambda}$, $CHI_{t-\lambda}$, $MR_{t-\lambda}$, $LOAN_{t-\lambda}$, and $LAPI_{t-\lambda}$. All variables measured in dollar values were converted into 1982 real or constant dollars using the LA metropolitan area consumer price index (CPI).

Stage II: Predicting the Explanatory Variables

In this stage, the forecast values of each explanatory variable (X_j) determined in the first stage are obtained employing two agents, GP and ANN. Agents responsible for obtaining forecasts of the X_j are each given a set of appropriate variables (Z_v) to explain variations in each previously identified X variable. Alternatively, $X_j = f(Z_{v\lambda})$, where $Z_{v\lambda}$ for $v = 1, ..., V$ explanatory variables and $\lambda = 3, .., L$ lags. Results from the two agents are compared and a decision is made on whether to take the better one or take their average. The two agents are assumed competitive if the results of one of them are deemed superior to the other.

Figure 2.

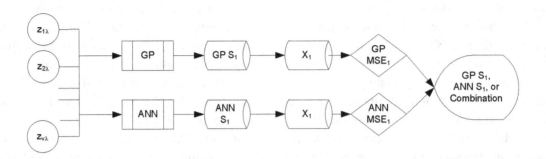

Superiority is determined according to training fitness (MSE). They are assumed cooperative if there is no difference between their performances and an average of them is taken. Figure 2 depicts the flow chart of the process used to obtain predictions for only one variable (X_1).

The same process is repeated for all the other explanatory variables ($X_2, ..., X_k$). The difference is in the variables used to capture the dynamics of each X_j and the final decision involving determining the best forecast after determining whether to employ a competitive or cooperative double-agent solution selection scheme. Given that two of the explanatory variables are identities (i.e., computed using other variables) and one is a dummy variable, the number of models to estimate is seven rather than ten. The specifications of the seven models identified as best in predicting the explanatory variables are:

$$COL_t = f(MR_{t-6}, ..., MR_{t-18}); \qquad (1)$$

$$\%D\_MR_t = f(\%D\_MR_{t-6}, ..., \%D\_MR_{t-18}); \quad (2)$$

$$FS_t = f(FS_{t-1,12}, COL_{t-1,12}, LPR_{t-1,12}); \qquad (3)$$

$$\%D\_SOLD_t = f(\%D\_MR_{t-6,12}, \%D\_LAPI_{t-1,12}); \qquad (4)$$

$$CMI_t = f(CMI_{t-3,12}); \qquad (5)$$

$$\%D\_LOAN_t = f(\%D\_MR_{t-3,18}); \qquad (6)$$

$$\%D\_LAPI_t = f(\%D\_LAPI_{t-3,6}, \%D\_CPI_{t-12,16}, \%D\_HWR_{t-12,16}). \qquad (7)$$

In the equations above, $\%D\_X = \%$ change in variable X and $X_{t-a,b} = X_{t-a}, X_{t-(a+1)}, ..., X_{t-b}$, where a and b are integer measures of temporal distances. Using lagged values helps capture the evolutionary

Table 1. Estimation statistics of fitted explanatory variables

Variable	MSE	R^2	Agent
%D_COL	7.39	0.94	ANN
%D_MR	6.10	0.26	GP
FS	2.98	0.61	ANN
%D_SOLD	18.81	0.50	ANN
CMI	9.84	0.96	GP
%D_LOAN	1.31	0.38	ANN
%D_LAPI	0.06	0.71	ANN

Figure 3. Flow chart starting with X_j explanitory input variables into each modeling technique. Solutions from the different models are identified by 'S' (GP S, ANN S, ..., etc.). Respective MSE computations determine the best model each technique produces. ANN is then used to estimate models that fit residuals output from GP and from RM.

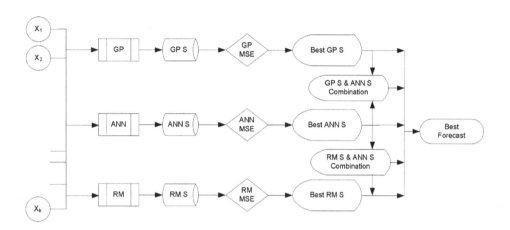

dynamics embedded in each variable's historical values. Two new variables were introduced in the equations above: LPR and HWR. HWR is the average monthly hourly wage in Los Angeles. Estimation statistics belonging to the models that generated forecasts of the seven explanatory variables (%D_COL, %D_MR, FS, %D_SOLD, CMI, %D_LOAN, %D_LAPI) are in Table 1. To produce lengthy forecasts, Equations (1) through (7) were used to forecast three months (the least lag in an equation) first. The three-month forecast was then used as input to solve each respective equation again to obtain the next three months forecast, and so on. Using this iterative process, a 24-month forecast of each of the explanatory variables was produced.

Stage III: Multi-Agent Modeling of %D_LXXR

In this stage, several agents are employed to produce forecasts of %D_LXXR and LXXR. Figure 3 portrays the flow of the implementation process. Historical as well as forecasted values of the explanatory variables are input to three techniques selected to produce %D_LXXR pre-

dictions. The three techniques are GP, ANN, and linear regression models (RM or OLS). All three are multivariate and they take exogenous variables as inputs. As shown in the figure, the X_j variables feed into the model generating techniques. Each technique acts as an agent whose task is to produce many solutions. The best solutions provided by the techniques then compete to determine the best forecast.

Solutions obtained using the different techniques remain competitive until all best solutions are identified. They act as cooperative agents to help deliver the best possible forecast in the final step. Cooperation is in two ways. The first involves fitting the residuals (= Actual − fitted) from one technique using a different technique. Because ANN produced the lowest MSE relative to GP and RM at the competitive level, ANN was used to fit residuals the other two produced. The idea assumes that whatever dynamics one technique missed (i.e., the residuals) may be captured using a different technique. The second cooperation involves using all outputs obtained from the different techniques as inputs to model some type of weight distribution between them and hopefully capture what may be the best forecast.

Figure 4.

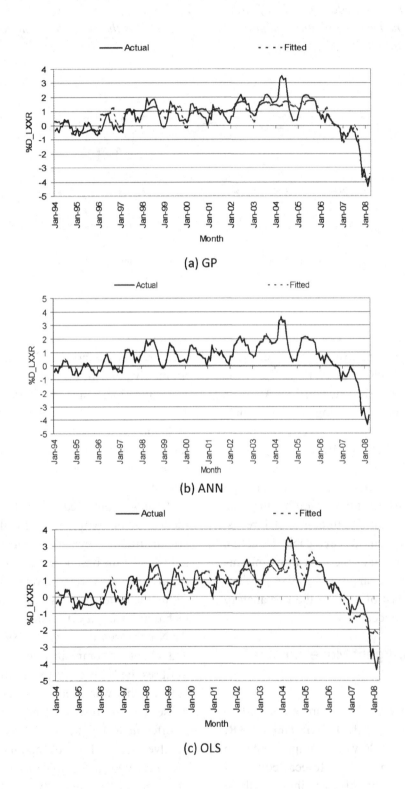

(a) GP

(b) ANN

(c) OLS

ESTIMATION RESULTS

Only the final estimation results and their statistics are presented here. Forecasts are presented in the next section. Using the same input variables, an extensive search is employed until the best possible output statistics are reached independently by each technique. ANN does not produce an estimated equation and only the best GP and RM-OLS estimated equations can be reported.

A search consisting of a total of 100 GP models was conducted. The best (lowest MSE) GP equation among the 100 attempted produced the fit shown in Figure 4 (a) and was as follows:

$$Y = \cos\{\sin[X1_{t\text{-}3} * \{\cos(X1_{t\text{-}6}) + X1_{t\text{-}3}\}] - X1_{t\text{-}6}\}$$
$$*[\sin[\sin(X2_{t\text{-}5}) + \{X3_{t\text{-}9} * \sin(X2_{t\text{-}5} + X1_{t\text{-}3})\}$$
$$] + \{ X3_{t\text{-}9} * [\{\sin(X2_{t\text{-}5}) + DV_t + X3_{t\text{-}9}\} + \{$$
$$\sin[X2_{t\text{-}5} + X1_{t\text{-}3} * \{ \cos(X1_{t\text{-}6}) + X1_{t\text{-}3}\}] -$$
$$\sin\{X1_{t\text{-}3} * [\cos\{\sin(X2_{t\text{-}5}) + DV_t\} + X1_{t\text{-}3}] \} +$$
$$X1_{t\text{-}6}\}] \}] - DV_t - 2 * X4_t \qquad (8)$$

where (and for aesthetic reasons), Y = %D_LXXR, X1 = CHI, X2 = %D_SOLD, X3 = COL, X4 = %D_MR, and ESDV = DV. The equation above had $R^2 = 0.79$ and MSE = 0.298.

As mentioned earlier, the ANN search involved trial-and-error routine to find the best fit. Ultimately, the best model was produced by a multilayered perceptron system with a layered feedforward network trained with static backpropagation. The best results were obtained using a single hidden layer with five input processing elements, a *TanhAxon* transfer function, used momentum learning rule (with step size = 1.0 and momentum = 0.7), with 6500 epochs, and was obtained after three training times. The best ANN fit obtained is shown in Figure 4 (b). It had $R^2 = 0.99$ and MSE = 0.006.

A trial-and-error procedure was also used to find the best OLS regression model. Explanatory variables were added and deleted and their lags varied until the best fit was found. Interestingly, only one extra explanatory variable (X5 = LOAN)

was added to the list the final GP model contained. The best OLS equation found (with $R^2 = 0.75$ and MSE = 0.36) was as follows:

$$Y = 7.49 - 39.41 X4_{t\text{-}6} - 0.74 X3_{t\text{-}5} + 0.55 X2_{t\text{-}11}$$
$$+ 3.40 \text{Ln}(X5_{t\text{-}9}) - 3.87 X5_{t\text{-}12}$$

(0.75) (14.3) (0.26) (0.18) (0.93) (0.77)

$$- 1.47 X1_{t\text{-}1} + 0.42 X1_{t\text{-}6}. \qquad (9)$$

(0.17) (0.22)

In (9), the figures in parentheses are the estimated coefficients' standard errors. These statistics suggest that all estimated coefficients are statistically different from zero at the 5% level of significance. Figure 4 (c) compares actual with OLS fitted values.

The results from the three techniques used suggest that ANN (with lowest MSE) should generate the best forecast. Logically then, GP and OLS residuals (Actual – Fitted) may be predictable using ANN. Thus, to capture the GP and OLS models unexplained variation in %D_LXXR, ANN was used to model them. The resulting fitted residuals were then added to originally obtained %D_LXXR fit to produce the new cooperative two-agent fits and forecasts. Two ANN systems were developed using the exact same explanatory variables with the dependent variable being GP residuals once and OLS residuals the second. Here are the results from estimating the two combinations performed:

ANN estimation of the GP residuals to obtain GP+ANN: $R^2 = 0.97$ MSE = 0.008

ANN estimation of the OLS residuals to obtain OLS+ANN: $R^2 = 0.95$ MSE = 0.02

Additional models were then obtained by reconciliatory cooperation. Reconciliatory cooperation entails employing all results obtained thus far as input variables to produce the final fits and forecasts. Alternatively, three agents (GP, ANN, and OLS) take as inputs the five solutions produced using GP, ANN, OLS, GP+ANN, and OLS+ANN

Table 2. %D_LXXR forecasts obtained by independent, complementing, and reconciliatory agents

	Independent			Complementing		Reconciliatory			
	ANN	GP	OLS	GP+ANN	OLS+ANN	GP_R	ANN_R	OLS_R	Ave
A-08	-3.96	-3.02	-3.29	-3.06	-4.68	-3.59	-3.44	-3.56	-3.53
M-08	-3.98	-1.09	-4.31	-1.84	-4.96	-3.08	-2.41	-3.00	-2.83
J-08	-3.66	-3.76	-3.95	-4.21	-3.70	-3.89	-3.60	-3.92	-3.81
J-08	-1.53	-2.45	-1.41	-3.10	-1.06	-2.19	-2.29	-2.25	-2.24
A-08	-0.67	-3.37	-0.26	-3.60	1.06	-1.90	-2.16	-2.02	-2.02
S-08	-2.32	-3.80	-0.49	-4.46	0.64	-3.22	-3.16	-3.32	-3.23
O-08	-3.71	-2.28	-1.30	-2.53	-0.36	-3.22	-2.42	-3.18	-2.94
N-08	-1.41	-3.04	-1.75	-2.53	-0.40	-1.88	-1.97	-1.93	-1.93
D-08	0.11	-2.96	-1.66	-1.66	-0.14	-0.63	-0.77	-0.70	-0.70
J-09	-0.89	-0.70	-0.96	0.36	-1.01	-0.37	-0.42	-0.31	-0.37
F-09	-0.79	-0.35	0.00	0.84	-1.26	-0.11	-0.23	-0.03	-0.12
M-09	0.76	-0.39	-0.04	0.40	-1.52	0.61	0.42	0.61	0.54
A-09	1.34	-0.63	-0.59	1.05	0.65	1.22	1.43	1.22	1.29
M-09	-0.27	-2.45	-1.33	-0.77	-0.55	-0.48	-0.58	-0.49	-0.52
J-09	-0.83	-0.98	-0.38	0.63	0.51	-0.22	-0.07	-0.15	-0.14
J-09	0.47	0.38	0.36	2.53	1.16	1.33	1.48	1.43	1.41
A-09	1.29	-1.16	0.48	0.65	0.93	1.02	1.33	1.01	1.12
S-09	1.19	0.10	0.02	0.89	0.20	1.07	1.10	1.07	1.08
O-09	-0.39	-0.96	-0.52	-0.06	-0.53	-0.25	-0.28	-0.23	-0.25
N-09	0.49	-0.47	-0.43	0.54	-0.51	0.51	0.48	0.53	0.51
D-09	0.86	-1.85	-0.78	0.12	-0.99	0.55	0.50	0.53	0.53
J-10	-1.02	-3.43	-1.31	-4.23	-1.54	-2.37	-3.06	-2.50	-2.64
F-10	-1.97	-2.04	-1.58	-3.19	-2.49	-2.48	-2.61	-2.53	-2.54
M-10	-1.74	-3.33	-1.51	-4.43	-2.47	-2.87	-3.41	-2.99	-3.09

modeling algorithms to produce new models and estimates. The final best GP- reconciliation model (GP_R) found is:

$$\%D\_LXXR = 0.4194 \, GPNN + 0.5806 \, NN \tag{10}$$

where GPNN = estimated values obtained from the GP+ANN combination and NN are the values produced using ANN alone. The equation above has only two right-hand-side variables because GP produced less desirable models otherwise

when other variables were selected. The OLS-reconciliation model (OLS_R) also produced the best outcome when the same two variables were employed. The best OLS_R model found is:

$$\%D\_LXXR = 0.011 + 0.463 \, GPNN + 0.543 \, NN. \tag{11}$$

(0.006) (0.053) (0.052)

The estimation MSE statistics of the cooperative models were almost identical: GP_R MSE = 0.005, OLS_R MSE = 0.004, ANN_R MSE =

Table 3. LXXR forecasts obtained by independent, complementing, and reconciliatory agents

	Independent			Complementing		Reconciliatory			
	ANN	GP	OLS	GP+ANN	OLS+ANN	GP_R	ANN_R	OLS_R	Ave
A-08	199.06	200.95	200.40	200.40	197.65	200.31	198.63	200.75	199.90
M-08	191.29	198.77	191.96	196.75	188.08	195.54	192.60	194.81	194.32
J-08	184.42	191.43	184.52	188.63	181.25	188.62	185.25	187.31	187.06
J-08	181.62	186.81	181.94	182.87	179.34	184.35	181.24	183.14	182.91
A-08	180.42	180.61	181.47	176.41	181.24	180.42	177.84	179.48	179.24
S-08	176.27	173.88	180.59	168.71	182.41	174.81	172.20	173.62	173.54
O-08	169.85	169.96	178.25	164.50	181.76	170.62	166.75	168.20	168.52
N-08	167.47	164.87	175.17	160.39	181.03	167.29	163.64	164.99	165.31
D-08	167.65	160.05	172.29	157.74	180.78	166.01	162.61	163.84	164.15
J-09	166.16	158.94	170.63	158.32	178.97	165.31	162.02	163.34	163.56
F-09	164.85	158.40	170.63	159.65	176.73	164.93	161.84	163.29	163.36
M-09	166.11	157.78	170.57	160.29	174.06	165.62	162.83	164.28	164.25
A-09	168.35	156.78	169.57	161.98	175.19	168.01	164.82	166.30	166.38
M-09	167.89	152.99	167.32	160.75	174.24	167.04	164.04	165.49	165.52
J-09	166.51	151.49	166.67	161.76	175.12	166.92	163.69	165.25	165.29
J-09	167.29	152.08	167.28	165.91	177.16	169.40	165.88	167.64	167.64
A-09	169.46	150.32	168.08	166.99	178.81	171.67	167.58	169.34	169.53
S-09	171.50	150.47	168.11	168.49	179.18	173.58	169.38	171.17	171.38
O-09	170.84	149.03	167.24	168.39	178.24	173.09	168.96	170.78	170.94
N-09	171.68	148.34	166.52	169.30	177.33	173.92	169.83	171.69	171.81
D-09	173.17	145.61	165.22	169.50	175.57	174.78	170.77	172.60	172.72
J-10	171.41	140.70	163.08	162.48	172.90	169.52	166.77	168.34	168.21
F-10	168.07	137.87	160.52	157.38	168.64	165.15	162.69	164.13	163.99
M-10	165.17	133.35	158.11	150.56	164.53	159.62	158.09	159.30	159.00

0.008, and their average MSE = 0.005. All four had $R^2 = 0.99$.

FORECAST RESULTS

Forecast results and their statistics are compared in this section. The three reconciliatory attempts produced similar forecasts. Table 2 shows the predicted monthly %D_LXXR produced by all agents and the average produced by the final three reconciliation agents. Although the rate of decline in prices is expected to decrease in 2008, prices start to increase marginally in 2009, but decline again early in 2010. Prices are expected to decrease by 7.5% in the third quarter of 2008 and decrease again by 5.56% in the fourth quarter. They are expected to increase by about 3% in third quarter of 2009, but decrease by 3.24% during the first quarter of 2010.

Table 3 presents forecasted LXXR values obtained using the %D_LXXR predictions where the predicted LXXR values are computed as follows:

Figure 5.

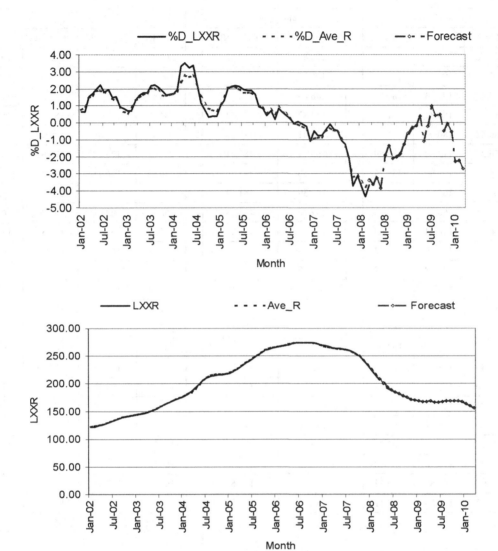

Figure 5. Plots of final predictions follow historical fitted values starting January 1992. The top plot shows the %D_LXXR fits and forecasts through March 2010. The bottom plot shows the LXXR fits and forecasts.

$$LXXR_t = \exp(\%D\_LXXR_t/100 + Ln(LXXR_{t-1}))$$
$$(12)$$

Given that the results of the agents are almost identical, Figure 5 presents the plots of predicted %D_LXXR and LXXR averages only. Predictions of housing price indexes in the Los Angeles metropolitan area throughout March of 2010 shown in Tables 2 and 3 are rather gloomy. Prices are expected to reverse to their levels during the third quarter of 2003 by the first quarter of 2010.

Efficacy of the forecast produced can be evaluated by comparing the forecast obtained here (that was available almost a year before the actual values materialized) with the actual LXXR values published after this research was completed. Of the 24-month period forecast values produced by the different agents as well as cooperation between them shown in Table 3, twelve-month of actual values became available prior to publishing this work. The prediction MAPE values for that period are 3.83%, 3.41%, 2.88%, 5.05%, 4.88%, 3.02%, 4.37%, 3.51%, and 3.55%. for GP, ANN, OLS, GP+ANN, OLS+ANN, GP_R, ANN_R, OLS_R, and the average reconciliated, respectively. Although the OLS independent forecasts appear best (with prediction MAPE = 2.8%), they are not statistically different from the GP_R forecasts (with prediction MAPE = 3%). Testing the null hypothesis that the mean difference between them is zero generated t-statistic = -0.18 and *p-value* = 0.86. This means that the null (i.e., difference is zero or the hypothesis that there is no difference between the two forecasts) cannot be rejected at any reasonable level of significance. Plot of actual materialized LXXR against that of the best forecast April 2008 through March 2009 are in Figure 6.

CONCLUSION

This chapter introduced novel thought on utilizing agent-based modeling to forecast the Los Angeles metropolitan area S&P/Case-Shiller Index for the 24 months: April 2008 through March 2010. The construction of agent-based modeling was based on progression defined in three stages. In the first, what may be perceived as the best variables that would explain variations in the monthly percentage change in the Los Angeles index were identified using a GP agent. In the second stage, GP and ANN agents were used to produce forecasts of the input variables identified in the first stage. In the third, agents competed and cooperated to produce a set of forecasts from which final forecast were obtained.

In the final analysis, the forecasts obtained by the employed agents were marginally different and can be viewed as too similar to select one as best. The average forecasts obtained using cooperation between the different agents employed is rather acceptable and as good as any. Actual index values published after the forecasts were completed (for April through March 2009) suggest that the OLS independent agent and the GP-reconciliation forecasts were more consistent with reality and are not statistically significantly different. Generally, the forecasts produced had a

Figure 6.

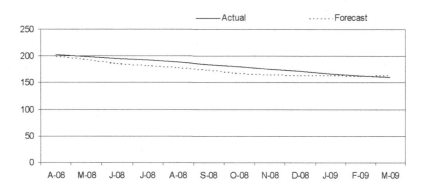

strong similarity between them. Given the strong similarity between them, it is easy to conclude that the Los Angeles metropolitan area housing market will remain depressed until the end of the forecast period considered in this research. Prices will stabilize in 2009 but resume their decline early in 2010.

REFERENCES

Appleton-Young, L. (2008). 2008 real estate market forecast. California Association of Realtors. Retrieved December 2008, from http://bayareahousingreview.com/wp-content/uploads/2008/02/ leslie_appleton_young _preso _read-only1.pdf.

Barto, A. (1996). Muti-agent reinforcement learning and adaptive neural networks. Retrieved December 2008, from http://stinet.dtic.mil/cgi-bin/GetTRDoc?AD=ADA315266&Location=U2&doc =GetTRDoc.pdf.

Bostonbubble.com. (2007). S&P/Case-Shiller Boston snapshot Q3 2007. Retrieved December 2008, from http://www.bostonbubble.com/forums/viewtopic.php?t=598.

C.A.R. (2008). U.S. economic outlook: 2008. Retrieved December 2008, from http://rodomino.realtor.org/Research.nsf/files/ currentforecast.pdf/$FILE/currentforecast.pdf.

Case, K., Glaeser, E., & Parker, J. (2000). Real estate and the macroeconomy. *Brookings Papers on Economic Activity, 2,* 119–162. doi:.doi:10.1353/eca.2000.0011

Case, K., & Shiller, R. (1989). The efficiency of the market for single-family homes. *The American Economic Review, 79,* 125–137.

Case, K., & Shiller, R. (1990). Forecasting prices and excess returns in the housing market. *American Real Estate and Urban Economics Association Journal, 18,* 263–273. doi:.doi:10.1111/1540-6229.00521

Case, K., & Shiller, R. (2003). Is there a bubble in the housing market? *Brookings Papers on Economic Activity, 1,* 299–342. doi:.doi:10.1353/eca.2004.0004

Chen, S., & Yeh, C. (1995). Predicting stock returns with genetic programming: Do the short-run nonlinear regularities exist? In D. Fisher (Ed.), *Proceedings of the Fifth International Workshop on Artificial Intelligence and Statistics* (pp. 95-101). Ft. Lauderdale, FL.

Chen, S., & Yeh, C. (1996). Genetic programming learning and the cobweb model . In Angeline, P. (Ed.), *Advances in Genetic Programming* (*Vol. 2,* pp. 443–466). Cambridge, MA: MIT Press.

Chen, X., & Tokinaga, S. (2006). Analysis of price fluctuation in double auction markets consisting of multi-agents using the genetic programming for learning. Retrieved from https://qir.kyushuu.ac.jp/dspace/bitstream /2324/8706/ 1/ p147-167.pdf.

China Bystanders. (2008). Bank profits trimmed by subprime losses. Retrieved from http://chinabystander. wordpress.com /2008/03/25/bank-profits-trimmed-by-subprime-losses/.

CME 2007. (n.d.). Retrieved December 2008, from http://www.cme.com/trading/prd/re/housing.html.

Commonweal of Australia. (2001). Economic Outlook. Retrieved December 2008, from http://www.budget.gov.au/2000-01/papers/ bp1/html/bs2.htm.

Economist.com. (2007). The world economy: Rocky terrain ahead. Retrieved December 2008, from http://www.economist.com/ daily/news/displaystory.cfm?storyid=9725432&top_story=1.

Fair, R., & Jaffee, D. (1972). Methods of estimation for markets in disequilibrium. *Econometrica*, *40*, 497–514. doi:.doi:10.2307/1913181

Freddie Mac. (2008a). CMHPI data. Retrieved December 2008, from http://www.freddiemac.com/finance/ cmhpi/#old.

Freddie Mac. (2008b). 30-year fixed rate historical Tables. Historical PMMS® Data. Retrieved December 2008, from http://www.freddiemac.com/pmms/pmms30.htm.

Group, C. M. E. (2007). S&P/Case-Shiller Price Index: Futures and options. Retrieved December 2008, from http://housingderivatives. typepad.com/housing_derivatives/files/cme_housing _fact_sheet.pdf.

Haji, K. (2007). Subprime mortgage crisis casts a global shadow – medium-term economic forecast (FY 2007~2017). Retrieved December 2008, from http://www.nli-research.co.jp/english/economics/2007/ eco071228.pdf.

Housing Predictor. (2008). Independent real estate housing forecast. Retrieved December 2008, from http://www.housingpredictor.com/california.html.

Iacono, T. (2008). Case-Shiller® Home Price Index forecasts: Exclusive house-price forecasts based on Fiserv's leading Case-Shiller Home Price Indexes. Retrieved December 2008, from http://www.economy.com/home/products/ case_shiller_indexes.asp.

Kaboudan, M. (2001). Genetically evolved models and normality of their residuals. *Journal of Economic Dynamics & Control, 25*, 1719–1749. doi:.doi:10.1016/S0165-1889(00)00004-X

Kaboudan, M. (2004). TSGP: A time series genetic programming software. Retrieved December 2008, from http://bulldog2.redlands.edu/ fac/mak_kaboudan/tsgp.

Koza, J. (1992). *Genetic programming*. Cambridge, MA: The MIT Press.

Ludwig, A., & Torsten, S. (2001). The impact of stock prices and house prices on consumption in OECD countries. Retrieved December 2008, from http://www.vwl.uni-mannheim.de/brownbag/ludwig.pdf.

Money, C. N. N. com (2008). World economy on thin ice - U.N.: The United Nations blames dire situation on the decline of the U.S. housing and financial sectors. Retrieved December 2008, from http://money.cnn.com/2008/05 /15/news/international/global_economy.ap/.

Moody's. Economy.com (2008). Case-Shiller® Home Price Index forecasts. Moody's Analytics, Inc. Retrieved December 2008, from http://www.economy.com/home/products/case_shiller_indexes.asp.

National Association of Home Builders, The Housing Policy Department. (2005). The local impact of home building in a typical metropolitan area: Income, jobs, and taxes generated. Retrieved December 2008, from http://www.nahb.org/fileUpload_details.aspx?contentTypeID=3&contentID=35601& subContentID=28002.

NeuroSolutions™ (2002). The Neural Network Simulation Environment. Version 3, NeuroDimensions, Inc., Gainesville, FL.

Principe, J., Euliano, N., & Lefebvre, C. (2000). *Neural and Adaptive Systems: Fundamentals through Simulations*. New York: John Wiley & Sons, Inc.

Standard & Poor's. (2008a). S&P/Case-Shiller® Home Price Indices Methodology. Standard & Poor's. Retrieved December 2008, from http://www2.standardandpoors.com/spf/pdf/index/ SP_CS_Home_ Price_Indices_ Methodology_ Web.pdf.

Standard & Poor's. (2008b). S&P/Case-Shiller Home Price Indices. Retrieved December 2008, from http://www2.standardandpoors.com/ portal/site/sp/en/us/page.topic/indices_csmahp/ 2,3,4,0,0,0,0,0,0,1,1,0,0,0,0,0.html.

Stark, T. (2008). Survey of professional forecasters: May 13, 2008. Federal Reserve Bank of Philadelphia. Retrieved December 2008, from http://www.philadelphiafed.org/files/spf/survq208.html

Tsang, E., Li, J., & Butler, J. (1998). EDDIE beats the bookies. *Int. J. Software. Practice and Experience, 28*, 1033–1043. doi:10.1002/(SICI)1097-024X(199808)28:10<1033::AID-SPE198>3.0.CO;2-1

U.S. Bureau of Economic Analysis. (2008). Regional economic accounts: State personal income. Retrieved December 2008, from http://www.bea.gov/regional/sqpi/default.cfm?sqtable=SQ1.

U.S. Census Bureau. (2008a). Housing vacancies and home ownership. Retrieved December 2008, from http://www.census.gov/hhes/ www/ histt10.html.

U.S. Census Bureau. (2008b). New residential construction. Retrieved December 2008, from http://www.census.gov/const/www/newresconstindex_excel.html.

Vanstone, B., & Finnie, G. (2007). An empirical methodology for developing stockmarket trading systems using artificial neural networks. Retrieved December 2008, from http://epublications.bond.edu.au/cgi/ viewcontent.cgi?article=1022&context=infotech_pubs.

Warren, M. (1994). Stock price prediction using genetic programming . In Koza, J. (Ed.), *Genetic Algorithms at Stanford 1994*. Stanford, CA: Stanford Bookstore.

Chapter 2
An Agent–Based Model for Portfolio Optimization Using Search Space Splitting

Yukiko Orito
Hiroshima University, Japan

Yasushi Kambayashi
Nippon Institute of Technology, Japan

Yasuhiro Tsujimura
Nippon Institute of Technology, Japan

Hisashi Yamamoto
Tokyo Metropolitan University, Japan

ABSTRACT

Portfolio optimization is the determination of the weights of assets to be included in a portfolio in order to achieve the investment objective. It can be viewed as a tight combinatorial optimization problem that has many solutions near the optimal solution in a narrow solution space. In order to solve such a tight problem, we introduce an Agent-based Model in this chapter. We continue to employ the Information Ratio, a well-known measure of the performance of actively managed portfolios, as an objective function. Our agent has one portfolio, the Information Ratio and its character as a set of properties. The evolution of agent properties splits the search space into a lot of small spaces. In a population of one small space, there is one leader agent and several follower agents. As the processing of the populations progresses, the agent properties change by the interaction between the leader and the follower, and when the iteration is over, we obtain one leader who has the highest Information Ratio.

INTRODUCTION

Portfolio optimization, based on the modern portfolio theory proposed by Markowitz (1952), determines the appropriate weights of assets included in a portfolio in order to achieve the investment objective. This optimization problem is one of the combinatorial optimization problems and the solution is the performance value obtained

DOI: 10.4018/978-1-60566-898-7.ch002

by a portfolio. When we attempt to solve such an optimization problem, we usually find candidates for the solutions that are better than others. The space of all feasible solutions is called the search space. There are a number of possible solutions in the search space and finding the best solution is thus equal to finding some extreme values, minimum or maximum. In the search space, we want to find the best solution, but it is hard to solve in reasonable time as the number of assets or the number of weights of each asset grows. Because there are many possible solutions in the search space, it is usually hard for us to know where to find a solution or where to start. In order to solve such a problem, many researchers use methods based on evolutional algorithms: for example, genetic algorithm (GA), simulated annealing, tabu search, some local searches and so on.

There are two investment objectives for portfolio management: active management and passive management. Active management is an investment strategy that seeks returns in excess of a given benchmark index. Passive management is an investment strategy that mirrors a given benchmark index. Thus, if you believe that it is possible to outperform the market, you should invest in an active portfolio. The Information Ratio and the Sharpe Ratio are well-known indices for active portfolio evaluation. On the other hand, if you think that it is not possible to outperform the market, you should invest in a passive portfolio. There are several reports that index funds that employ passive management show better performance than other mutual funds (e.g. see Elton et. al, 1996; Gruber, 1996; Malkiel, 1995). The correlation between the portfolio price and the benchmark index and Beta are famous indices that are used to evaluate the passive portfolio.

This optimization problem can be viewed as a discrete combinatorial problem regardless of the index we choose to evaluate the performance of active or passive portfolios. Hence, this optimization problem has two subproblems. The first one is that portfolios consisting of quite different weights

of assets have similar performance values. The second problem is that there are many solutions near the optimal solution. It is hard to solve such a tight optimization problem even with strong evolutionary algorithms.

In this chapter, we propose an Agent-based Model in order to solve this tight optimization problem. In general, agent-based models describe interactions and dynamics of a group of traders in the artificial financial market (LeBaron, 2000). Our Agent-based Model is implemented as a global and local search method for the portfolio optimization problem. Our agent has a set of properties: its own portfolio, a performance value obtained by the portfolio and its character. In the starting population, there is one leader agent, and there are many follower agents. The follower agents are categorized into three groups, namely obedient group, disobedient group, and indifferent group. In the first group, the followers obediently follow the leader's behaviors. In the second group, the followers are disobedient and adopt behaviors opposite to that of the leader. In the third group, the followers determine their behaviors quite independently. As processing of the population proceeds through search space splitting, the agent properties change through the interaction between the leader and the followers, and gradually a best performing agent (the leader agent) with the highest performance value emerges as the optimal solution. Hence, our Agent-based Model has the advantage that our model searches solutions in global space as well as local spaces for this tight optimization problem, because plural leader agents appear and disappear during search space splitting.

The structure of the balance of this chapter is as follows: Section 2 describes related works. Section 3 defines the portfolio optimization problem and describes the performance value used to evaluate the portfolio as the objective function. In Section 4, we propose an Agent-based Model in order to optimize the portfolios. Section 5 shows the results of numerical experiments obtained by the simulation of the Agent-based Model.

RELATED WORKS

Markowitz (1987) proposed the mean-variance methodology for portfolio optimization problems. The objective function in his methodology is to minimize the variance of the expected returns as the investment risk under the expected return as the investment return. Many researchers have extended his methodology to practical formulae and applications, and have tackled this problem by using methods based on evolutionary algorithms. Xia et al. (2000) proposed a new mean-variance model with an order of the expected returns of securities, and applied a GA to optimize the portfolios. Chang et al. (2000) proposed the extended mean-variance models, and applied various evolution algorithms, such as GA, simulated annealing and tabu search. Lin and Liu (2008) proposed the extended mean-variance model with minimum transaction lots, and applied GA to optimize the practical portfolios. Streichert & Tanaka-Yamawaki (2006) evaluated a multi-objective evolutionary algorithm with a quadratic programming local search for the multi-objective constrained or unconstrained portfolio optimization problems. For passive portfolio optimization problems, Oh et al. (2005) showed the effectiveness of index funds optimized by a GA on the Korean Stock Exchange. Their objective function was a function based on a beta which is a measure of correlation between the fund's price and the benchmark index. Orito & Yamamoto (2007) and Orito et al. (2009) proposed GA methods with a heuristic local search and optimized the index funds on the Tokyo Stock Exchange. Their objective functions were correlation coefficients between the fund's return rates and the changing rates of benchmark indices. Aranha & Iba (2007) also proposed a similar but different method on the Tokyo Stock Exchange.

On the other hand, agent-based models for artificial financial markets have recently been popular in research. The agent properties change by the interaction between agents, and gradually a best performing agent emerges as an optimal solution (e.g. see LeBaron et al., 1999; Chen & Yeh, 2002). Based on the idea of agent-based models, we propose an Agent-based Model for the portfolio optimizations in this chapter. In general, many agent-based models describe interactions and dynamics in a group of traders in an artificial financial market (LeBaron, 2000). Our Agent-based Model is implemented as a global and local search method for portfolio optimization.

PORTFOLIO OPTIMIZATION PROBLEM

First, we define the following notations for the portfolio optimization problem.

N: total number of all the assets included in a portfolio.

i: Asset i, $i = 1,\ldots, N$.

$F_{index}\left(t\right)$: the value of the given benchmark index at t.

$\mathbf{P}_{index}$: the sequence of the rates of changes of benchmark index over $t = 1,\ldots,T$. That is the vector $\mathbf{P}_{index} = \left(P_{index}\left(1\right),\cdots, P_{index}\left(T\right)\right)$ whose $P_{index}\left(t\right)$ is defined as $P_{index}\left(t\right) = \left(F_{index}\left(t+1\right) - F_{index}\left(t\right)\right)\big/F_{index}\left(t\right)$.

$F_i(t)$: the price of Asset i at t.

P_i: the sequence of the return rates of Asset i over $t = 1,\ldots, T$. That is the vector $P_i=(P_i(1),\ldots,P_i(T))$hose $P_i(t)$ is defined as $P_i(t) = (F_i(t+1)-F_i(t))/F_i(t)$.

M: total number of all the units of investment.

M_i: the unit of investment for Asset i. That is an integer such that $\sum M_i = M$.

w_i: the weight of Asset i included in the portfolio. That is a real number $w_i = M_i/M$ ($0 \leq w_i \leq 1$. Note that we do not discuss the short sale here.

G_k: the portfolio of k-th agent. That is the vector $G_k = w_1, \ldots, w_N)$ such that $\sum_{i=1}^{N} w_i = 1$.

$\mathbf{P}_{G_j}$: the sequence of the return rates of portfolio G_k over $t = 1, \ldots, T$. That is the vector $\mathbf{P}_{G_k} = \left(P_{G_k}(1), \cdots, P_{G_k}(T) \right)$ whose $P_{G_k}(t) = \sum_{i=1}^{N} w_i \cdot P_i(t)$.

It is well known that the Information Ratio, which is built on the modern portfolio theory, is an index for the evaluation of active portfolios. It is defined as the active return divided by the tracking error. The active return means the amount of performance over or under a given benchmark index. The tracking error is the standard deviation of the active returns. Therefore, it is desirable to achieve a high Information Ratio.

In this chapter, we define the Information Ratio of portfolio G_k as the objective function for the portfolio optimization problem to be as follows:

$$ \max \quad IR_{G_k} = \frac{E\left[\mathbf{P}_{G_k} - \mathbf{P}_{index}\right]}{\sqrt{\mathrm{var}\left[\mathbf{P}_{G_k} - \mathbf{P}_{index}\right]}}, \qquad (1) $$

where $E\left[\mathbf{P}_{G_k} - \mathbf{P}_{index}\right]$ is the expected value of the historical data of portfolio's return rates over or under the rates of changes of benchmark index, and $\sqrt{\mathrm{var}\left[\mathbf{P}_{G_k} - \mathbf{P}_{index}\right]}$ is the standard deviation from the same data.

In this chapter, we optimize portfolios that consist of the combination of weights of N assets with M elements in order to maximize the Information Ratio given by Equation (1). To obtain one portfolio $G_k = (w_1, \ldots, w_N)$ means to obtain its Information Ration as one solution. The number of combinations, i.e. the number of solutions, is given by $(M + N - 1)!/N!(M-1)!$.

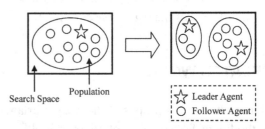

Figure 1. Re-composition of the populations

Search Space   Population

Leader Agent
Follower Agent

AGENT-BASED MODEL

In our Agent-based Model, the search space is split into several small spaces. Agents in one population search for solutions in one small space. One population consists of one leader agent and several follower agents. As the agents' properties through evolve, a new leader appears, and then all populations in the search space are re-composed. Such re-composition of populations represents the search space splitting as shown in Figure 1.

In this section, we describe our Agent-based Model. Section 1 defines the agent. Section 2 defines the evolutional behavior of the agents in a population. Section 3 describes the outcome of processing of the population as the search space splitting.

Agents

Figure 2 shows a typical agent in our Agent-based Model. Each agent has its own portfolio, a unit of investment of the portfolio, the Information Ratio, and its character as the properties. The character of agent is either "obedient", "disobedient", or "independent." The agent's evolutional behavior depending on its character is described in Section 2.

Evolutional Behavior of Agents in a Population

For our Agent-based Model, let s be the number of iterations of the evolutional process. In the first ($s = 1$) iteration, we set W agents as the initial

Figure 2. Agent properties

Agent k	Portfolio: $\mathbf{G}_k = \left(w_1, \cdots, w_N\right)$
	Unit of investment of Portfolio: $\left(M_1, \cdots, M_N\right)$
	Information Ratio: $IR_{\mathbf{G}_k}$
	Character: Obedient/Disobedient/Independent

Figure 3. Behavior of the obedient agent

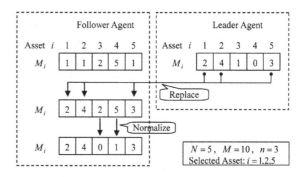

population. For the unit of investment of Asset i for Agent k, an integer M_i ($i = 1,…, N$) is randomly given. Then Agent k has the portfolio G_k and its Information Ratio. In addition, Agent k is randomly given one of the following characters: obedient, disobedient or independent. The rates of obedient agent, disobedient agent and independent agent in the initial population are given by α_{obd}, α_{dob} and α_{idp}, respectively.

An agent whose Information Ratio is the highest of all the agents in the population becomes the leader agent in its population and the others remain as followers. On the s-th process, the follower agents' properties change through the interaction between the leader and the follower agents. On the other hand, the leader agent's properties change by its own rule. The evolutional behaviors of each kind of follower and the leader are described below.

Behavior of Follower Agent

The follower agents, depending on their character: obedient, disobedient or independent, change their properties through the interaction with the leader agent, and evolve into new agents.

Obedient Follower

If Agent k is a follower agent and its character is obedient, Agent k is an agent that imitates a part of the leader's portfolio. The procedure of the obedient follower's evolutionary behavior is as follows:

1. We randomly select n assets from N assets to put in the portfolio. For each of these n assets, we replace the follower's investment units M_i with those of the leader's.
2. We randomly repeat the selection of one asset from $N - n$ assets, which are not chosen in step 1 and increase or decrease one investment unit of it until $\sum M_i = M$. We call this operation "normalization" in this chapter.

    For example, Figure 3 shows an evolutional behavior of an obedient follower agent with $N = 5$, $M = 10$ and $n = 3$.

Disobedient Follower

If Agent k is a follower agent and its character is disobedient, Agent k is a follower agent that does not imitate a part of the leader's portfolio. The procedure of the disobedient follower's evolutional behavior is as follows:

1. We randomly select n assets from N assets to put in the portfolio. For each of these n assets, we subtract the follower's investment units M_i from those of the leader. If the selected asset's investment units becomes 0, the subtraction of the asset stops.
2. For normalization, we randomly repeat the selection of one asset from $N - n$ assets not chosen in step 1 and increase one investment unit of it until $\sum M_i = M$.

Figure 4. [Behavior of the disobedient agent]

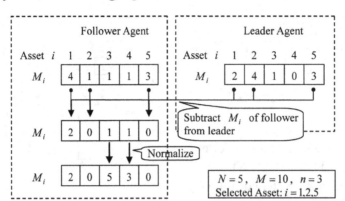

For example, Figure 4 shows an evolutional behavior of a disobedient agent with $N = 5$, $M = 10$ and $n = 3$.

Independent Follower

The independent follower is an agent whose behavior is not influenced by the leader, and changes its portfolio by its own independent rule defined as Equation (2). A randomly chosen integer M_i is the set for the unit of investment.

$$M_i \in [0, V] \ (i = 1, \ldots, N), \tag{2}$$

where V is a given integer parameter as a upper bound of the number of investment units. The new investments units are to be normalized.

By performing the evolutionary behaviors of obedient, disobedient and independent followers according to their assigned characters, all the follower agents have new investment units ($M_1, \ldots, M_N$). Hence the new portfolio $G_k = (g_1, \ldots, g_N)$ is calculated again and the Information Ratio of the portfolio is updated in order to complete the evolutions of all the properties of each follower.

Behavior of the Leader Agent

The leader in a population is an agent whose Information Ratio is the highest of all agents in the search space when the population is generated. The leader agent's properties change by its own rule. The procedure for the evolutionary behavior of the leader is as follows. We note that the leader's Information Ratio in the re-composed population is not the highest of all agents in the search space. We describe the outcome of processing of the population in the Section 3.

1. We randomly select n assets from N assets to put in the portfolio. We express these n assets as $\{i_1, \cdots, i_n\}$ and call it Group k_1. As a similar manner, we express the $N - n$ assets as $\{i_{n+1}, \cdots, i_N\}$ and call it Group k_2. We define the expected value of the historical data of each group's return rates over or under the rates of changes of benchmark index as follows;

$$\begin{cases} E\left[\mathbf{P}_{\mathbf{G}_{k_1}} - \mathbf{P}_{index}\right] & \left(\mathbf{G}_{k_1} = \left(w_{i_1}, \cdots, w_{i_n}\right)\right) \\ E\left[\mathbf{P}_{\mathbf{G}_{k_2}} - \mathbf{P}_{index}\right] & \left(\mathbf{G}_{k_2} = \left(w_{i_{n+1}}, \cdots, w_{i_N}\right)\right) \end{cases} \tag{3}$$

Figure 5. Behavior of the leader agent

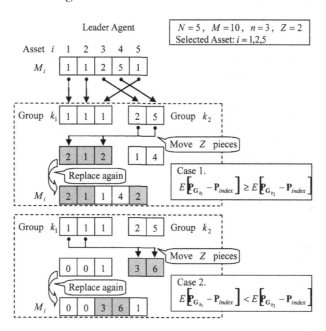

2. If $E\left[\mathbf{P}_{G_{k_1}} - \mathbf{P}_{index}\right] \geq E\left[\mathbf{P}_{G_{k_2}} - \mathbf{P}_{index}\right]$ in Equation (3) is satisfied, we move z investment units for the randomly selected assets from Group k_2 to Group k_1. On the other hand, if

$$E\left[\mathbf{P}_{G_{k_1}} - \mathbf{P}_{index}\right] < E\left[\mathbf{P}_{G_{k_2}} - \mathbf{P}_{index}\right]$$ is

satisfied, we move z investment units for the randomly selected assets from Group k_1 to Group k_2.

For example, Figure 5 shows an evolutional behavior of the leader agent with $N = 5$, $M = 10$, $n = 3$ and $z = 2$.

Upon execution of this behavior, the leader agent has the new investment units $(M_1, ..., M_N$. Hence the new portfolio $G_k = (g_1, ..., g_N)$ is calculated again and the Information Ratio of the portfolio is updated in order to complete the evolution of all the properties.

Processing of the Population

In this section, we describe the outcome of processing the population. On the s-th process, the properties of the leader and the follower agents are updated by their evolutionary behaviors described in the section on the evolutional behavior of agents. Hence, the new leader agent who has the highest Information Ratio appears. The new population, with its new leader agent, is generated on the s +1st process. In other words, the progression of the population in our Agent-based Model repeats the generation of the new leader agent and the division of current populations of the s-th process. Populations with no follower agents, which are caused by repeated splitting of search spaces, disappear. The leader of such a population becomes a follower agent in one of other populations. Figure 6 shows such a series of processes as an example. We note that, in our model, it is possible that an agent who has lower Information Ratio than the new leader's ratio and has higher Information Ratio than the current leader's ratio remains as a follower agent in the current leader's population.

As shown in Figure 6, one population consists of one leader agent and its follower agents. When a population is split, all follower agents select the current leader in the current population or the new leader whose Information Ratio is the highest in the search space as their own leader on the next process. The criterion for this selection is defined as follows.

1. When the new leader appears in the population to which some of the current follower agents belong, the rate β_{same} is used to select the new followers in the population of the new leader agent on the next process. On the other hand, the ratio of $1 - \beta_{same}$ followers remain to be followers of the current leader.

2. When the new leader appears in the population to which the follower agents do not belong, the rate β_{diff} is used to select the followers in the population of the new leader agent on the next process. On the other hand, the ratio of $1 - \beta_{diff}$ followers remain to be followers of the current leader.

After repeating these processes, we select the portfolio whose Information Ratio is the highest in the search space as our optimal portfolio.

SIMULATION RESULTS

We have applied our Agent-based Model to each of 12 data periods on the First Section of Tokyo Stock Exchange from Jan. 6, 1997 to Oct. 2, 2006. Each data period is 100 days, and is shifted every 200 days from Jan. 6, 1997. It is called from Period 1 (Jan. 6, 1997 - May. 30, 1997) to Period 12 (May. 12, 2006 - Oct. 2, 2006), respectively. The dataset consists of assets in the order of high turnover average as a subset of TOPIX. The TOPIX, a well

Figure 6. Progression populations during search space splitting

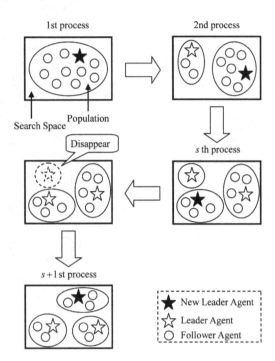

known benchmark index in the First Section of Tokyo Stock Exchange, represents the increase or decrease in stock values of all companies listed on the market. For this, these assets are typical assets that have a big influence on the value of TOPIX.

Information Ratio as a Function of Parameters in an Agent-Based Model

The Information Ratios of the optimal portfolios obtained by our Agent-based Model depend on the starting parameters. For the combinatorial optimization problem with the total number of all the assets included in a portfolio $N = 200$ and the total number of all the units of investment $M = N \times 100$, we applied our Agent-based Model using parameters given by Table 1 for the numerical experiments.

For $n = 20 (=N \times 0.1)$ and $n = 40 (=N \times 0.2)$, the Information Ratios of the optimal portfolios obtained by our Agent-based Model for the twelve

Table 1. Parameters in Agent-based Model

The total number of all the assets included in a portfolio: N	200
The total number of all the units of investment: M	20000 ($=N \times 100$)
The total number of agents in the solution space	100
The number of iterations of the evolutionary process: K	100,200,300,400,500
The number of selected assets for the interaction between the leader and the follower: n	$N \times c (c = 0.0, 0.1, 0.2,...,1)$ 0,20,40,60,80,100,120,140,160,180,200
The ratio of obedient, disobedient and independent agent in the solution space: $\left\{ \alpha_{abd}, \alpha_{dob}, \alpha_{idp} \right\}$	{1,0,0},{0.8,0.2,0},{0.8,0,0.2},{0.6,0.4,0}, {0.6,0.2,0.2},{0.6,0,0.4},{0.4,0.6,0},{0.4,0.4,0.2}, {0.4,0.2,0.4},{0.4,0,0.6},{0.2,0.8,0},{0.2,0.6,0.2}, {0.2,0.4,0.4},{0.2,0.2,0.6},{0.2,0,0.8},{0,1,0}, {0,0.8,0.2},{0,0.6,0.4},{0,0.4,0.6},{0,0.2,0.8},{0,0,1}
The ratio of follower agent who moves to the new leader's population from the same and the different population to which the current leader belong: $\left\{ \beta_{same}, \beta_{diff} \right\}$	{1,1},{1,0.75},{1,0.5},{1,0.25},{1,0},{0.75,1}, {0.75,0.75},{0.75,0.5},{0.75,0.25},{0.75,0}, {0.5,1},{0.5,0.75},{0.5,0.5},{0.5,0.25},{0.5,0}, {0.25,1},{0.25,0.75},{0.25,0.5},{0.25,0.25}, {0.25,0},{0,1},{0,0.75},{0,0.5},{0,0.25},{0,0}
The number of investment units moved between Groups k_1 and k_2: z	2000 ($=M \times 0.1$)

periods as a function of the repetition of the progresses process are shown in Figure 7 (a) and (b), respectively. We repeated the evolution process up to 500 times. Here, we set the parameters $\left\{ \alpha_{abd}, \alpha_{dob}, \alpha_{idp} \right\}$ and $\left\{ \beta_{same}, \beta_{diff} \right\}$ to $\left\{ \alpha_{abd}, \alpha_{dob}, \alpha_{idp} \right\} = \left\{ 0.6, 0.2, 0.2 \right\}$ and $\left\{ \beta_{same}, \beta_{diff} \right\} = \left\{ 0.5, 0.5 \right\}$.

(a) n=20

(b) n=40

From Figure 7 (a) and (b), we can observe that the Information Ratio becomes higher as the number of repetitions of the process increases. Therefore, we set the number of repetitions to be $K = 500$ for further experiments.

Next, the Information Ratio average of all periods as a function of the number of selected assets for the interaction between the leader and the follower n is shown in Figure 8. As well as Figure 7, we set the parameters $\left\{ \alpha_{abd}, \alpha_{dob}, \alpha_{idp} \right\}$ and $\left\{ \beta_{same}, \beta_{diff} \right\}$ to

$\left\{ \alpha_{abd}, \alpha_{dob}, \alpha_{idp} \right\} = \left\{ 0.6, 0.2, 0.2 \right\}$ and $\left\{ \beta_{same}, \beta_{diff} \right\} = \left\{ 0.5, 0.5 \right\}$.

From Figure 8, we can observe that the Information Ratio is high when n is small except for the case of $n = 0$. As defined in the behavior of follower agent section, n is the number of assets for which an obedient follower matches the leader's weight or that a disobedient follower subtracts from its portfolio contrary to the leader's weight of that asset in the leader's portfolio. In this context, we can conclude that Figure 8 suggests that good agents do not appear when the interaction between the leader and the follower is too strong. Therefore, we set as $n = 40$ for further experiments, because its Information Ratio is the highest of all. This means that the followers interact with 20% of leader's assets.

Next, we discuss the influence of parameters $\left\{ \alpha_{abd}, \alpha_{dob}, \alpha_{idp} \right\}$ and $\left\{ \beta_{same}, \beta_{diff} \right\}$ on the portfolio optimization problem. The Information Ratio average of all periods as a function of $\left\{ \alpha_{abd}, \alpha_{dob}, \alpha_{idp} \right\}$ and a function of $\left\{ \beta_{same}, \beta_{diff} \right\}$

Figure 7. Information Ratio as a function of the number of iterations of the evolutionary process

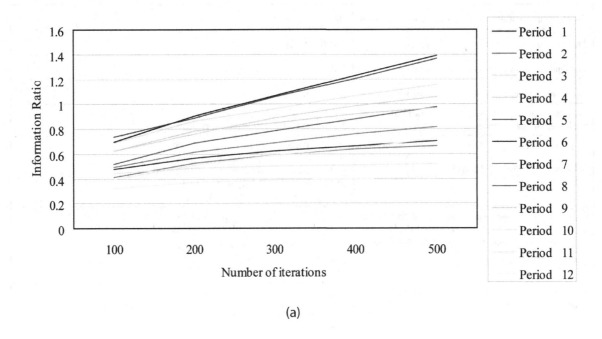

(a)

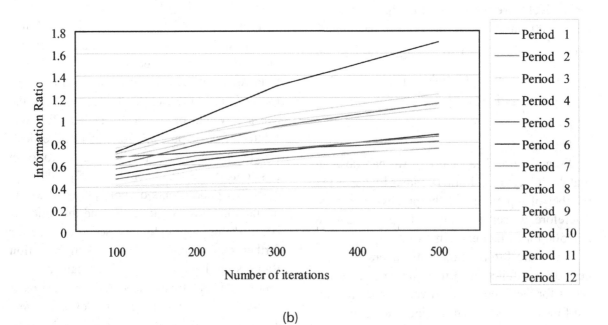

(b)

Figure 8. Information Ratio as a function of the number of selected assets for the interaction between the leader and the follower

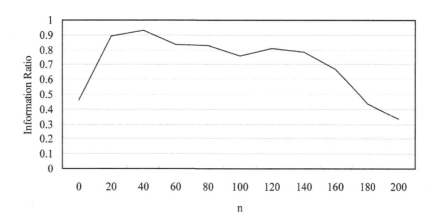

is shown in Figures 9 and 10, respectively. Note that we set $\left\{\beta_{same}, \beta_{diff}\right\} = \left\{0.5, 0.5\right\}$ for Figure 9 and $\left\{\alpha_{abd}, \alpha_{dob}, \alpha_{idp}\right\} = \left\{0.6, 0.2, 0.2\right\}$ for Figure 10.

From Figure 9, the Information Ratio with $\left\{\alpha_{abd}, \alpha_{dob}, \alpha_{idp}\right\} = \left\{0.8, 0, 0.2\right\}$ is the highest. We can conclude that the Information Ratio is high when the proportion of obedient followers in the solution space is large. This means that the effective population should consist of many obedient agents and few disobedient and independent agents. Therefore, we set $\left\{\alpha_{abd}, \alpha_{dob}, \alpha_{idp}\right\}$ to be {0.8, 0, 0.2} for experiments in the next section.

On the other hand, from Figure 10, we can observe that almost of all the Information Ratios are similar. The exceptions are the cases when each of β_{same} and β_{diff} is set to 0 or 1. This means

Figure 9. Information Ratio as a function of the ratio {alpha_obd,alpha_dob,alpha_idp}

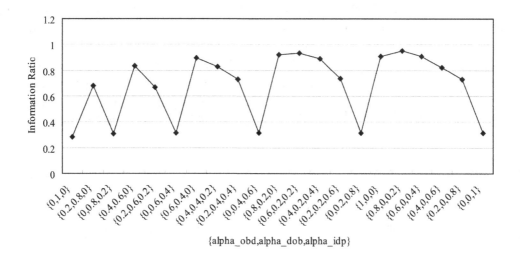

Figure 10. Information Ratio as a function of the ratio {beta_same,beta_diff}

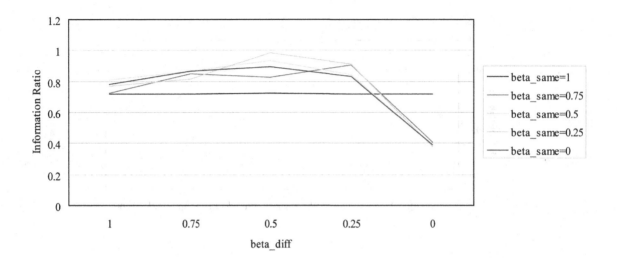

that the ratio of the follower agents who moves to the new leader's population from the same or the different population to which the current leader belong does not affect the results in our Agent-based Model.

Comparison of Agent-Based Model and GA

We compare our Agent-based Model with GA for the combinatorial optimization problem with the total number of assets $N = 200$ and the total number of units of investment $M = 20000$. Therefore, the number of combinations is given by $(20000 + 200 - 1)!/200! (20000 - 1)!$. For our model, we set parameters as $K = 500$, $n = 40$, $\left\{\alpha_{abd}, \alpha_{dob}, \alpha_{idp}\right\} = \left\{0.8, 0, 0.2\right\}$ and $\left\{\beta_{same}, \beta_{diff}\right\} = \left\{0.5, 0.5\right\}$. On the other hand, it

is well known that GA is a useful stochastic search method for such optimization problems (for this, see e.g. Holland, 1975; Goldberg, 1989). The genetic representation of our GA is shown in Figure 11. A gene represents a weight of an asset w_i and a chromosome represents a portfolio G_k. The fitness value of GA is defined as the Information Ratio.

On the first generation of the GA, we randomly generate the initial population. We apply the uniform crossover for exchanging the partial structure between the two chromosomes and repair to a probability distribution via renormalization. We also apply the uniform mutation for replacing the partial structure of the selected chromosomes with a new random value in [0, 1] and repair to a probability distribution via renormalization. After making offspring, we apply a roulette wheel selection and an elitism method of 10% chromosomes

Figure 11. Genetic representation

	Asset 1	Asset 2	...	Asset N
Chromosome $\mathbf{G}_k$	w_1	w_2	...	w_N

based on the fitness value. As a termination criterion of GA, we apply the generation size. Note that the population size, the generation size, the crossover rate and the mutation rate are set to the similar values of our Agent-based Model, 100 (the same value of the total number of assets in the solution space), 500 (the same value of K), 0.8 (the same value of the ratio of obedient follower in the solution space) and 0.2 (the same value of the ratio of independent follower in the solution space), respectively. The Information Ratio average of 20 simulations obtained by our Agent-based Model and GA are shown in Table 2.

From Table 2, we can observe that the Information Ratios of portfolios obtained by our Agent-based Model are higher than those of GA for all the periods. In almost all the periods, the Information Ratios obtained by our model exceeds the final results of GA within 50 iterations of evolutionary process.

Therefore, we can conclude that our Agent-based Model using search space splitting produces more optimal portfolios than simple GA. However, the results obtained by our Agent-based Model depend on the parameters; the number of selected assets for the interaction between the leader and the n followers, and the ratio of the obedient, disobedient and independent agents in the solution space $\left\{ \alpha_{abd}, \alpha_{dob}, \alpha_{idp} \right\}$.

CONCLUSION

In this chapter, we proposed an Agent-based Model to solve the portfolio optimization problem which is a tight optimization problem. The most notable advantage of our Agent-based Model is that our model searches solutions in global space as well as local spaces.

From the numerical experiments, we conclude that our Agent-based Model using search space splitting produces more optimal portfolios than simple GA. However, the results obtained by our model depend on the parameters; the number of selected assets for the interaction between the leader and the n followers, and the ratio of the obedient, disobedient and independent agents in the solution space $\left\{ \alpha_{abd}, \alpha_{dob}, \alpha_{idp} \right\}$.

Sometimes, portfolios consisting of quite different weights of assets have the similar Information Ratios to the ratios of other portfolios. We do not have reasonable explanations for this fact.

Table 2. Information Ratio average

Period No.	Agent-based Model	GA
1	0.886301	0.359296
2	0.713341	0.263974
3	1.353191	0.470981
4	1.384440	0.488758
5	1.673957	0.542885
6	0.792513	0.445230
7	1.002361	0.350776
8	0.765283	0.401970
9	1.129314	0.462260
10	0.869310	0.309170
11	0.514538	0.332241
12	0.364809	0.226260

For this, it would be beneficial for us to visualize the landscape partially and improve our Agent-based Model on the partial landscape. It is hard to visualize the landscape of solutions, however, because this problem can be viewed as a discrete combinatorial problem. In addition, we need to rebalance the portfolios in the future period in order to maintain their performance. These issues are reserved for our future work.

ACKNOWLEDGMENT

The authors are grateful to Kimiko Gosney who gave us useful comments. The first author acknowledges partial financial support by Grant #20710119, Grant-in-Aid for Young Scientists (B) from JSPS, (2008-).

REFERENCES

Aranha, C., & Iba, H. (2007). Portfolio Management by Genetic Algorithms with Error Modeling. In *JCIS Online Proceedings of International Conference on Computational Intelligence in Economics & Finance.*

Chang, T. J., Meade, N., Beasley, J. E., & Sharaiha, Y. M. (2000). Heuristics for Cardinality Constrained Portfolio Optimization. *Computers & Operations Research, 27,* 1271–1302. doi:10.1016/S0305-0548(99)00074-X

Chen, S. H., & Yeh, C. H. (2002). On the Emergent Properties of Artificial Stock Markets: The Efficient Market Hypothesis and the Rational Expectations Hypothesis. *Journal of Behavior & Organization, 49,* 217–239. doi:10.1016/S0167-2681(02)00068-9

Elton, E., Gruber, G., & Blake, C. (1996). Survivorship Bias and Mutual Fund Performance. *Review of Financial Studies, 9,* 1097–1120. doi:10.1093/rfs/9.4.1097

Goldberg, D. E. (1989). *Genetic Algorithms in Search, Optimization and Machine Learning.* Addison-Wesley.

Gruber, M. J. (1996). Another Puzzle: The Growth in Actively Managed Mutual Funds. *The Journal of Finance, 51*(3), 783–810. doi:10.2307/2329222

Holland, J. H. (1975). *Adaptation in Natural and Artificial Systems.* University of Michigan Press.

LeBaron, B. (2000). Agent-based Computational Finance: Suggested Readings and Early Research. *Journal of Economics & Control, 24,* 679–702. doi:10.1016/S0165-1889(99)00022-6

LeBaron, B., Arthur, W. B., & Palmer, R. (1999). Time Series Properties of an Artificial Stock Market. *Journal of Economics & Control, 23,* 1487–1516. doi:10.1016/S0165-1889(98)00081-5

Lin, C. C., & Liu, Y. T. (2008). Genetic Algorithms for Portfolio Selection Problems with Minimum Transaction Lots. *European Journal of Operational Research, 185*(1), 393–404. doi:10.1016/j.ejor.2006.12.024

Malkiel, B. (1995). Returns from Investing in Equity Mutual Funds 1971 to 1991. *The Journal of Finance, 50,* 549–572. doi:10.2307/2329419

Markowitz, H. (1952). Portfolio Selection. *The Journal of Finance, 7,* 77–91. doi:10.2307/2975974

Markowitz, H. (1987). *Mean-Variance Analysis in Portfolio Choice and Capital Market.* New York: Basil Blackwell.

Oh, K. J., Kim, T. Y., & Min, S. (2005). Using Genetic Algorithm to Support Portfolio Optimization for Index Fund Management. *Expert Systems with Applications, 28,* 371–379. doi:10.1016/j.eswa.2004.10.014

Orito, Y., Takeda, M., & Yamamoto, H. (2009). Index Fund Optimization Using Genetic Algorithm and Scatter Diagram Based on Coefficients of Determination. *Studies in Computational Intelligence: Intelligent and Evolutionary Systems*, *187*, 1–11.

Orito, Y., & Yamamoto, H. (2007). Index Fund Optimization Using a Genetic Algorithm and a Heuristic Local Search Algorithm on Scatter Diagrams. In *Proceedings of 2007 IEEE Congress on Evolutionary Computation* (pp. 2562-2568).

Streichert, F., & Tanaka-Yamawaki, M. (2006). The Effect of Local Search on the Constrained Portfolio Selection Problem. In *Proceedings of 2006 IEEE Congress on Evolutionary Computation* (pp. 2368-2374).

Xia, Y., Liu, B., Wang, S., & Lai, K. K. (2000). A Model for Portfolio Selection with Order of Expected Returns. *Computers & Operations Research*, *27*, 409–422. doi:10.1016/S0305-0548(99)00059-3

KEY TERMS AND DEFINITIONS

Portfolio Optimization: A combinatorial optimization problem that determines proportion-weighted combination in a portfolio in order to achieve an investment objective.

Information Ratio: A well-known measure of performance of actively managed portfolios.

Agent Property: An agent has one portfolio, its Information Ratio and a character as a set of properties.

Leader Agent: An agent whose Information Ratio is the highest of all agents in search space.

Follower Agent: An agent is categorized into any of three groups, namely obedient group, disobedient group, and indifferent group. An obedient agent is an agent that imitates a part of the leader's portfolio. A disobedient agent is an agent that does not imitate a part of the leader's portfolio. An independent agent is an agent whose behavior is not influenced by the actions of the leader agent.

Search Space Splitting: Re-composition of populations. One population consists of one leader agent and several follower agents. As agents' properties evolve, a new leader appears, and then all populations in a search space are re-composed.

Section 2
Neuro-Inspired Agents

Chapter 3
Neuroeconomics:
A Viewpoint from Agent–Based Computational Economics

Shu-Heng Chen
National Chengchi University, Taiwan

Shu G. Wang
National Chengchi University, Taiwan

ABSTRACT

Recently, the relation between neuroeconomics and agent-based computational economics (ACE) has become an issue concerning the agent-based economics community. Neuroeconomics can interest agent-based economists when they are inquiring for the foundation or the principle of the software-agent design, normally known as agent engineering. It has been shown in many studies that the design of software agents is non-trivial and can determine what will emerge from the bottom. Therefore, it has been quested for rather a period regarding whether we can sensibly design these software agents, including both the choice of software agent models, such as reinforcement learning, and the parameter setting associated with the chosen model, such as risk attitude. In this chapter, we shall start a formal inquiry by focusing on examining the models and parameters used to build software agents.

NEUROECONOMICS: AN ACE VIEWPOINT

From the perspective of agent-based computational economics (ACE), our interest in neuroeconomics is different from that of general psychologists and neural scientists. Agent-based computational economics advocates a *bottom-up research paradigm* for economics. This paradigm does not treat micro and macro as two separate entities and work with each of them separately; instead, it studies the relationship between the two in a coherent framework. Therefore, given the bottom-up manner, we pay more attention to the micro details, and always start the modeling at the level of agents. This *methodological individualism* drives us to incorporate the psychological, cognitive, and neural attributes of human beings into the study of economics. What causes ACE to differ from these behavioral sciences is the scope of the research questions; therefore, while ACE cares about the fundamental cause (the

DOI: 10.4018/978-1-60566-898-7.ch003

neural cause) of the cognitive biases, it is more concerned with the implications of these cognitive biases for any possible emergent mesoscopic or macroscopic phenomena. Furthermore, ACE researchers do not regard the behavioral factors as given (exogenous); they also study the feedback from the aggregate level (social outcome) to the bottom level (individual behavior). [1]

Given what has been said above, we believe that unless neuroeconomics can provide some important lessons for agent-based computational economists, its significance may hardly go far beyond neural science, and would not draw much attention from economists. This, therefore, motivates us to ask: *Does neuroeconomics provide some important lessons for agent-based economic modeling?* It is this question that this chapter would like to address.

In the following, we will review the recent progresses in neuroeconomics in light of its contributions to different aspects of agent engineering. We start from the most fundamental part of agents, i.e., preferences (Section 2), which points to two foundational issues in economics, namely, the *measurement* or *representation* of preference and the *formation* of preference. Some recent advances in the study of these two issues may lead to new insights in the future of agent engineering with regard to *preference development*. We then move to the immediate issue after preferences, i.e., *choices*, or, more precisely, value-based choices (Section 3), and further specify the *intertemporal choice* (Section 3.1), where we can see how the *discount rate* should be more carefully designed. We then focus more on two behavioral aspects pertaining to the design of financial agents, namely, *risk perception* (Section 3.2.1) and *risk preference* (Section 3.2.2). The neural mechanism regarding learning or adaptation is given in Section 4. Finally, the chapter ends with a final remark that connects the relationships among behavioral economics, neural economics and agent-based economics, which is a continuation of the points made earlier (Chen, 2008).

PREFERENCE

"The nature of wealth and value is explained by the consideration of an infinitely small amount of pleasure and pain, just as the theory of statics is made to rest upon the equality of indefinitely small amounts of energy. (Jevons, 1879, p. 44; Italics, added)"

Standard economic theory takes individual preferences as given and fixed over the course of the individual's lifetime. It would be hard to imagine how economic models can stand still by giving up preferences or utility functions. They serve as the very foundation of economics just as we quoted above from William Stanley Jevons (1835-1882). Without preference or utility, it will no longer be clear what we mean by welfare, and hence we make welfare-enhancing policy ill-defined. Nevertheless, preference is now in a troubling moment in the development of economics. Even though its existence has been questioned, the development of neuroeconomics may further deepen this turbulent situation.

The Brain as a Multi-Agent System

The recent progress in neural science provides economists with some foundational issues of economic theory. Some of its findings may lend support to many heated discussions which are unfortunately neglected by mainstream economics. The most important series of questions is that pertaining to *preference*. While its existence, formalization (construction), measurement, consistency and stability has long been discussed outside mainstream economics, particularly in the realm of behavioral economics, neuroeconomics provides us with solid ground to tackle these issues.[2]

To see how neuroscience can inform economists, it is important to perceive that the *brain is a multi-agent system*. For example, consider the *Triune Brain Model* proposed by Maclean (1990).

The brain is composed of three major parts: the reptilian brain (the brainstem), the mammalian brain (the limbic system), and the hominid brain (the cerebral cortex). Each of the three is associated with different cognitive functions, while receiving and processing different signals. The three parts also have various interactions (competition or cooperation) with the embedded network. The three "agents'" and their interactions, therefore, constitute the very basis of this multi-agent system.

This multi-agent system (MAS) approach to the brain compels us to think hard on what would be a *neural representation of preference*. Preference is unlikely to be represented by a signal neuron or a single part of the brain, but by an emergent phenomenon from the interactions of many agents. Hence, many agents of the brain can contribute to part of the representation. So, when asked what the preference for commodity *A* is and its relative comparison for *B*, many agents of the brain work together either in a synchronous or asynchronous manner to generate a representation, the utility of *A* and *B*, say *U(A)* and *U(B)*.

During the process, some agents retrieve the past experiences (memory) of consuming *A* and *B*, and some agents aggregate this information. These processes can be collaborative or competitive; it is likely that some agents inhibit other agents to function. As a result, the memory can be partial, which, depending on the external elicitation and other conditions, can vary from time to time.

This rough but simple picture of the multi-agent neurodynamics may indicate why a steady preference conventionally assumed in economics may not be there. The alternative is that people do not have given unchanging preferences, but rather their preferences are constructed to fit the situations they face. Herbert Simon is one of the precursors of the idea of preference construction (Simon, 1955, 1956).

Preference Construction

"On the contrary, we approach choice within specific, quite *narrow frames of reference* that continually shift with the circumstances in which we find ourselves and with the thoughts that are evoked in our minds by these particular circumstances. Thus, in any given choice situation, we evoke and make use only a small part even of the limited information, knowledge and reasoning skills that we have stored in our memory, and these memory contents, even if fully evoked, would give us only a pale and highly inexact picture of the world in which we live." (Simon, 2005, p. 93, *Italics* added)

The MAS approach to the study of the brain may connect us to the literature on preference construction for real human beings (Fischhoff, 1991; Slovic, 1995; Lichtenstein and Slovic, 2006), and, in particular, the role of *experiences* and *imagination* in preference formation. In the following, we would like to exemplify a few psychological studies which shed light on the *experience-based* or *imagination-based* preferences.

Adaptive Decision Makers (Payne, Bettman, and Johnson, 1993) The *effort-accuracy* framework proposed by Payne, Bettman, and Johnson (1993) represents an attempt to shift the research agenda from demonstrations of irrationality in the form of heuristics and biases to an understanding of the causal mechanisms underlying the behavior. It has considerable merit as a model of how decision makers cope with cognitive limitations. The *adaptive decision maker* is a person whose repertoire of strategies may depend upon many factors, such as cognitive development, experience, and more formal training and education. Payne, Bettman, and Johnson (1993) suggest that decision-making behavior is a *highly contingent form* of information processing and is highly sensitive to task factors and context factors. They consider that the cognitive effort required to make a decision can be usefully measured in terms of the total number of basic information processes needed to solve a particular problem using a specific decision strategy. In addition, they state that individual differences in decision behavior may be related to differences in how much effort the various elementary information processes the individuals are required to make.

Hedonic Psychology Hedonic psychology is the study of what makes experiences and life pleasant or unpleasant (Kahneman, Diener, and Schwarz, 2003). It is concerned with feelings of pleasure and pain, of interest and boredom, of joy and sorrow, and of satisfaction and dissatisfaction. All decisions involve *predictions of future tastes or feelings*. Getting married involves a prediction of one's long-term feelings towards one's spouse; returning to school for an advanced degree involves predictions about how it will feel to be a student as well as predictions of long-term career preferences; buying a car involves a prediction of how it would feel to drive around in different cars. In each of these examples, the quality of the decision depends critically on the accuracy of the prediction; errors in predicting feelings are measured in units of divorce, dropout, career burnout and consumer dissatisfaction (Loewenstein and Schkade, 2003).

Empathy Gaps People are often incorrect about what determines happiness, leading to prediction errors. In particular, the well-known *empathy gaps*, i.e., the inability to imagine opposite feelings when experiencing heightened emotion, be it happy or sad, lead to errors in predicting both feelings and behavior (Loewenstein, 2005). So, people seem to think that if disaster strikes it will take longer to recover emotionally than it actually does. Conversely, if a happy event occurs, people overestimate how long they will emotionally benefit from it.

Psychological Immune System The cognitive bias above also indicates that agents may underestimate the proper function of their psychological immune systems. The psychological immune system is a system which helps fight off bad feelings that result from unpleasant situations

(Kagan, 2006). This system is activated when humans are faced with potential or actual negative events in their life. The system functions to assist in protecting humans from extreme reactions to those negative events. Sharot, De Martino and Dolan (2008) studied how hedonic psychology affects our choices from a neural perspective. They combined participants' estimations of the pleasure they will derive from future events with fMRI data recorded *while they imagined those events*, both before, and after making choices. It was found that activity in the *caudate nucleus* predicted the choice agents made when forced to choose between two alternatives they had previously rated equally. Moreover, post choice the selected alternatives were valued more strongly than pre-choice, while discarded ones were valued less. This *post-choice preference change* was mirrored in the caudate nucleus response. The choice-sensitive preference observed above is similar to behavior driven by reinforcement learning.

VALUE AND CHOICE

"Neuroeconomics is a relatively new discipline that studies the computations that the brain carries out in order to make value-based decisions, as well as the neural implementation of those computations. It seeks to build a biologically sound theory of how humans make decisions that can be applied in both the natural and the social sciences." (Rangel, Camerer, and Montague, 2008)

"In a choice situation, we usually look at a few alternatives, sometimes including a small number that we generate for the purpose but more often limiting ourselves to those that are already known and available. These alternatives are generated or evoked in response to specific goals or drives (i.e. specific components of the utility function), so that different alternatives are generated when we are hungry from when we are thirsty; when we are thinking about our science from when we are thinking about our children." (Simon, 2005, p. 93)

The very basic economics starts with value assignment and choice making. However, traditional economics makes little effort to understand the cognitive and computation loading involved in this very fundamental economic activity. A number of recent studies have challenged the view that what we used to be taught may be misplaced when we take into account the value-assignment problem more seriously (Iyengar and Lepper, 2000; Schwartz, 2003). These studies lead us to question the impact of the dimensionality of choice space upon our behavior of value assignment and choice making. It seems that when the number of choices increases, the ability to make the best choice becomes problematic.

Going one step further, Louie, Grattan, and Glimcher(2008) attempt to theorize this paradox of choice by exploring the neural mechanism underlying value representation during decision-making and how such a mechanism influences choice behavior in the presence of alternative options. In their analysis, value assignment is relatively normalized when new alternatives are presented. The linear proportionate normalization is a simple example. Because value is relatively coded rather than absolutely coded, the value differences between two alternatives may become narrow when more alternatives are presented.

Intertemporal Choice

Agent-based economic models are dynamic. Time is an inevitable element, and the *time preference* becomes another important setting for agents in the agent-based models. However, in mainstream economic theory, the time preference has been largely standardized as an exponential discounting with a time-invariant discount rate. However, recent studies have found that people discount future outcomes more steeply when they have the opportunity for immediate gratification than when all outcomes occur in the future. This has led to the modification of the declining discount rates or *hyperbolic-discounting* (Laibson, 1997).

Frederick, Loewenstein, and O'Donoghue (2002) provided an extensive survey on the empirical studies showing that the observed discount rates are not constant over time, but appear to decline.

Loewenstein (1988) has further demonstrated that discount rates can be dramatically affected by whether the change in delivery time of an outcome is framed as an *acceleration* or a *delay* from some temporal reference point. So, when asked whether they would be willing to wait for a month to receive $110 instead of receiving $100 today, most people choose $100 today. By contrast, when asked whether they would prefer to speed up the receipt of $110 in a month by receiving $100 today instead, most people exhibit patience and take the $110 in a month. This phenomenon has been used as evidence for the *gain-loss asymmetry* or the *prospect theory*. It has also been connected to the *endowment effect*, which predicts that people tend to value objects more highly after they come to feel that they own them (Kahneman, Knetsch and Thaler, 1990; Kahneman, 1991). The endowment effect explains the reluctance of people to part with assets that belong to their endowment. Nonetheless, Lerner, Small and Loewenstein (2004) show that the agents' *mood*, sad or neutral, can affect the appearance of this effect.

Query Theory Recently, query theory, proposed by Johnson, Haeubl and Keinan (2007), has been used to explain this and other similar choice inconsistencies. Query theory assumes that preferences, like all knowledge, are subject to the processes and dynamics of memory encoding and retrieval, and explores whether *memory and attentional processes* can explain observed anomalies in evaluation and choice. Weber et al. (2007) showed that the directional asymmetry in discounting is caused by the different order in which memory is queried for reasons favoring immediate versus future consumption, with earlier queries resulting in a richer set of responses, and reasons favoring immediate consumption being generated earlier for delay vs. acceleration decisions.

Neural Representation of Hyperbolic Discounting McClure et al. (2004) investigate the neural systems that underlie discounting the value of rewards based on the delay until the time of delivery. They test the theory that hyperbolic discounting results from the combined function of *two separate brain systems*}. The *beta system* is hypothesized to place special weight on immediate outcomes, while the *delta system* is hypothesized to exert a more consistent weighting across time. They further hypothesize that *beta* is mediated by limbic structures and *delta* by the lateral prefrontal cortex and associated structures supporting higher cognitive functions. Extending McClure et al. (2004), Finger et al. (2008) conducted an fMRI study investigating participants' neural activation underlying acceleration vs. delay decisions. They found hyperbolic discounting only in the delay, but not the acceleration, function.

Risk

Risk preference plays an important role in many agent-based economic models, in particular agent-based financial models. The frequently used assumptions are CARA (Constant Absolute Risk Aversion), CRRA (Constant Relative Risk Aversion), HARA (Hyperbolic Absolute Risk Aversion), and mean-variance, but, so far, few have ever justified the use of any of these with a neural foundation. This question can be particularly hard because, with the recent development of neuroscience, we are inevitably pushed to ask a deeper question: what the risk is. How does the agent recognize the risk involved in his or her decision making? What may cause the perceived risk to deviate from the real risk? Is there any particular region in our brain which corresponds to a different order of *moments*, the statistics used to summarize the probabilistic uncertainty?

Neural Representation of Risk

One of the main issues currently discussed in neuroeconomics is the neural representation of risk.

Through a large variety of risk experiments, it can be shown that many different parts of the brain are involved in decisions under risk, and they vary with experimental designs. Based on the activated areas of the brain, one may define a neural representation of the risk associated with a given experiment. Different kinds of risks may be differentiated by their different neural representations, and different risk-related concepts may also be distinguished in this way. For example, the famous Knight's distinction between uncertainty and risk can now be, through delicate experimental designs, actually distinguished from their associated neural representations. Using the famous Iowa Gambling Task, Lin et al. (2008) show that uncertainty is represented by the brain areas closely pertaining to emotion, whereas risk is associated with the prefrontal cortex. In this vein, Pushkarskaya et al. (2008) distinguishes ambiguity from conflicts, and Mohr et al. (2008) separate behavioral risk from reward risk.

Identifying the neural representations of different risks may also shed light on the observed deviations of human behavior based on probability-based predictions. For example, a number of experiments, such as Feldman's Experiment (Feldman, 1962) or the Iowa Gambling Task (Lin, 2008), have indicated that even though subjects are given a risk environment, they may still behave as if they are in a uncertain environment. It is left for further study as to what are the neural processes behind this pattern recognition test which may inhibit or enhance the discovery of the underlying well-defined probabilistic environment.

Risk Preference

Different assumptions of risk preference, such as the mean-variance, CARA, CRRA, or HARA, are used in economic theory, usually in an arbitrary way. While agent-based modeling relies heavily on the idea of heterogeneity, preference or risk preference in most studies is normally assumed to be homogeneous. Little has been explored on the aggregate dynamics generated by a society of agents with heterogeneous risk preference.[3] Nevertheless, it seems to be quite normal to see agents with heterogeneous risk preferences in neuroeconomic experiments (Paulsen et al., 2008).

Genetics have contributed in accounting for the difference in risk preference. Kuhnen and Chiao (2008) showed that several genes previously linked to emotional behavior and addiction are also found to be correlated with risk-taking investment decisions. They found that 5HTLPR ss allele carriers are more risk averse than those carrying the sl or ll alleles of the gene. D4DR 7-repeat allele carriers are more risk seeking than individuals without the 7-repeat allele. Individuals with the D2DR A1/A1 genotype have more stable risk preferences than those with the A1/A2 or A2/A2 genotype, while those with D4DR 4-repeat allele have less stable preferences than people who do not have the 4-repeat allele.

One of the essential developments in neuroeconomics is to provide neural foundations of the risk preferences. It is assumed that the human brain actually follows the finance approach, encoding the various statistical inputs needed for the effective evaluation of the desirability of risky gambles. In particular, neurons in parts of the brain respond immediately (with minimal delay) to changes in expected rewards and with a short delay (about 1 to 2 seconds) to risk, as measured by the payoff variance (Preuschoff, Bossaerts and Quartz, 2006). Whether one can find evidence of higher-order risk (skewness aversion, for instance) remains an interesting issue.

Some initial studies indicate that risk preference may be *context-dependent* or *event-driven*, which, to some extent, can be triggered by how the risky environment is presented. d'Acremont and Bossaerts (2008) show that the dominance of mean-variance preference over the expected utility depends on the number of states. When the number of states increases, it is more likely that the mean-variance preference may fit the data better than the expected utility.

LEARNING AND THE DRPE HYPOTHESIS

One essential element of agent-based computational economics is the notion of *autonomous agents*, i.e, the agents who are able to learn and adapt on their own. It would have been a big surprise to us if neuroscience had not cared about learning. However, it will also be a surprise to us if the learning algorithms which we commonly use for the software agents can actually have their neural representations. Nonetheless, a few recent studies have pointed in this direction.

Studies start with how the brain encodes the prediction error, and how other neural modules react to these errors. The most famous hypothesis in this area is the *Dopaminergic reward prediction error* (DRPE) *hypothesis*. This hypothesis states that neurons that contain the neurotransmitter release dopamine in proportion to the difference between the *predicted reward* and the *experienced reward* of a particular event. Recent theoretical and experimental work on dopamine release has focused on the role that this neurotransmitter plays in learning and the resulting choice behavior. Neuroscientists have hypothesized that the role of dopamine is to update the *value* that humans and animals attach to different actions and stimuli, which in turn affects the probability that such an action will be chosen. If true, this theory suggests that a deeper understanding of dopamine will expand economists' understanding of how beliefs and preferences are formed, how they evolve, and how they play out in the act of choice.

Caplin and Dean (2008) formulate the DRPE hypothesis in axiomatic terms. Their treatment has precisely the *revealed preference* characteristic of identifying any possible reward function directly from the observables. They discuss the potential for measured dopamine release to provide insight into belief formation in repeated games and to learning theory, e.g., *reinforcement learning*. Their axiomatic model specifies three easily testable conditions for the *entire class of reward prediction error (RPE) models*. Briefly, the axioms will be satisfied if activity is (1) increase wit prize magnitude (2) decreasing with lottery expected value and (3) equivalent for outcomes from all lotteries with a single possible outcome. These three conditions are both necessary and sufficient for any RPE signal. If they hold, there is a way of defining experienced and predicted reward such that the signal encodes RPE with respect to those definitions. Rutledge et al. (2008) used the BOLD responses at the outcome time to test whether activity in the nucleus accumbens satisfies the axioms of the RPE model.

Klucharev et al. (2008) show that a deviation from the group opinion is detected by neural activity in the rostral cingular zone (RCZ) and ventral striatum. These regions produce a neural signal similar to the prediction error signal in reinforcement learning that indicates a need for social conformity: a strong conflict-related signal in the RCZ and NAc trigger adjustment of judgments in line with group opinion. Using an olfactory categorization task performed by rats, Kepecs, Uchida, and Mainen (2008) attempt to obtain evidence for quantitative measurements of learning increments and test the hypothesis implied by the reinforcement learning, i.e., one should learn more when uncertain and less when certain.

Studies also try to find the neural representation of different learning algorithms. The commonly used reinforcement learning and Bayesian learning is compared in Bossaerts et al. (2008) where they address the existence of the dual system.[4] They consider the *reflective system* and the *reflexive system* as the neural representation of Bayesian learning and reinforcement learning, respectively. Using the trust game, they were able to stratify subjects into two groups. One group used well-

adapted strategies. EEG recordings revealed activation of a reflective (conflict-resolution) system, evidently to inhibit impulsive emotional reactions after disappointing outcomes. Pearson et al. (2008) initiated another interesting line of research, i.e., the neural representations which distinguish *exploration* from *exploitation*, the two fundamental search strategies frequently used in various intelligent algorithms, say, genetic algorithms.

DUAL SYSTEM CONJECTURE

The dual system conjecture generally refers to the hypothesis that human thinking and decision-making are governed by two different but interacting systems. This conjecture has been increasingly recognized as being influential in psychology (Kahneman, Diener, and Schwarz, 2003), neural science (McClure, 2004), and economics. The two systems are an *affective system* and a deliberative system (Loewenstein and O'Donoghue, 2005) or a *reflexive system* and a *reflective system* (Lieberman, 2003). The affective system is considered to be myopic, activated by environmental stimuli, and primarily driven by affective states. The deliberative system is generally described as being goal-oriented and forward-looking. The former is associated with the areas of the brain that we have labeled the ventral striatum (nucleus accumbens, ventral caudate, and ventral putamen), the right striatum, neostriatum and amygdala, among others, whereas the latter is associated with the areas of the brain that we have labeled the ventromedial and dorsolateral prefrontal and anterior cingulate, among others.

The dual system of the brain has become the neuroeconomic area which economic theorists take the most seriously. This has also helped with the formation of the new field known as *neuroeconomic theory*. A number of dual-process models have been proposed in economics with applications to *intertemporal choice* (Loewenstein and O'Donoghue, 2005; Fudenberg and Levin, 2006; Brocas and Carrillo, 2008), *risk preferences* (Loewenstein and O'Donoghue, 2005), and *social preferences* (Loewenstein and O'Donoghue, 2005). All these models view economic behavior as being determined by the interaction between two different systems.

The application of the dual system conjecture to learning is just the beginning. Earlier, we have mentioned the cognitive loading between different learning algorithms, such as reinforcement learning vs. Bayesian learning (see Section 4). This issue has been recently discussed in experimental economics (Charness and Levin, 2005), and now also in neuroeconomics (Bossaerts et al.,2008).

Software Agents with Neurocognitive Dual Systems

While agents with dual systems have been considered to be a new research direction in neuroeconomic theory (Brocas and Carrillo, 2008a, Brocas and Carrillo, 2008b), software agents or autonomous agents in agent-based modeling mostly follow a single system. However, the dual system interpretation exists for many agent-based economic models. Consider the fundamentalist-chartist model as an example, where the fundamentalist's and chartist's behavior can be differentiated by the associated neural systems, say, assuming the former is associated with a deliberative system while the latter is associated with the affective system.

Another example is the *individual learning* vs. *social learning*. These two learning schemes have been frequently applied to model the learning behavior in experiments and their fit to the experimental data are different (Hanaki, 2005). Agent-based simulation has also shown that their emergent patterns are different. For example, in the context of an artificial stock market, Yeh and Chen (2001) show that agents using individual learning behave differently from agents using social learning in terms of market efficiency, price dynamics

and trading volume. If individual learning can be associated with, say, the deliberative system, and social learning can be connected to the affective system, then the dual system can also be applied to agent-based modeling. This issue opens the future to collaboration between agent-based economics and neuroeconomics.

FROM MODULAR MIND/BRAIN TO MODULAR PREFERENCE

At present, modularity (Simon, 1965) is still not a part of agent-based economic modeling. This absence is a little disappointing since ACE is regarded as a complement to mainstream economics in terms of articulating the mechanism of evolution and automatic discovery. One way of making progress is to enable autonomous agents to discover the modular structure of their surroundings, and hence they can adapt by using modules. This is almost equivalent to causing their "brain" or "mind" to be designed in a modular way as well.

The only available work in agent-based economic modeling which incorporates the idea of modularity is that related to the agent-based models of innovation initiated by Chen and Chie (2004). They proposed a *modular economy* whose demand side and supply side both have a decomposable structure. While the decomposability of the supply side, i.e., production, has already received intensive treatment in the literature, the demand side has not. Inspired by the study of *neurocognitive modularity*, Chen and Chie (2004) assume that the preference of *consumers can be decomposable*.[5] In this way, the demand side of the modular economy corresponds to a market composed of a set of consumers with *modular preference*.

In the modular economy, the assumption of modular preference is made in the form of a dual relationship with the assumption of modular production. Nevertheless, whether in reality the two can have a nice mapping, e.g., a one-to-one relationship, is an issue related to the distinction between *structural modularity* and *functional modularity*. While in the literature this distinction has been well noticed and discussed, "recent progress in developmental genetics has led to remarkable insights into the molecular mechanisms of morphogenesis, but has at the same time blurred the clear distinction between structure and function." (Callebaut and Rasskin-Gutman, 2005, p. 10)

The modular economy considered by Chen and Chie (2004) does not distinguish between the two kinds of modularity, and they are assumed to be identical. One may argue that the notion of modularity that is suitable for preference is structural, i.e., *what it is*, whereas the one that is suitable for production is process, i.e., *what is does*. However, this understanding may be partial. Using the LISP (List Programming) parse-tree representation, Chen and Chie (2004) have actually integrated the two kinds of modularity. Therefore, consider drinking coffee with sugar as an example. Coffee and sugar are modules for both production and consumption. Nevertheless, for the former, producers add sugar to coffee to deliver the final product, whereas for the latter, the consumers drink the mixture knowing of the existence of both components or by "seeing" the development of the product.

Chen and Chie (2007) tested the idea of augmented genetic programming (augmented with automatically defined terminals) in a modular economy. Chen and Chie (2007)considered an economy with two oligopolistic firms. While both of these firms are autonomous, they are designed differently. One firm is designed with simple GP (SGP), whereas the other firm is designed with augmented GP (AGP). These two different designs match the two watchmakers considered by Simon (1965). The modular preferences of consumers not only define the search space for firms, but also a search space with different hierarchies. While it is easier to meet consumers' needs with very low-end products, the resulting profits are negligible. To gain higher profits, firms have to satisfy consum-

ers up to higher hierarchies. However, consumers become more and more heterogeneous when their preferences are compared at higher and higher hierarchies, which calls for a greater diversity of products.[6] It can then be shown that the firm using a modular design performs better than the firm not using a modular design, as Simon predicted.

CONCLUDING REMARKS: AGENT BASED OR BRAIN BASED?

Can we relate agent-based economics to brain-based economics (neuroeconomics)? Can we use the knowledge which we obtain from neuroeconomics to design software agents? One of the features of agent-based economics is the emphasis on the *heterogeneity* of agents. This heterogeneity may come from behavioral genetics. Research has shown that genetics has an effect on our risk preference. Kuhnen and Chiao (2008), Jamison et al. (2008), and Weber et al. (2008) show that preferences are affected by the genes and/or education (environment). With the knowledge of genetics and neuroeconomics, the question is: How much more heterogeneity do we want to include in agent-based modeling? Does it really matter?

Heterogeneity may also result from age. The neuroeconomics evidence shows that certain functions of the brain will age. The consequence is that elderly people will make some systematic errors more often than young people, and, age will affect financial decisions as well (Samanez Larkin, Kuhnen, and Knutson, 2008). Thus the same question arises: when engaging in agent-based modeling, should we take age heterogeneity into account? So, when a society ages, should we constantly adjust our agent-based model so that it can match the empirical age distribution of the society? So far we have not seen any agent-based modeling that features the aspect of aging.

Neuroeconomics does encourage the modular design of agents, because our brain is a modular structure. Many different modules in the brain have been identified. Some modules are related to emotion, some are related to cognition, and some are related to self-control. When human agents are presented with different experimental settings, we often see different combinations of these modules.

ACKNOWLEDGMENT

The author is grateful for the financial support provided by the NCCU Top University Program.

NSC research grant No. 95-2415-H-004-002-MY3 is also gratefully acknowledged.

REFERENCES

Baldassarre, G. (2007, June). *Research on brain and behaviour, and agent-based modelling, will deeply impact investigations on well-being (and theoretical economics)*. Paper presented at International Conference on Policies for Happiness, Certosa di Pontignano, Siena, Italy.

Bossaerts, P., Beierholm, U., Anen, C., Tzieropoulos, H., Quartz, S., de Peralta, R., & Gonzalez, S. (2008, September). *Neurobiological foundations for "dual system" theory in decision making under uncertainty: fMRI and EEG evidence*. Paper presented at Annual Conference on Neuroeconomics, Park City, Utah.

Brocas, I., & Carrillo, J. (2008a). The brain as a hierarchical organization. *The American Economic Review*, *98*(4), 1312–1346. doi:10.1257/aer.98.4.1312

Brocas, I., & Carrillo, J. (2008b). Theories of the mind. *American Economic Review: Papers\& Proceedings*, *98*(2), 175-180.

Callebaut, W., & Rasskin-Gutman, D. (Eds.). (2005). *Modularity: Understanding the development and evolution of natural complex systems*. MA: MIT Press.

Caplin, A., & Dean, M. (2008). Economic insights from ``neuroeconomic'' data. *The American Economic Review*, *98*(2), 169–174. doi:10.1257/aer.98.2.169

Charness, G., & Levin, D. (2005). When optimal choices feel wrong: A laboratory study of Bayesian updating, complexity, and affect. *The American Economic Review*, *95*(4), 1300–1309. doi:10.1257/0002828054825583

Chen, S.-H. (2008). Software-agent designs in economics: An interdisciplinary framework. *IEEE Computational Intelligence Magazine*, *3*(4), 18–22. doi:10.1109/MCI.2008.929844

Chen, S.-H., & Chie, B.-T. (2004). Agent-based economic modeling of the evolution of technology: The relevance of functional modularity and genetic programming. *International Journal of Modern Physics B*, *18*(17-19), 2376–2386. doi:10.1142/S0217979204025403

Chen, S.-H., & Chie, B.-T. (2007). Modularity, product innovation, and consumer satisfaction: An agent-based approach . In Yin, H., Tino, P., Corchado, E., Byrne, W., & Yao, X. (Eds.), *Intelligent Data Engineering and Automated Learning* (pp. 1053–1062). Heidelberg, Germany: Springer. doi:10.1007/978-3-540-77226-2_105

Chen, S.-H., & Huang, Y.-C. (2008). Risk preference, forecasting accuracy and survival dynamics: Simulations based on a multi-asset agent-based artificial stock market. *Journal of Economic Behavior & Organization*, *67*(3), 702–717. doi:10.1016/j.jebo.2006.11.006

d'Acremont, M., & Bossaerts, P. (2008, September). *Grasping the fundamental difference between expected utility and mean-variance theories*. Paper presented at Annual Conference on Neuroeconomics, Park City, Utah.

Feldman, J. (1962). Computer simulation of cognitive processes . In Broko, H. (Ed.), *Computer applications in the behavioral sciences*. Upper Saddle River, NJ: Prentice Hall.

Figner, B., Johnson, E., Lai, G., Krosch, A., Steffener, J., & Weber, E. (2008, September). *Asymmetries in intertemporal discounting: Neural systems and the directional evaluation of immediate vs future rewards*. Paper presented at Annual Conference on Neuroeconomics, Park City, Utah.

Fischhoff, B. (1991). Value elicitation: Is there anything in there? *The American Psychologist, 46*, 835–847. doi:10.1037/0003-066X.46.8.835

Frederick, S., Loewenstein, G., & O'Donoghue, T. (2002). Time discounting and time preference: A critical review. *Journal of Economic Literature, XL*, 351–401. doi:10.1257/002205102320161311

Fudenberg, D., & Levine, D. (2006). A dual-self model of impulse control. *The American Economic Review, 96*(5), 1449–1476. doi:10.1257/aer.96.5.1449

Hanaki, N. (2005). Individual and social learning. *Computational Economics, 26*, 213–232. doi:10.1007/s10614-005-9003-5

Iyengar, S., & Lepper, M. (2000). When choice is demotivating: Can one desire too much of a good thing? *Journal of Personality and Social Psychology, 79*(6), 995–1006. doi:10.1037/0022-3514.79.6.995

Jamison, J., Saxton, K., Aungle, P., & Francis, D. (2008). *The development of preferences in rat pups*. Paper presented at Annual Conference on Neuroeconomics, Park City, Utah.

Jevons, W. (1879). *The Theory of Political Economy, 2nd Edtion. Edited and introduced by R. Black (1970)*. Harmondsworth: Penguin.

Johnson, E., Haeubl, G., & Keinan, A. (2007). Aspects of endowment: A query theory account of loss aversion for simple objects. *Journal of Experimental Psychology. Learning, Memory, and Cognition, 33*, 461–474. doi:10.1037/0278-7393.33.3.461

Kagan, H. (2006). *The Psychological Immune System: A New Look at Protection and Survival*. Bloomington, IN: AuthorHouse.

Kahneman, D., Diener, E., & Schwarz, N. (Eds.). (2003). *Well-Being: The Foundations of Hedonic Psychology*. New York, NY: Russell Sage Foundation.

Kahneman, D., Knetsch, J., & Thaler, R. (1990). Experimental tests of the endowment effect and the Coase theorem. *The Journal of Political Economy, 98*, 1325–1348. doi:10.1086/261737

Kahneman, D., Knetsch, J., & Thaler, R. (1991). Anomalies: The endowment effect, loss aversion, and status quo bias. *The Journal of Economic Perspectives, 5*(1), 193–206.

Kahneman, D., Ritov, I., & Schkade, D. (1999). Economic preferences or attitude expressions? An analysis of dollar responses to public issues. *Journal of Risk and Uncertainty, 19*, 203–235. doi:10.1023/A:1007835629236

Kepecs, A., Uchida, N., & Mainen, Z. (2008, September). *How uncertainty boosts learning: Dynamic updating of decision strategies*. Paper presented at Annual Conference on Neuroeconomics, Park City, Utah.

Klucharev, V., Hytonen, K., Rijpkema, M., Smidts, A., & Fernandez, G. (2008, September). *Neural mechanisms of social decisions*. Paper presented at Annual Conference on Neuroeconomics, Park City, Utah.

Kuhnen, C., & Chiao, J. (2008, September). *Genetic determinants of financial risk taking.* Paper presented at Annual Conference on Neuroeconomics, Park City, Utah.

Laibson, D. (1997). Golden eggs and hyperbolic discounting. *The Quarterly Journal of Economics, 12*(2), 443–477. doi:10.1162/003355397555253

Lerner, J., Small, D., & Loewenstein, G. (2004). Heart strings and purse strings: Carry-over effects of emotions on economic transactions. *Psychological Science, 15,* 337–341. doi:10.1111/j.0956-7976.2004.00679.x

Lichtenstein, S., & Slovic, P. (Eds.). (2006). *The Construction of Preference.* Cambridge, UK: Cambridge University Press. doi:10.1017/CBO9780511618031

Lieberman, M. (2003). Reflective and reflexive judgment processes: A social cognitive neuroscience approach . In Forgas, J., Williams, K., & von Hippel, W. (Eds.), *Social Judgments: Explicit and Implicit Processes* (pp. 44–67). New York, NY: Cambridge University Press.

Lin, C.-H., Chiu, Y.-C., Lin, Y.-K., & Hsieh, J.-C. (2008, September). *Brain maps of Soochow Gambling Task.* Paper presented at Annual Conference on Neuroeconomics, Park City, Utah.

Lo, A. (2005). Reconciling efficient markets with behavioral finance: The adaptive market hypothesis. *The Journal of Investment Consulting, 7*(2), 21–44.

Loewenstein, G. (1988). Frames of mind in intertemporal choice. *Management Science, 34,* 200–214. doi:10.1287/mnsc.34.2.200

Loewenstein, G. (2005). Hot-cold empathy gaps and medical decision making. *Health Psychology, 24*(4), S49–S56. doi:10.1037/0278-6133.24.4.S49

Loewenstein, G., & O'Donoghue, T. (2005). Animal spirits: Affective and deliberative processes in economic behavior. Working Paper. Carnegie Mellon University, Pittsburgh.

Loewenstein, G., & Schkade, D. (2003). Wouldn't it be nice?: Predicting future feelings . In Kahneman, D., Diener, E., & Schwartz, N. (Eds.), *Hedonic Psychology: The Foundations of Hedonic Psychology* (pp. 85–105). New York, NY: Russell Sage Foundation.

Louie, K., Grattan, L., & Glimcher, P. (2008). Value-based gain control: Relative reward normalization in parietal cortex. Paper presented at Annual Conference on Neuroeconomics, Park City, Utah.

MacLean, P. (1990). *The Triune Brain in Evolution: Role in Paleocerebral Function.* New York, NY: Plenum Press.

McClure, S., Laibson, D., Loewenstein, G., & Cohen, J. (2004). Separate neural systems value immediate and delayed monetary rewards. *Science, 306,* 503–507. doi:10.1126/science.1100907

Mohr, P., Biele, G., & Heekeren, H. (2008, September). *Distinct neural representations of behavioral risk and reward risk.* Paper presented at Annual Conference on Neuroeconomics, Park City, Utah.

Paulsen, D., Huettel, S., Platt, M., & Brannon, E. (2008, September). *Heterogeneity in risky decision making in 6-to-7-year-old children.* Paper presented at Annual Conference on Neuroeconomics, Park City, Utah.

Payne, J., Bettman, J., & Johnson, E. (1993). *The adaptive decision maker.* New York, NY: Cambridge University Press.

Pearson, J., Hayden, B., Raghavachari, S., & Platt, M. (2008) *Firing rates of neurons in posterior cingulate cortex predict strategy-switching in a k-armed bandit task.* Paper presented at Annual Conference on Neuroeconomics, Park City, Utah.

Preuschoff, K., Bossaerts, P., & Quartz, S. (2006). Neural Differentiation of Expected Reward and Risk in Human Subcortical Structures. *Neuron, 51*(3), 381–390. doi:10.1016/j.neuron.2006.06.024

Pushkarskaya, H., Liu, X., Smithson, M., & Joseph, J. (2008, September). *Neurobiological responses in individuals making choices in uncertain environments*: Ambiguity and conflict. Paper presented at Annual Conference on Neuroeconomics, Park City, Utah.

Rangel, A., Camerer, C., & Montague, R. (2008). A framework for studying the neurobiology of value-based decision making. *Nature Reviews. Neuroscience, 9*, 545–556. doi:10.1038/nrn2357

Rutledge, R., Dean, M., Caplin, A., & Glimcher, P. (2008, September). *A neural representation of reward prediction error identified using an axiomatic model*. Paper presented at Annual Conference on Neuroeconomics, Park City, Utah.

Samanez Larkin, G., Kuhnen, C., & Knutson, B. (2008). *Financial decision making across the adult life span*. Paper presented at Annual Conference on Neuroeconomics, Park City, Utah.

Schwartz, B. (2003). *The Paradox of Choice: Why More Is Less*. New York, NY: Harper Perennial.

Sharot, T., De Martino, B., & Dolan, R. (2008, September) *Choice shapes, and reflects, expected hedonic outcome*. Paper presented at Annual Conference on Neuroeconomics, Park City, Utah.

Simon, H. (1955). A behavioral model of rational choice. *The Quarterly Journal of Economics, 69*, 99–118. doi:10.2307/1884852

Simon, H. (1956). Rational choice and the structure of the environment. *Psychological Review, 63*, 129–138. doi:10.1037/h0042769

Simon, H. (1965). The architecture of complexity. *General Systems, 10*, 63–76.

Simon, H. (2005). Darwinism, altruism and economics. In: K. Dopfer (Ed.), *The Evolutionary Foundations of Economics* (89-104), Cambridge, UK: Cambridge University Press.

Slovic, P. (1995). The construction of preference. *The American Psychologist, 50*, 364–371. doi:10.1037/0003-066X.50.5.364

Weber, B., Schupp, J., Reuter, M., Montag, C., Siegel, N., Dohmen, T., et al. (2008). *Combining panel data and genetics: Proof of principle and first results*. Paper presented at Annual Conference on Neuroeconomics, Park City, Utah.

Weber, E., Johnson, E., Milch, K., Chang, H., Brodscholl, J., & Goldstein, D. (2007). Asymmetric discounting in intertemporal choice: A query-theory account. *Psychological Science, 18*, 516–523. doi:10.1111/j.1467-9280.2007.01932.x

Yeh, C.-H., & Chen, S.-H. (2001). Market diversity and market efficiency: The approach based on genetic programming. *Journal of Artificial Simulation of Adaptive Behavior, 1*(1), 147–165.

ENDNOTES

[1]   See also Baldassarre (2007). While it has a sharp focus on the *economics of happiness*, the idea of building economic agents upon the empirical findings of psychology and neuroscience and placing these agents in an agent-based computational framework is the same as what we argue here. From Baldassarre (2007), the reader may also find a historical development of the *cardinal utility* and *ordinal utility* in economics. It has been a while since economists first considered that utility is a very subjective thing which cannot be measured in a scientific way, so that interpersonal comparison of utility is impossible, which further causes any redistribution policy to lose its ground.

2   It is not clear where preferences come from, i.e., their formation and development process, nor by when in time they come to their steady state and become fixed. Some recent behavioral studies have even asserted that people do not have preferences, in the sense in which that term is used in economic theory (Kahneman, Ritov, and Schkade, 1999).

3   For an exception, see Chen and Huang (2008).

4   See Section 5 for the dual system conjecture.

5   Whether one can build preference modules upon the brain/mind modules is of course an issue deserving further attention.

6   If the consumers' preferences are randomly generated, then it is easy to see this property through the combinatoric mathematics. On the other hand, in the parlance of economics, moving along the hierarchical preferences means traveling through different regimes, from a primitive manufacturing economy to a quality service economy, from the mass production of homogeneous goods to the limited production of massive quantities of heterogeneous customized products.

Chapter 4
Agents in Quantum and Neural Uncertainty

Germano Resconi
Catholic University Brescia, Italy

Boris Kovalerchuk
Central Washington University, USA

ABSTRACT

This chapter models quantum and neural uncertainty using a concept of the Agent–based Uncertainty Theory (AUT). The AUT is based on complex fusion of crisp (non-fuzzy) conflicting judgments of agents. It provides a uniform representation and an operational empirical interpretation for several uncertainty theories such as rough set theory, fuzzy sets theory, evidence theory, and probability theory. The AUT models conflicting evaluations that are fused in the same evaluation context. This agent approach gives also a novel definition of the quantum uncertainty and quantum computations for quantum gates that are realized by unitary transformations of the state. In the AUT approach, unitary matrices are interpreted as logic operations in logic computations. We show that by using permutation operators any type of complex classical logic expression can be generated. With the quantum gate, we introduce classical logic into the quantum domain. This chapter connects the intrinsic irrationality of the quantum system and the non-classical quantum logic with the agents. We argue that AUT can help to find meaning for quantum superposition of non-consistent states. Next, this chapter shows that the neural fusion at the synapse can be modeled by the AUT in the same fashion. The neuron is modeled as an operator that transforms classical logic expressions into many-valued logic expressions. The motivation for such neural network is to provide high flexibility and logic adaptation of the brain model.

INTRODUCTION

We model quantum and neural uncertainty using the Agent–based Uncertainty Theory (AUT) that

DOI: 10.4018/978-1-60566-898-7.ch004

uses complex fusion of crisp conflicting judgments of agents. AUT represents and interprets uniformly several uncertainty theories such as rough set theory, fuzzy sets theory, evidence theory, and probability theory. AUT exploits the fact that agents as independent entities can give conflicting evaluations of the same attribute. It models conflicting evaluations that are fused in the same evaluation context. If only one evaluation is allowed for each statement in each context (world) as in the modal logic then there is no logical uncertainty. The situation that the AUT models is inconsistent (fuzzy) and is very far from the situation that modeled by the traditional logic that assumes consistency. We argue that the AUT by incorporating such inconsistent statements is able to model different types of conflicts and their fusion known in many-valued logics, fuzzy logic, probability theory and other theories.

This chapter shows how the agent approach can be used to give a novel definition of the quantum uncertainty and quantum computations for quantum gates that are realized by unitary transformations of the state. In the AUT approach, unitary matrices are interpreted as logic operations in logic computations. It is shown, that by using permutation operators that are unitary matrixes any type of complex classical logic expression can be generated. The classical logic has well-known difficulties in quantum mechanics. Now with the quantum gate we introduce classical logic into the quantum domain. We connect the intrinsic irrationality of the quantum system and the non-classical quantum logic with the agents. We argue that Agent-based uncertainty theory (AUT) can help to find meaning for quantum superposition of non-consistent states for which one particle can be at the different points in the same time or the same particle can have spin up and down in the same time.

Next, this chapter shows that the neural fusion at the synapse can be modeled by the AUT. Agents in the neural network are represented by logic input values in the neuron itself. In the ordinary neural networks any neuron is a processor that models a Boolean function. We change the point of view and consider a neuron as an operator that transforms classical logic expressions into many-valued logic

expressions or in other words, changes crisp sets into fuzzy sets. This neural network consists of neurons at two layers. At the first one, neurons or agents implement the classical logic operations. At the second layer neurons or nagents (neuron agents) compute the same logic expression with different results. These are many-valued neurons that fuse results provided by different agents at the first layer. They fuse conflicting or inconsistent situations. The network is based on use of the logic of the uncertainty instead of the classical logic. The motivation for such neural network is to provide high flexibility and logic adaptation of the brain model. In this brain model, communication among agents is specified by the fusion process in the neural elaboration.

The probability calculus does not incorporate explicitly the concepts of irrationality or logic conflict of agent's state. It misses structural information at the level of individual objects, but preserves global information at the level of a set of objects. Given a dice the probability theory studies frequencies of the different faces $E=\{e\}$ as independent (elementary) events. This set of elementary events E has *no structure*. It is only required that elements of E are *mutually exclusive* and *complete*, that is no other alternative is possible. The order of its elements is irrelevant to probabilities of each element of E. No irrationality or conflict is allowed in this definition relative to mutual exclusion. The classical probability calculus does not provide a mechanism for modelling uncertainty when agents communicate (collaborates or conflict). Recent work by Halpern (2005) is an important attempt to fill this gap.

This chapter is organized as follows: Sections 2 and 3 provide a summary of the AUT starting from concepts and definitions. Section 4 presents links between quantum mechanics and first order conflicts in the AUT. Section 5 discusses the neural images of the AUT. Section 6 concludes this chapter.

CONCEPTS AND DEFINITIONS

Now we will provide more formal definition of AUT concepts. It is done first for individual agents then for sets of agents. Consider a set of agents $G=\{g_1, g_2,....,g_n\}$. Each agent g_k assigns binary true/false value $v\in\{True, false\}$ to proposition p. To show that v was assigned by the agent g_k we use notation $g_k(p) = v_k$.

Definition. A triple $g = <N, A_R, A_A >$ is called an *agent* g if N is label that is interpreted as agent's name and A_R is as set of truth-evaluation actions and A_A is a set of non- truth-evaluation actions associated with name N. For instance agent g called Professor has a set of truth-evaluation actions A_R such as grading students' answer, while delivering a lecture is in another category A_A.

Definition. An agent g is called a *reasoning agent* if g assigns a truth-value v(p) to any proposition p from a set of propositions S and any logical formula based on S.

The actions of a general agent may or may not include truth-evaluation actions. From a mathematical viewpoint a reasoning agent g serves as a mapping,

g: p → v(p)

While natural agents have both AR, and AA, artificial software agents and robots may have only one of them.

Definition. A set of reasoning agents G is called *totally consistent* for proposition p if any agent g from {g} always provides the same truth value for p.

In other words, all agents in G are in concord or logical coherence for the same proposition. Thus, changing the agent does not change logic value v(p) for a totally consistent set of agents. Here v(p) is *global* for G and has *no local variability* that is independent on the individual agent's evaluation. The classical logic is applicable for such set of consistent (rational) agents.

Definition. A set of reasoning agents G is called *inconsistent for* proposition p if there are two subset of agents G_1, G_2 such that agents from them provides different truth values for p.

Multiple reasons lead to agents' inconsistency. The most general one is the context (or hidden variables) in which each agent is evaluating statement p. Even in a relatively well formalized environment, context is not fully defined. Agent g_1 may evaluate p in context "abc", but agent g_2 in context "ab". Agent (robot) g_2 may have no sensor to obtain signal c or abilities to process and reason correctly with input from that sensor (e.g., color blind agent with a panchromatic camera).

Definition. Let S be a set propositions, $S=\{p_1, p_2, ...,p_n\}$ then set $\neg S = \{\neg p_1, \neg p_2, ..., \neg p_n\}$ is called a *complementary set* of S.

Definition. A set of reasoning agents G is called *S-only-consistent* if agents {g} are consistent only for propositions in $S=\{p_1, p_2, ...,p_n\}$ and are inconsistent in the complimentary set $\neg S$.

The evaluations of p is a vector-function $v(p)=(v_1(p), v_2(p)...,v_n(p))$ for a set of agents G that we will represent as follows:

$$\mu(p \vee q) = \frac{w_1(v_1(p) \vee v_1(q)) + .. + w_N(v_N(p) \vee v_N(q))}{N}$$

(1)

An example of the logic evaluation by five agent is shown below for $p = $"$A>B$"

$$\mathbf{v}(p) = \begin{pmatrix} g_1 & g_2 & \cdots & g_{n-1} & g_n \\ v_1 & v_2 & \cdots & v_{n-1} & v_n \end{pmatrix}$$

Here A > B is true for agents $g_1,g_3,$ and g_5 and it is false for the agents $g_2,$ and g_4.

Kolmogorov's axioms of the probability theory are based on a totally consistent set of agents (for a set of statements) on mutual exclusion of elementary events. It follows from the definitions below for a set of events $E=\{e_1,e_2,...,e_n\}$.

<u>Definition</u>. Set E={$e_1.e_2,...,e_n$} is called a set of *elementary events* (or mutually exclusive events) for predicate El if

$\forall$ e_i e_j $\in$ A El(e_i) $\vee$ El(e_j) =True and
El($e_{i)}$$\wedge$El($e_j$)=False.

In other words, event e_i is an *elementary event* (El(e_i)=true) if for any j, j≠i events e_i and e_j cannot happen simultaneously, that is probability P($e_i \wedge e_j$) = 0 and P($e_i \vee e_j$)= P(e_i)+P(e_j). Property P($e_i \wedge e_j$)=0 is the *mutual exclusion axiom* (ME- axiom).

Let S={$p_1, p_2,..., p_n$} be a set of statements, where p_i=p(e_i)=True if and only if event e_i is an elementary event. In probability theory, p(e_i) is not associated with **any** specific agent. It is assumed to be a global property (applicable to all agents). In other words, statements p(e_i) are totally consistent for all agents.

<u>Definition</u>. A set of agent S = {$g_1.g_2,...,g_n$} is called a *ME-rational* set of agents if

$\forall$ e_i e_j $\in$ A, $\forall$ g_i $\in$ S, p(e_i) $\vee$ p(e_j) =True and
p($e_{i)}$$\wedge$p($e_j$)=False.

In other words, these agents are *totally consistent* or *rational on Mutual Exclusion.* Previously we assumed that each agent assigns value v(p) and we did not model this process explicitly. Now we introduce a set of criteria C={$C_1, C_2,...,C_m$} by which an agent can decide if a proposition p is true or false, i.e., now v(p) = v(p, C_i), which means that p is evaluated by using the criterion C_i.

<u>Definition</u> Given a set of criteria C, agent g is in a *self-conflicting state* if

$\exists$ C_i, C_j (C_i, C_j $\in$ C) & v(p,C_i) $\neq$ v(p,C_j)

In other words, an agent is in a self–conflicting state if two criteria exist such that p is true for one of them and false for another one. With the explicit set of criteria, the logic evaluation function v(p) is not vector-function any more, but it is expanded to be a matrix function as shown below:

$$f(p) = \begin{pmatrix} g_1 & g_2 & g_3 & g_4 & g_5 \\ true & false & true & false & true \end{pmatrix}$$

(2)

For example, four agents using four criteria can produce v(p) as follows:

$$\mathbf{v}(p) = \begin{bmatrix} & g_1 & g_2 & \cdots & g_n \\ C_1 & v_{1,1} & v_{1,2} & \cdots & v_{1,n} \\ C_2 & v_{2,1} & v_{2,2} & \cdots & v_{2,n} \\ \cdots & \cdots & \cdots & \cdots & \cdots \\ C_m & v_{m,1} & v_{m,2} & \cdots & v_{m,n} \end{bmatrix}$$

Note that introduction of a set of criteria C explains self-conflict, but does not remove it. The agent still needs to resolve the ultimate preference contradiction having a goal, say, to buy only one and better car. It also does not resolve conflict among different agents if they need to buy a car jointly.

If agent g can modify criteria in C making them consistent for p then g can resolves self-conflict. The agent can be in a logic conflict state because of inability to understand the complex *context* and to evaluate *criteria*. For example, in the stock market environment, some traders quite often do not understand a complex context and rational criteria of trading. These traders can be in logic conflict exhibiting chaotic, random, and impulsive behavior. They can sell and buy stocks, exhibiting logic conflicting states "sell" = p and "buy" = ¬p in the same market situation that appears as irrational behavior, which means that the statement p $\wedge$ ¬p can be true.

A logical structure of self-conflicting states and agents is much more complex than it is without self-conflict. For m binary criteria C_1, C_2,...C_m that evaluate the logic value for the same attribute, there are 2^m possible states and only two of them (all true or all false values) do not exhibit conflict between criteria.

FRAMEWORK OF FIRST ORDER OF CONFLICT LOGIC STATE

<u>Definition.</u> A set of agents G is in a first order of conflicting logic state (*first order conflict*, for short) if

$$\exists\ g_i,\ g_j\ (g_i,\ g_j \in G)\ \&\ v(p,g_i) \neq v(p,g_j).$$

In other words, there are agents g_i and g_j in G for which exist different values v_i, v_j in

$$\mathbf{v}(p) = \begin{bmatrix} & g_1 & g_2 & g_3 & g_4 \\ C_1 & true & false & false & true \\ C_2 & false & false & true & true \\ C_3 & true & false & true & true \\ C_4 & false & false & true & true \end{bmatrix}$$

A set of agents G is in the *first order of conflict* if

$$G(A{>}B) \cap G(A{<}B) = \emptyset\ ,\ G(A{>}B) \neq \emptyset,\ G(A{>}B)$$
$$\cup\ G(A{<}B) = G.$$

The following definition presents this idea in general terms.

<u>Definition.</u> A set of agents G is in a **First Order Conflict (FOC)** for proposition p if

$$G(p) \cap G(\neg p) = \emptyset,\ and\ G(p) \neq \emptyset,\ G(p) \cup$$
$$G(\neg p) = G.$$

Figure 1 shows a set of 20 agents in the logic conflicting state, where 7 white agents are in the state True and 13 black agents are in the state False for the same proposition p = "A > B".

Below we show that at the first order of conflicts, AND and OR operations should differ from the classical logic operations and should be **vector operations** in the space of the agents' evaluations (*agents space*). The vector operations reflect a structure of logic conflict among coherent individual agent evaluations.

Figure 1. A set of 20 agents in the first order of logic conflict

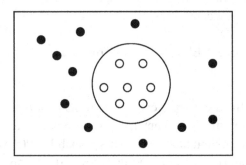

Fusion Process

If a single decision must be made at the first order of conflict, then we must introduce a **fusion process** of the logic values of proposition p given by all agents. A basic way to do this is to compute the weighted frequency of logic value given by all agents:

$$\mathbf{v}(p) = \begin{pmatrix} g_1 & g_2 & \cdots & g_{n-1} & g_n \\ v_1 & v_2 & \cdots & v_{n-1} & v_n \end{pmatrix} \qquad (3)$$

where

$$\mu(p) = w_1 v_1(p) + \ldots + w_n v_n(p) = \begin{bmatrix} v_1(p) \\ v_2(p) \\ \ldots \\ v_n(p) \end{bmatrix}^T \begin{bmatrix} w_1 \\ w_2 \\ \ldots \\ w_n \end{bmatrix}$$

are two vectors in the space of the agents' evaluations. The first vector contains all logic states (True/False) for all agents, the second vector (with

property $\begin{bmatrix} v_1(p) \\ v_2(p) \\ \ldots \\ v_n(p) \end{bmatrix}$ and $\begin{bmatrix} w_1 \\ w_2 \\ \ldots \\ w_n \end{bmatrix}$) contains non-negative

weights (utilities) that are given to each agent in the fusion process. In a simple frequency case, each weight is equal to 1/n. At first glance, $\mu(p)$ is the same as used in the probability and utility theories. However, classical axioms of the probability theory have no references to agents producing initial uncertainty values and do not violate the mutual exclusion. Below we define **vector logic operations** for the first order of conflict logic states $\mathbf{v}(p)$.

Definition

$$\sum_{k=1}^{n} w_k = 1$$

$$\mathbf{v}(p \wedge q) = v_1(p) \wedge v_1(q), \ldots, v_n(p) \wedge v_n(q),$$

$$\mathbf{v}(p \vee q) = v_1(p) \vee v_1(q), \ldots, v_n(p) \vee v_n(q)$$

where the symbols $\wedge$, $\vee$, $\neg$ in the right side of the equations are the classical AND, OR, and NOT operations.

Below these operations are written with explicit indication of agents (in the first row):

$$\mathbf{v}(\neg p) = \neg v_1(p), \ldots, \neg v_n(p),$$

Below we present the important properties of sets of conflicting agents at the first order of conflicts. Let $|G(x)|$ be the numbers of agents for which proposition x is true.

Statement 1 sets up properties of the AND and OR operations for nested sets of conflicting agents.

Statement 1 (general non min/max properties of $\wedge$ and $\vee$ operations)

If G is a set of agents at the first order of conflicts and

$$|G(q)| \leq |G(p)|$$

then $\wedge$ and $\vee$ logic operations satisfy the following properties

$$|G(p \wedge q)| = \min(|G(p)|, |G(q)|) - |G(\neg p \wedge q)|,$$
$$(4)$$

$$|G(p \vee q)| = \max(|G(p)|, |G(q)|) + |G(\neg p \wedge q)|.$$
$$(5)$$

If G is a set of agents at the first order of conflicts and

$$|G(p)| \leq |G(q)|$$

then $\wedge$ and $\vee$ logic operations satisfy the following properties

$$|G(p \wedge q)| = \min(|G(p)|, |G(q)|) - |G(\neg q \wedge p)|,$$

$$|G(p \vee q)| = \max(|G(p)|, |G(q)|) + |G(\neg q \wedge p)|.$$

and also

$$|G(p \vee q)| = |G(p) \cup G(q)|, \quad G(p \wedge q) = |G(p) \cap G(q)|$$

Corollary 1 (min/max properties of $\wedge$ and $\vee$ operations for nested sets of agents)

If G is a set of agents at the first order of conflicts such that $G(q) \subset G(p)$ or $G(p) \subset G(q)$ then

$$G(\neg p \wedge q)| = \emptyset \text{ or } G(\neg q \wedge p)| = \emptyset$$

$$|G(p \wedge q)| = \min(|G(p)|, |G(q)|)$$

$$|G(p \vee q)| = \max(|G(p)|, |G(q)|)$$

This follows from the statement 1. The corollary presents a well-known condition when the use of min, max operations has the clear justification.

Let $G^c(p)$ is a *complement* of $G(p)$ in G: $G^c(p) = G \setminus G(p)$, $G = G(p) \cup G^c(p)$.

Statement 2. $G = G(p) \cup G^c(p) = G(p) \cup G(\neg p)$.

Corollary 2. $G(\neg p) = G^c(p)$

It follows directly from Statement 2.

Statement 3. If G is a set of agents at the first order of conflicts then

$$G(p \vee \neg p) = G(p) \cup G(\neg p) = G(p) \cup G^c(p) = G$$

Figure 2. A set of total 10 agents with two different splits to G(p) and G(q) subsets (a) and (b)

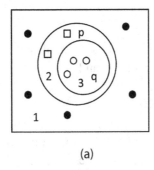 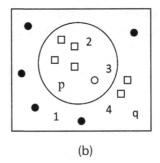

(a)   (b)

G (p ∧ ¬ p) = G(p) ∩ G(¬ p) = G(p) ∩ Gᶜ(p) = ∅

It follows from the definition of the first order of conflict and statement 2. In other words, G (p ∧ ¬ p) = ∅ corresponds to the contradiction p ∧ ¬ p, that is always false and G (p ∨ ¬ p)= G corresponds to the tautology p ∨ ¬ p, that is always true in the first order conflict.

Let $G_1 \oplus G_2$ be a *symmetric difference* of sets of agents G_1 and G_2,

$G_1 \oplus G_2 = (G_1 \cap G_2^c) \cup (G_1^c \cap G_2)$ and let p⊕q be the *exclusive or* of propositions p and q,

p⊕q = (p ∧ ¬q) ∨ (¬ p ∧ q).

Consider, a set of agents G(p⊕q). It consists of agents for which values of p and q differ from each other, that is

G(p⊕q)= G ((p ∧ ¬q) ∨ (¬ p ∧ q)).

Below we use the number of agents in set G(p⊕q) to define a measure of difference between statements p and q and a measure of difference between sets of agents G(p) a G(q).

Definition. A measure of difference D(p,q) between statements p and q and a measure of difference D(G(p),G(q)) between sets of agents G(p) a G(q) are defined as follows:

D(p,q) = D(G(p),G(q))= |G(p)⊕ G(q)|

Statement 4. D(p,q) = D(G(p),G(q)) is a distance, i.e., it satisfies distance axioms

D(p,q) ≥ 0

D(p,q) = D(q,p)

D(p,q)+D(q,h) ≥ D(p,h).

This follows from the properties of the symmetric difference ⊕ (e.g.., Flament 1963).

Figure 2 illustrates a set of agents G(p) for which p is true and a set of agents G(q) for which q is true. In Figure 2(a) the number of agents for which truth values of p and q are different, (¬ p ∧ q) ∨ (p ∧ ¬q), is equal to 2. These agents are represented by white squares. Therefore the distance between G(p) and G(q) is 2. Figure 2(b) shows other G(p) and G(q) sets with the number of the agents for which ¬ p ∧ q) ∨ (p ∧ ¬q is true equal to 6 (agents shown as white squares and squares with the grid). Thus, the distance between the two sets is 6.

In Figure 2(a), set 2 consists of 2 agents |G((p ∧ ¬ q)| = 2 and set 4 is empty, |G((¬ p ∧ ¬ q)| = 0, thus D(Set2, Set4)=2. This emptiness means that a set of agents with true p includes a set of agents with true q.

In Figure 2(b), set 2 consists of 4 agents |G((p ∧ ¬ q)| = 4 and set 4 consists of 2 agents, |G((¬ p ∧ ¬ q)| = 2, thus D(Set2,Set4)=6.

These splits produce different distances between G(p) and G(q). The distance in the case (a) is equal to 2; the distance in the case (b) is equal to 6. Set 1 (black circles) consists of agents for which both p and q are false, Set 2 (white squares) consists of agents for which p is true but q is false. Set 3 (white circles) consists of agents for which p and q are true, and Set 4 (squares with grids) consists of agents for which p is false and q is true,

Figure 3. Graph of the distances for three sentences p_1, p_2, and p_3

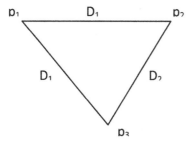

For complex computations of logic values provided by the agents we can use a *graph of the distances* among the sentences p_1, p_2, ...,p_N. For example, for three sentences p_1, p_2, and p_3, we have a graph of the distances shown in Figure 3.

This graph has can be represented by a distance matrix

$$\mathbf{v}(p \wedge q) = \begin{pmatrix} g_1 & g_2 & ... & g_{n-1} & g_n \\ v_1(p) \wedge v_1(q) & v_2(p) \wedge v_2(q) & ... & v_{n-1}(p) \wedge v_{n-1}(q) & v_n(p) \wedge v_n(q) \end{pmatrix}$$

$$\mathbf{v}(p \vee q) = \begin{pmatrix} g_1 & g_2 & ... & g_{n-1} & g_n \\ v_1(p) \vee v_1(q) & v_2(p) \vee v_2(q) & ... & v_{n-1}(p) \vee v_{n-1}(q) & v_n(p) \vee v_n(q) \end{pmatrix}$$

$$\mathbf{v}(\neg p) = \begin{pmatrix} g_1 & g_2 & ... & g_{n-1} & g_n \\ v_1(\neg p) & v_2(\neg p) & ... & v_{n-1}(\neg p) & v_n(\neg p) \end{pmatrix}$$

which has a general form of a symmetric matrix,

$$D = \begin{bmatrix} 0 & D_{1,2} & D_{1,3} \\ D_{1,2} & 0 & D_{2,3} \\ D_{1,3} & D_{2,3} & 0 \end{bmatrix}$$

Having distances D_{ij} between propositions we can use them to compute complex expressions in agents' logic operation. For instance, using $0 \leq D(p, q) \leq G(p) + G(q)$ and

$$G(p \wedge q) \equiv G(p \vee q) - G((\neg p \wedge q) \vee (p \wedge \neg q))$$
$$\equiv G(p \vee q) - D(p, q)$$

Evidence Theory and Rough Set Theory with Agents

In the evidence theory any subset $A \subseteq U$ of the universe U is associated with a value m(A) called a basic assignment probability such that

$$D = \begin{bmatrix} 0 & D_{1,2} & ... & D_{1,N} \\ D_{1,2} & 0 & ... & D_{2,N} \\ ... & ... & ... & .. \\ D_{1,N} & D_{2,N} & ... & 0 \end{bmatrix}$$

Respectively, the belief Bel(Ω) and plausibility Pl(Ω) measures for any set Ω are defined as

$$\sum_{k=1}^{2^N} m(A_k) = 1$$

Thus, Bel measure includes all subsets that inside Ω, The Pl measure includes all sets with non-empty intersection with Ω. In the evidence theory as in the probability theory we associate one and only one agent with an element.

Figure 4 shows set A at the border of the set Ω, which is divided in two parts: one inside Ω and another one outside it. For the belief measure, we exclude the set A, thus we exclude the false state (f,f) and a logical self-conflicting state (t,f). But for the plausibility measure we accept the (t, t) and the self-conflicting state (f,t). In the belief measure, we exclude any possible self-conflicting state, but in the plausibility measure we accept self-conflicting states.

There are two different criteria to compute the belief and plausibility measures in the evidence theory. For belief, C_1 criterion is related to set Ω, thus C_1 is true for the cases inside Ω. The second criterion C_2 is related to set A, thus C_2 is false for the cases inside A. Now we can put in evidence a logically self-conflicting state (t, f), where we eliminate A also if it is inside Ω. The same is applied to the plausibility, where C_2 is true for cases inside A. In this situation, we accept a

Figure 4. Example of irrational agents for the belief theory

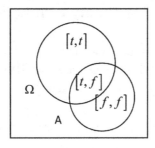
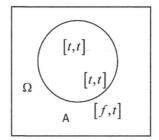

Belief measure   Plausibility measure

Figure 5. Example of irrational agents for the Rough sets theory

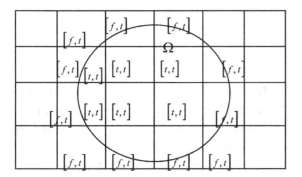

logically self-conflicting situation (*f, t*), where cases in Ω are false but cases in A are true. We also use the logical self-conflict to study the roughs set as shown in Figure 5.

Figure 5 shows set Ω that includes the logically self-conflicting states (f, t).

The internal and external part of Ω is defined by a non empty frontier, where the couples [t, t]are present and as well as the self-conflicting couples [f, t].

QUANTUM MECHANICS: SUPERPOSITION AND FIRST ORDER CONFLICTS

The Mutual Exclusive (ME) principle states that an object cannot be in two different locations at

the same time. We call this locality phenomenon. In quantum mechanics, we have *non-locality phenomenon* for which the same particle can be in many different positions at the same time. This is a clear violation of the ME principle for location. The non-locality is essential for quantum phenomena. Given proposition p = "The particle is in position x in the space", agent g_i can say that p is true, but agent g_j can say that the same p is false.

Individual states of the particle include its position, momentum and others. The *complete state* of the particle is a *superposition* of the quantum states of the particle w_i at different positions x_i in the space, This superposition can be a global wave function:

$$Bel(\Omega) = \sum_{A_k \subseteq \Omega} m(A_k), \quad Pl(\Omega) = \sum_{A_k \cap \Omega \neq \emptyset} m(A_k)$$

where $\psi = w_1 | x_1 \rangle + w_2 | x_2 \rangle + ... + w_n | x_n \rangle$ denotes the state at the position x_i (D'Espagnat, 1999). If we limit our consideration by an individual quantum state of the particle and ignore other states then we are in Mutual Exclusion situation of the classical logic (the particle is at x_i or not). In the same way in AUT, when we observe only one agent at the time the situation collapses to the classical logic that may not adequately represent many-valued logic properties. Having multiple states associated with the particle, we use AUT multi-valued logic as a way to model the multiplicity of states. This situation is very complex because measuring position x_i of the particle changes the superposition value ψ. However, this situation can be modelled by the first order conflict, because each individual agent has no self-conflict.

Quantum Computer

The modern concept of the quantum computer is based on following statements (Abbott, Doering, Caves, Lidar, Brandt, Hamilton, Ferry, Gea-Banacloche, Bezrukov, & Kish, 2003; DiVincenzo, 2000; DiVincenzo, 1995; Feynman, 1982; Jaeger, 2006; Nielsen & Chuang, 2000; Benenti, 2004; Stolze & Suter, 2004; Vandersypen, Yannoni, & Chuang, 2000; Hiroshi & Masahito, 2006):

1. Any state denoted as $|x_i\rangle$ is a field of complex numbers on the reference space (position and time). In our notation, the quantum state is a column of values for different points (objects).
2. Any combination of states $|n\rangle$ is a product of fields.

   Given two atoms with independent states of energy $|nm\rangle = |n\rangle|m\rangle$ the two atoms have the state $|n\rangle$ and $|m\rangle$ that is the product of the separate state for atom 1 and atom 2.

3. The space H is the space which dimension is the number of elementary fields or states.
4. In quantum mechanics, we compute the probability that the qubit takes values 1 or 0 in the all space time.

   A qubit has some similarities to a classical bit, but is overall very different. Like a bit, a qubit can have two possible values—normally a 0 or a 1. The difference is that whereas a bit *must* be either 0 or 1, a qubit can be 0, 1, or a superposition of both.

Explanation:

The classical digital computer operates with bits that is with 1 and 0. The quantum computer operates with two states $|nm\rangle = |n\rangle|m\rangle$ In quantum mechanics, the state $|1\rangle, |0\rangle$ is associated with a field of probability that the bit assumes the value 1 in the space time. Thus, in the classical sense the bit is 1 in a given place and time. In quantum mechanics, we have the distribution of the probability that a bit assumes the value 1 in the space time. Therefore we have a field of uncertainty for value 1 as well as a field of uncertainty for value 0. This leads to the concept of the qubit.

5. Any space of $2^m - 1$ dimensions is represented by H and UH is the unitary transformation of H by which computations are made in quantum physics.

Explanation:

H is a matrix of qubits that can assume different values. For simplicity we use only the numbers 1 and 0, where 1 means qubit that assume the value 1 in the all space time and the same for 0.

In quantum mechanics, the qubit value can be changed only with particular transformations U or unitary transformations. Only the unitary transformations can be applied. Given the unitary transformation

$$|1\rangle$$

such that $U^T U = I$ and

$$U = \begin{bmatrix} 1 & 0 & 0 & 0 \\ 0 & 1 & 0 & 0 \\ 0 & 0 & 0 & 1 \\ 0 & 0 & 1 & 0 \end{bmatrix}$$

with $2^2 = 4$ qubits we have

$$H = \begin{bmatrix} 0 & 0 \\ 0 & 1 \\ 1 & 0 \\ 1 & 1 \end{bmatrix} \quad (1)$$

Also

$$UH = \begin{bmatrix} 1 & 0 & 0 & 0 \\ 0 & 1 & 0 & 0 \\ 0 & 0 & 0 & 1 \\ 0 & 0 & 1 & 0 \end{bmatrix} \begin{bmatrix} 0 & 0 \\ 0 & 1 \\ 1 & 0 \\ 1 & 1 \end{bmatrix} = \begin{bmatrix} 0 & 0 \\ 0 & 1 \\ 1 & 1 \\ 1 & 0 \end{bmatrix}$$

is the XOR that can be written as c = a $\oplus$ b.

In quantum mechanics, (1) is represented by the Deutch circuit shown in Figure 6

Now we want to clarify and extend the Deutch's interpretation of the quantum computer to introduce not only a Boolean algebra in the quantum computer, but also a many-valued logic. This is more in line with the quantum mechanics intrinsic inconsistency due to the superposition of mutually exclusive states. In quantum mechanics a particle can have two opposite spins at the same time and different positions again at the same time. Here we cannot apply the mutual exclusive states of the classical physics where we have logical consistency that is any particle has one and only one position.

Permutations, Unitary Transformation and Boolean Algebra

In quantum mechanics any physical transformation is governed by special transformations of the Hilbert space of the states. For the unitary transformation U the probability of the quantum system is invariant. The unitary transformation has property $U\,U^T = 1$ and is extensively used in the quantum computer. Now we establish a

Figure 6. Quantum Computer circuit that represents the transformation

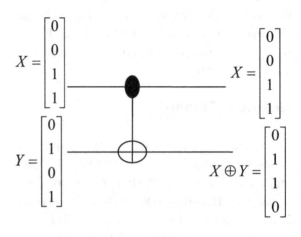

connection between unitary transformation and permutation. In fact we have

$$\begin{bmatrix} 0 & 0 \\ 0 & 1 \\ 1 & 0 \\ 1 & 1 \end{bmatrix} \Rightarrow \begin{bmatrix} 0 \\ 1 \\ 1 \\ 0 \end{bmatrix}$$

where

$$Q = UH = \begin{bmatrix} 1 & 0 & 0 & 0 \\ 0 & 1 & 0 & 0 \\ 0 & 0 & 0 & 1 \\ 0 & 0 & 1 & 0 \end{bmatrix} \begin{bmatrix} 0 & 0 \\ 0 & 1 \\ 1 & 0 \\ 1 & 1 \end{bmatrix} = \begin{bmatrix} 0 & 0 \\ 0 & 1 \\ 1 & 1 \\ 1 & 0 \end{bmatrix}$$

and

$$U = \begin{pmatrix} P_1 & P_2 & P_3 & P_4 \\ P_1 & P_2 & P_4 & P_3 \end{pmatrix}$$

Now for

$$(P_1 , P_2 , P_3 , P_4) = ((0,0) , (0,1) , (1,0) , (1,1))$$

and

$$Q_1(X,Y) = \begin{bmatrix} 0 \\ 0 \\ 1 \\ 1 \end{bmatrix} = X \ , Q_2(X,Y) = \begin{bmatrix} 0 \\ 1 \\ 1 \\ 0 \end{bmatrix}$$

Similarly, for the negation we have

$$Q_2(X,1) = \begin{bmatrix} 1 \\ 0 \end{bmatrix} = \neg X$$

$$\neg X = \begin{pmatrix} P_1 & P_2 & P_3 & P_4 \\ P_1 & P_2 & P_4 & P_3 \end{pmatrix} = (3,4)$$

and

$$Q = UH = \begin{bmatrix} 0 & 1 & 0 & 0 \\ 1 & 0 & 0 & 0 \\ 0 & 0 & 1 & 0 \\ 0 & 0 & 0 & 1 \end{bmatrix} \begin{bmatrix} 0 & 0 \\ 0 & 1 \\ 1 & 0 \\ 1 & 1 \end{bmatrix} = \begin{bmatrix} 0 & 1 \\ 0 & 0 \\ 1 & 0 \\ 1 & 1 \end{bmatrix}$$

and

$$Q_2(X,1) = \begin{bmatrix} 0 \\ 1 \end{bmatrix} = X$$

$$X = \begin{pmatrix} P_1 & P_2 & P_3 & P_4 \\ P_2 & P_1 & P_3 & P_4 \end{pmatrix} = (1,2)$$

for

$$\neg X = \left\{ (1,2) \right\}^C = \left\{ (3,4) \right\}$$

Therefore,

$$Q = UH = \begin{bmatrix} 0 & 1 & 0 & 0 \\ 1 & 0 & 0 & 0 \\ 0 & 0 & 0 & 1 \\ 0 & 0 & 1 & 0 \end{bmatrix} \begin{bmatrix} 0 & 0 \\ 0 & 1 \\ 1 & 0 \\ 1 & 1 \end{bmatrix} = \begin{bmatrix} 0 & 1 \\ 0 & 0 \\ 1 & 1 \\ 1 & 0 \end{bmatrix}$$

$$Q_2(X,1) = \begin{bmatrix} 0 \\ 0 \end{bmatrix} = false$$

$$\neg X \wedge X = \begin{pmatrix} P_1 & P_2 & P_3 & P_4 \\ P_2 & P_1 & P_4 & P_3 \end{pmatrix} = (1,2)(3,4)$$

and

$$false = \neg X \wedge X = \left\{ (1,2) \right\} \cup \left\{ (3,4) \right\} = \left\{ (1,2),(3,4) \right\} = Universe$$

Next we have

$$Q = UH = \begin{bmatrix} 1 & 0 & 0 & 0 \\ 0 & 1 & 0 & 0 \\ 0 & 0 & 1 & 0 \\ 0 & 0 & 0 & 1 \end{bmatrix} \begin{bmatrix} 0 & 0 \\ 0 & 1 \\ 1 & 0 \\ 1 & 1 \end{bmatrix} = \begin{bmatrix} 0 & 0 \\ 0 & 1 \\ 1 & 0 \\ 1 & 1 \end{bmatrix}$$

$$Q_2(X,1) = \begin{bmatrix} 1 \\ 1 \end{bmatrix} = true$$

and

$$\neg X \vee X = \begin{pmatrix} P_1 & P_2 & P_3 & P_4 \\ P_1 & P_2 & P_3 & P_4 \end{pmatrix} = \varnothing$$

The quantum computer circuit can be explained as follows. Having

$c = X \oplus Y = (X \wedge \neg Y) \vee (\neg X \wedge Y)$ we infer that

$$Y = 0 \Rightarrow X \oplus Y = X$$

$$Y = 1 \Rightarrow X \oplus Y = \neg X$$

Thus, we have what is shown in Figure 7

Now we use a common symbolic representation of elementary states $false = \neg X \vee X = \left\{ (1,2) \right\} \cap \left\{ (3,4) \right\} = \left\{ \ \right\} = \varnothing$ $|0\rangle$ as unitary vectors in 2-D Hilbert space

$|1\rangle$

Figure 7. Quantum Computer circuit by NOT operation or "¬ "

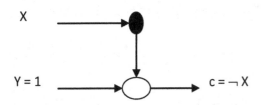

Any vector in this 2-D space is a superposition

$$\left|1\right\rangle = \begin{bmatrix} 1 \\ 0 \end{bmatrix} \ , \ \left|0\right\rangle = \begin{bmatrix} 0 \\ 1 \end{bmatrix}$$

that is symbolically represented as

$$\psi = \alpha\left|1\right\rangle + \beta\left|0\right\rangle$$

Thus,

$$\psi = \begin{bmatrix} \alpha \\ \beta \end{bmatrix}$$

Points in 2-D Hilbert space for the four states are

$$\psi = \left|1\right\rangle = \begin{bmatrix} 1 \\ 0 \end{bmatrix} \ , \ \psi = \left|0\right\rangle = \begin{bmatrix} 0 \\ 1 \end{bmatrix}$$

and

$$H = \begin{bmatrix} \left|0\right\rangle\left|0\right\rangle \\ \left|0\right\rangle\left|1\right\rangle \\ \left|1\right\rangle\left|0\right\rangle \\ \left|1\right\rangle\left|1\right\rangle \end{bmatrix} = \begin{bmatrix} 0 & 0 \\ 0 & 1 \\ 1 & 0 \\ 1 & 1 \end{bmatrix}$$

In the classical Boolean algebra, complex expressions are obtained by composition using the elementary operations (NOT, AND, and OR). In quantum computer, we cannot use the same elementary operations and must change our point of view. The quantum mechanics is based on the idea of states, superposition of states, product of states (entanglement or relationship among states) and the Hilbert space of the states with unitary transformation. Thus, the mathematical formalism is closer to the vector space than to the Boolean algebra. We remark that permutations can be taken as generators of mathematical groups. This sets up an interesting link between *groups of symmetry* and *Boolean algebra*, which can lead to a deeper connection with the Boolean logic. In fact, given the permutation with a set of points P in 2-D space, every permutation of the states can generate only the following set of Boolean functions

$$\left|11\right\rangle = \left|1\right\rangle\left|1\right\rangle = \begin{bmatrix} 1 \\ 0 \end{bmatrix} \otimes \begin{bmatrix} 1 \\ 0 \end{bmatrix} = \begin{bmatrix} 1\begin{bmatrix}1\\0\end{bmatrix} \\ 0\begin{bmatrix}1\\0\end{bmatrix} \end{bmatrix} = \begin{bmatrix} 1 \\ 0 \\ 0 \\ 0 \end{bmatrix} \ , \ \left|10\right\rangle = \left|1\right\rangle\left|0\right\rangle = \begin{bmatrix} 1 \\ 0 \end{bmatrix} \otimes \begin{bmatrix} 0 \\ 1 \end{bmatrix} = \begin{bmatrix} 1\begin{bmatrix}0\\1\end{bmatrix} \\ 0\begin{bmatrix}0\\1\end{bmatrix} \end{bmatrix} = \begin{bmatrix} 0 \\ 1 \\ 0 \\ 0 \end{bmatrix} \ ,$$

$$\left|01\right\rangle = \left|0\right\rangle\left|1\right\rangle = \begin{bmatrix} 0 \\ 1 \end{bmatrix} \otimes \begin{bmatrix} 1 \\ 0 \end{bmatrix} = \begin{bmatrix} 0\begin{bmatrix}1\\0\end{bmatrix} \\ 1\begin{bmatrix}1\\0\end{bmatrix} \end{bmatrix} = \begin{bmatrix} 0 \\ 0 \\ 1 \\ 0 \end{bmatrix} \ , \ \left|00\right\rangle = \left|0\right\rangle\left|0\right\rangle = \begin{bmatrix} 0 \\ 1 \end{bmatrix} \otimes \begin{bmatrix} 0 \\ 1 \end{bmatrix} = \begin{bmatrix} 0\begin{bmatrix}0\\1\end{bmatrix} \\ 1\begin{bmatrix}0\\1\end{bmatrix} \end{bmatrix} = \begin{bmatrix} 0 \\ 0 \\ 0 \\ 1 \end{bmatrix}$$

To obtain the Boolean function NAND we must extend the space of the objects and introduce a 3-D attribute space. Thus, we have the Toffoli transformation and the permutation

$$F = \left\{ X, \neg X, \neg X \wedge X, \neg X \vee X) \right\}$$

For

$$U = \begin{pmatrix} P_1 & P_2 & P_3 & P_4 & P_5 & P_6 & P_7 & P_8 \\ P_1 & P_2 & P_3 & P_4 & P_5 & P_6 & P_8 & P_7 \end{pmatrix} = (7,8)$$

and

$$U = \begin{bmatrix} 1 & 0 & 0 & 0 & 0 & 0 & 0 & 0 \\ 0 & 1 & 0 & 0 & 0 & 0 & 0 & 0 \\ 0 & 0 & 1 & 0 & 0 & 0 & 0 & 0 \\ 0 & 0 & 0 & 1 & 0 & 0 & 0 & 0 \\ 0 & 0 & 0 & 0 & 1 & 0 & 0 & 0 \\ 0 & 0 & 0 & 0 & 0 & 1 & 0 & 0 \\ 0 & 0 & 0 & 0 & 0 & 0 & 0 & 1 \\ 0 & 0 & 0 & 0 & 0 & 0 & 1 & 0 \end{bmatrix}$$

and

and

$$X = \begin{bmatrix} 0 \\ 0 \\ 0 \\ 0 \\ 1 \\ 1 \\ 1 \\ 1 \end{bmatrix}, Y = \begin{bmatrix} 0 \\ 0 \\ 1 \\ 1 \\ 0 \\ 0 \\ 1 \\ 1 \end{bmatrix}, Z = \begin{bmatrix} 0 \\ 1 \\ 0 \\ 1 \\ 0 \\ 1 \\ 0 \\ 1 \end{bmatrix}, H = \begin{bmatrix} 0 & 0 & 0 \\ 0 & 0 & 1 \\ 0 & 1 & 0 \\ 0 & 1 & 1 \\ 1 & 0 & 0 \\ 1 & 0 & 1 \\ 1 & 1 & 0 \\ 1 & 1 & 1 \end{bmatrix} = \begin{bmatrix} X & Y & Z \end{bmatrix}$$

$$Y = \begin{pmatrix} P_1 & P_2 & P_3 & P_4 & P_5 & P_6 & P_7 & P_8 \\ P_2 & P_1 & P_4 & P_3 & P_5 & P_6 & P_7 & P_8 \end{pmatrix} = (1,2)(3,4)$$

or

Now we put

$$UP = \begin{bmatrix} 0 & 1 & 0 & 0 & 0 & 0 & 0 & 0 \\ 1 & 0 & 0 & 0 & 0 & 0 & 0 & 0 \\ 0 & 0 & 0 & 1 & 0 & 0 & 0 & 0 \\ 0 & 0 & 1 & 0 & 0 & 0 & 0 & 0 \\ 0 & 0 & 0 & 0 & 1 & 0 & 0 & 0 \\ 0 & 0 & 0 & 0 & 0 & 1 & 0 & 0 \\ 0 & 0 & 0 & 0 & 0 & 0 & 1 & 0 \\ 0 & 0 & 0 & 0 & 0 & 0 & 0 & 1 \end{bmatrix} \begin{bmatrix} 0 & 0 & 0 \\ 0 & 0 & 1 \\ 0 & 1 & 0 \\ 0 & 1 & 1 \\ 1 & 0 & 0 \\ 1 & 0 & 1 \\ 1 & 1 & 0 \\ 1 & 1 & 1 \end{bmatrix} = \begin{bmatrix} 0 & 0 & 1 \\ 0 & 0 & 0 \\ 0 & 1 & 1 \\ 0 & 1 & 0 \\ 1 & 0 & 0 \\ 1 & 0 & 1 \\ 1 & 1 & 0 \\ 1 & 1 & 1 \end{bmatrix}$$

$$Q(X,Y,Z) = UH = \begin{bmatrix} 1 & 0 & 0 & 0 & 0 & 0 & 0 & 0 \\ 0 & 1 & 0 & 0 & 0 & 0 & 0 & 0 \\ 0 & 0 & 1 & 0 & 0 & 0 & 0 & 0 \\ 0 & 0 & 0 & 1 & 0 & 0 & 0 & 0 \\ 0 & 0 & 0 & 0 & 1 & 0 & 0 & 0 \\ 0 & 0 & 0 & 0 & 0 & 1 & 0 & 0 \\ 0 & 0 & 0 & 0 & 0 & 0 & 0 & 1 \\ 0 & 0 & 0 & 0 & 0 & 0 & 1 & 0 \end{bmatrix} \begin{bmatrix} 0 & 0 & 0 \\ 0 & 0 & 1 \\ 0 & 1 & 0 \\ 0 & 1 & 1 \\ 1 & 0 & 0 \\ 1 & 0 & 1 \\ 1 & 1 & 0 \\ 1 & 1 & 1 \end{bmatrix} = \begin{bmatrix} 0 & 0 & 0 \\ 0 & 0 & 1 \\ 0 & 1 & 0 \\ 0 & 1 & 1 \\ 1 & 0 & 0 \\ 1 & 0 & 1 \\ 1 & 1 & 1 \\ 1 & 1 & 0 \end{bmatrix}$$

Also

$$Q_3 \begin{bmatrix} X=0 & Y=0 & Z=1 \\ X=1 & Y=0 & Z=1 \\ X=0 & Y=1 & Z=1 \\ X=1 & Y=1 & Z=1 \end{bmatrix} = \begin{bmatrix} 0 \\ 0 \\ 1 \\ 1 \end{bmatrix} = Y$$

Thus,

$$Q_1(X,Y,Z) = \begin{bmatrix} 0 \\ 0 \\ 0 \\ 0 \\ 1 \\ 1 \\ 1 \\ 1 \end{bmatrix} = X, \quad Q_2(X,Y,Z) = \begin{bmatrix} 0 \\ 0 \\ 1 \\ 1 \\ 0 \\ 0 \\ 1 \\ 1 \end{bmatrix} = Y, \quad Q_3(X,Y,Z) = \begin{bmatrix} 0 \\ 1 \\ 0 \\ 0 \\ 1 \\ 1 \\ 1 \\ 0 \end{bmatrix}$$

or

$$X = \begin{pmatrix} P_1 & P_2 & P_3 & P_4 & P_5 & P_6 & P_7 & P_8 \\ P_2 & P_1 & P_3 & P_4 & P_6 & P_5 & P_7 & P_8 \end{pmatrix} = (1,2)(5,6)$$

and

Therefore,

$$Q_3 \begin{bmatrix} X=0 & Y=0 & Z=1 \\ X=1 & Y=0 & Z=1 \\ X=0 & Y=1 & Z=1 \\ X=1 & Y=1 & Z=1 \end{bmatrix} = \begin{bmatrix} 1 \\ 1 \\ 1 \\ 0 \end{bmatrix} = \neg(X \wedge Y) = X\ NAND\ Y$$

$$UP = \begin{bmatrix} 0 & 1 & 0 & 0 & 0 & 0 & 0 & 0 \\ 1 & 0 & 0 & 0 & 0 & 0 & 0 & 0 \\ 0 & 0 & 1 & 0 & 0 & 0 & 0 & 0 \\ 0 & 0 & 0 & 1 & 0 & 0 & 0 & 0 \\ 0 & 0 & 0 & 0 & 0 & 1 & 0 & 0 \\ 0 & 0 & 0 & 0 & 1 & 0 & 0 & 0 \\ 0 & 0 & 0 & 0 & 0 & 0 & 1 & 0 \\ 0 & 0 & 0 & 0 & 0 & 0 & 0 & 1 \end{bmatrix} \begin{bmatrix} 0 & 0 & 0 \\ 0 & 0 & 1 \\ 0 & 1 & 0 \\ 0 & 1 & 1 \\ 1 & 0 & 0 \\ 1 & 0 & 1 \\ 1 & 1 & 0 \\ 1 & 1 & 1 \end{bmatrix} = \begin{bmatrix} 0 & 0 & 1 \\ 0 & 0 & 0 \\ 0 & 1 & 0 \\ 0 & 1 & 1 \\ 1 & 0 & 1 \\ 1 & 0 & 0 \\ 1 & 1 & 0 \\ 1 & 1 & 1 \end{bmatrix}$$

The permutation of (7,8) introduces a zero, this $Q_3(1, 1, 1) = 0$.

For the permutation we have

$$Q_3(X,Y,1) = \neg(X \wedge Y) = X\ NAND\ Y$$

For the operation $X \wedge Y$, we have the permutation

$$Q_3 \begin{bmatrix} X=0 & Y=0 & Z=1 \\ X=1 & Y=0 & Z=1 \\ X=0 & Y=1 & Z=1 \\ X=1 & Y=1 & Z=1 \end{bmatrix} = \begin{bmatrix} 0 \\ 1 \\ 0 \\ 1 \end{bmatrix} = X$$

because

$$X \wedge Y = \begin{pmatrix} P_1 & P_2 & P_3 & P_4 & P_5 & P_6 & P_7 & P_8 \\ P_2 & P_1 & P_4 & P_3 & P_6 & P_5 & P_7 & P_8 \end{pmatrix} = (1,2)(3,4)(5,6)$$

or

$X = (1,2)(5,6)$ and $Y = (1,2)(3,4)$
$X \wedge Y = \{(1,2),(5,6)\} \cup \{(1,2),(3,4)\} = \{(1,2),(3,4),(5,6)\}$

and

$$UP = \begin{bmatrix} 0&1&0&0&0&0&0&0 \\ 1&0&0&0&0&0&0&0 \\ 0&0&0&1&0&0&0&0 \\ 0&0&1&0&0&0&0&0 \\ 0&0&0&0&0&1&0&0 \\ 0&0&0&0&1&0&0&0 \\ 0&0&0&0&0&0&1&0 \\ 0&0&0&0&0&0&0&1 \end{bmatrix} \begin{bmatrix} 0&0&0 \\ 0&0&1 \\ 0&1&0 \\ 0&1&1 \\ 1&0&0 \\ 1&0&1 \\ 1&1&0 \\ 1&1&1 \end{bmatrix} = \begin{bmatrix} 0&0&1 \\ 0&0&0 \\ 0&1&1 \\ 0&1&0 \\ 1&0&1 \\ 1&0&0 \\ 1&1&0 \\ 1&1&1 \end{bmatrix}$$

For the operation X ∨ Y, we have the permutation

$$Q_3 \begin{bmatrix} X=0 & Y=0 & Z=1 \\ X=1 & Y=0 & Z=1 \\ X=0 & Y=1 & Z=1 \\ X=1 & Y=1 & Z=1 \end{bmatrix} = \begin{bmatrix} 0 \\ 0 \\ 0 \\ 1 \end{bmatrix} = X \wedge Y$$

and because

$$X \vee Y = \begin{pmatrix} P_1 & P_2 & P_3 & P_4 & P_5 & P_6 & P_7 & P_8 \\ P_2 & P_1 & P_3 & P_4 & P_5 & P_6 & P_7 & P_8 \end{pmatrix} = (1,2)$$

and

$X = \{(1,2),(5,6)\}, X = \{(1,2),(3,4)\}$
$X \vee Y = \{(1,2),(5,6)\} \cap \{(1,2),(3,4)\} = \{(1,2)\}$
or

$$UP = \begin{bmatrix} 0&1&0&0&0&0&0&0 \\ 1&0&0&0&0&0&0&0 \\ 0&0&1&0&0&0&0&0 \\ 0&0&0&1&0&0&0&0 \\ 0&0&0&0&1&0&0&0 \\ 0&0&0&0&0&1&0&0 \\ 0&0&0&0&0&0&1&0 \\ 0&0&0&0&0&0&0&1 \end{bmatrix} \begin{bmatrix} 0&0&0 \\ 0&0&1 \\ 0&1&0 \\ 0&1&1 \\ 1&0&0 \\ 1&0&1 \\ 1&1&0 \\ 1&1&1 \end{bmatrix} = \begin{bmatrix} 0&0&1 \\ 0&0&0 \\ 0&1&0 \\ 0&1&1 \\ 1&0&0 \\ 1&0&1 \\ 1&1&0 \\ 1&1&1 \end{bmatrix}$$

For the negation operation, we generate a complementary set of permutations and

$$Q_3 \begin{bmatrix} X=0 & Y=0 & Z=1 \\ X=1 & Y=0 & Z=1 \\ X=0 & Y=1 & Z=1 \\ X=1 & Y=1 & Z=1 \end{bmatrix} = \begin{bmatrix} 0 \\ 1 \\ 1 \\ 1 \end{bmatrix} = X \vee Y$$

or

$$\neg X = \begin{pmatrix} P_1 & P_2 & P_3 & P_4 & P_5 & P_6 & P_7 & P_8 \\ P_1 & P_2 & P_4 & P_3 & P_5 & P_6 & P_8 & P_7 \end{pmatrix} = \{(3,4)(7,8)\}$$

$X = \{(1,2),(5,6)\}$
$\neg X = \{(1,2),(5,6)\}^c = \{(3,4),(7,8)\}$

$$UP = \begin{bmatrix} 1&0&0&0&0&0&0&0 \\ 0&1&0&0&0&0&0&0 \\ 0&0&0&1&0&0&0&0 \\ 0&0&1&0&0&0&0&0 \\ 0&0&0&0&1&0&0&0 \\ 0&0&0&0&0&1&0&0 \\ 0&0&0&0&0&0&0&1 \\ 0&0&0&0&0&0&1&0 \end{bmatrix} \begin{bmatrix} 0&0&0 \\ 0&0&1 \\ 0&1&0 \\ 0&1&1 \\ 1&0&0 \\ 1&0&1 \\ 1&1&0 \\ 1&1&1 \end{bmatrix} = \begin{bmatrix} 0&0&0 \\ 0&0&1 \\ 0&1&1 \\ 0&1&0 \\ 1&0&0 \\ 1&0&1 \\ 1&1&1 \\ 1&1&0 \end{bmatrix}$$

For contradiction we have

$$Q_3 \begin{bmatrix} X=0 & Y=0 & Z=1 \\ X=1 & Y=0 & Z=1 \\ X=0 & Y=1 & Z=1 \\ X=1 & Y=1 & Z=1 \end{bmatrix} = \begin{bmatrix} 1 \\ 0 \\ 1 \\ 0 \end{bmatrix} = \neg X$$

Thus, we have the Unitary transformation

$$Q_3 \begin{bmatrix} X=0 & Y=0 & Z=1 \\ X=1 & Y=0 & Z=1 \\ X=0 & Y=1 & Z=1 \\ X=1 & Y=1 & Z=1 \end{bmatrix} = \begin{bmatrix} 0 \\ 0 \\ 0 \\ 0 \end{bmatrix} = \neg X \wedge X$$

and

$$UP = \begin{bmatrix} 0&1&0&0&0&0&0&0 \\ 1&0&0&0&0&0&0&0 \\ 0&0&0&1&0&0&0&0 \\ 0&0&1&0&0&0&0&0 \\ 0&0&0&0&0&1&0&0 \\ 0&0&0&0&1&0&0&0 \\ 0&0&0&0&0&0&0&1 \\ 0&0&0&0&0&0&1&0 \end{bmatrix} \begin{bmatrix} 0&0&0 \\ 0&0&1 \\ 0&1&0 \\ 0&1&1 \\ 1&0&0 \\ 1&0&1 \\ 1&1&0 \\ 1&1&1 \end{bmatrix} = \begin{bmatrix} 0&0&1 \\ 0&0&0 \\ 0&1&1 \\ 0&1&0 \\ 1&0&1 \\ 1&0&0 \\ 1&1&1 \\ 1&1&0 \end{bmatrix}$$

This is due to

$$X \wedge \neg X = \begin{pmatrix} P_1 & P_2 & P_3 & P_4 & P_5 & P_6 & P_7 & P_8 \\ P_2 & P_1 & P_3 & P_3 & P_6 & P_5 & P_8 & P_7 \end{pmatrix} = (1,2)(3,4)(5,6)(7,8)$$

and

$$X = \big\{(1,2),(5,6)\big\}, \qquad \neg X = \big\{(3,4),(7,8)\big\}$$

Next by using the De Morgan rule for negation

$$X \wedge \neg X = \big\{(1,2),(5,6)\big\} \cup \big\{(3,4),(7,8)\big\} = \big\{(1,2),(3,4),(5,6),(7,8)\big\}$$

we get

$$\neg(X \wedge Y) = (\neg X \vee \neg Y)$$

Thus,

$$\neg(X \wedge Y) = \begin{pmatrix} P_1 & P_2 & P_3 & P_4 & P_5 & P_6 & P_7 & P_8 \\ P_2 & P_1 & P_4 & P_3 & P_6 & P_5 & P_7 & P_8 \end{pmatrix} = (7,8) = X\,NAND\,Y$$

$$\neg(X \wedge Y) = \big\{(1,2),(3,4),(5,6)\big\}^C = \big\{(7,8)\big\}$$

or

$$UP = \begin{bmatrix} 1&0&0&0&0&0&0&0 \\ 0&1&0&0&0&0&0&0 \\ 0&0&1&0&0&0&0&0 \\ 0&0&0&1&0&0&0&0 \\ 0&0&0&0&1&0&0&0 \\ 0&0&0&0&0&1&0&0 \\ 0&0&0&0&0&0&0&1 \\ 0&0&0&0&0&0&1&0 \end{bmatrix} \begin{bmatrix} 0&0&0 \\ 0&0&1 \\ 0&1&0 \\ 0&1&1 \\ 1&0&0 \\ 1&0&1 \\ 1&1&0 \\ 1&1&1 \end{bmatrix} = \begin{bmatrix} 0&0&0 \\ 0&0&1 \\ 0&1&0 \\ 0&1&1 \\ 1&0&0 \\ 1&0&1 \\ 1&1&1 \\ 1&1&0 \end{bmatrix}$$

Now having permutations and rules to compute AND, OR and NOT we can generate any function with two variables X and Y. For example,

$$\neg X = \begin{pmatrix} P_1 & P_2 & P_3 & P_4 & P_5 & P_6 & P_7 & P_8 \\ P_1 & P_2 & P_4 & P_3 & P_5 & P_6 & P_8 & P_7 \end{pmatrix} = \big\{(3,4),(7,8)\big\}$$

$$\neg Y = \begin{pmatrix} P_1 & P_2 & P_3 & P_4 & P_5 & P_6 & P_7 & P_8 \\ P_1 & P_2 & P_3 & P_4 & P_6 & P_5 & P_7 & P_8 \end{pmatrix} = \big\{(5,6),(7,8)\big\}$$

$$\neg X \vee \neg Y = \big\{(3,4),(7,8)\big\} \cap \big\{(5,6)(7,8)\big\} = \big\{(7,8)\big\} = \neg(X \wedge Y)$$

Finally in quantum computer, given set of elementary permutation

$$\Gamma = \{(1,2),\ (3,4),\ (5,6),\ (7,8)\}$$

any Boolean function of two variables is a subset of Γ.

Agents theory AUT in the quantum interpretation of logic. For the quantum mechanics the true and false value or qbit is given by the elementary vectors

$$\neg X = \begin{pmatrix} P_1 & P_2 & P_3 & P_4 & P_5 & P_6 & P_7 & P_8 \\ P_1 & P_2 & P_4 & P_3 & P_5 & P_6 & P_8 & P_7 \end{pmatrix} = \big\{(3,4),(7,8)\big\}$$

$$Y = \begin{pmatrix} P_1 & P_2 & P_3 & P_4 & P_5 & P_6 & P_7 & P_8 \\ P_1 & P_2 & P_3 & P_4 & P_6 & P_5 & P_7 & P_8 \end{pmatrix} = \big\{(1,2),(3,4)\big\}$$

$$X \rightarrow Y = \neg X \vee Y = \big\{(3,4),(7,8)\big\} \cap \big\{(1,2)(3,4)\big\} = \big\{(3,4)\big\}$$

where

$$true = \begin{bmatrix} 1 \\ 0 \end{bmatrix}, false = \begin{bmatrix} 0 \\ 1 \end{bmatrix} = \begin{bmatrix} 0 & 1 \\ 1 & 0 \end{bmatrix} \begin{bmatrix} 1 \\ 0 \end{bmatrix}$$

We define a quantum agent as any entity that can define the quantum logic values $|\psi\rangle = \alpha|true\rangle + \beta|false\rangle$

now $|\psi\rangle$ is function of the parameters $\begin{bmatrix} \alpha \\ \beta \end{bmatrix}$

If these parameters are equal to $\begin{bmatrix} 1 \\ 0 \end{bmatrix}$ then ,

$$|\psi\rangle = |true\rangle$$

If these parameters are equal to $\begin{bmatrix} 0 \\ 1 \end{bmatrix}$ then

$$|\psi\rangle = |false\rangle$$

where false can be spin down and true can be spin up. Now for different particles a set of logic values v(p) can be defined by using quantum agents,

$$|\psi\rangle = |true\rangle \ , \ |\psi\rangle = |false\rangle$$

and p is a logic proposition, that is p = " the particle is in the state

$$v(p) = \begin{bmatrix} v_1 \\ v_2 \\ ... \\ v_n \end{bmatrix} \text{ where } \nu_k \in \{true, false\} = \left\{ \begin{bmatrix} 1 \\ 0 \end{bmatrix}, \begin{bmatrix} 0 \\ 1 \end{bmatrix} \right\}$$

". For example the particle can be in the state up for the spin or the particle has the velocity v and so on.

Definition. A qagent is an agent that performs a quantum measure.

The quantum agents or "qagent" and the evaluation vector can be written in the language of AUT in this usual way

$$|\psi\rangle$$

where

$$v(p) = \begin{bmatrix} qagent_1 & qagent_2 & .... & qagent_n \\ v_1 & v_2 & .... & v_n \end{bmatrix}$$

$$v_k \in \{(1,2),(5,6)\}$$

and

$$v(q) = \begin{bmatrix} qagent_1 & qagent_2 & .... & qagent_n \\ \eta_1 & \eta_2 & .... & \eta_n \end{bmatrix}$$

Now we use the previous definition of the logic in quantum mechanics, thus

$$\eta_k \in \{(1,2),(3,4)\}$$

where

$$v(p \wedge q) = \begin{bmatrix} qagent_1 & qagent_2 & .... & qagent_n \\ v_1 \wedge \eta_1 & v_2 \wedge \eta_2 & .... & v_n \wedge \eta_n \end{bmatrix}$$

Now for

$$v_k \wedge \eta_k \in \{(1,2),(5,6)\} \cup \{(1,2),(3,4)\} = \{(1,2),(3,4),(5,6)\}$$

where

$$v(p \vee q) = \begin{bmatrix} qagent_1 & qagent_2 & .... & qagent_n \\ v_1 \vee \eta_1 & v_2 \vee \eta_2 & .... & v_n \vee \eta_n \end{bmatrix}$$

Finally, we have the logic vector

Table 1. Logic operation AND by qagents with quantum states 1 and 0

$p \wedge q$	$\lvert 1\wedge 1\rangle, \lvert 0\wedge 0\rangle, \lvert 1\wedge 0\rangle, \lvert 0\wedge 1\rangle$	$\begin{pmatrix} qagent_1 & qagent_2 \\ \nu_1=\lvert 1\rangle & \nu_2=\lvert 1\rangle \end{pmatrix}$	$\begin{pmatrix} qagent_1 & qagent_2 \\ \nu_1=\lvert 1\rangle & \nu_2=\lvert 0\rangle \end{pmatrix}$	$\begin{pmatrix} qagent_1 & qagent_2 \\ \nu_1=\lvert 0\rangle & \nu_2=\lvert 1\rangle \end{pmatrix}$
$\begin{pmatrix} qagent_1 & qagent_2 \\ \nu_1=\lvert 0\rangle & \nu_2=\lvert 0\rangle \end{pmatrix}$	$\begin{pmatrix} qagent_1 & qagent_2 \\ \nu_1=\lvert 1\rangle & \nu_2=\lvert 1\rangle \end{pmatrix}$	$\begin{pmatrix} qagent_1 & qagent_2 \\ \nu_1=\lvert 1\rangle & \nu_2=\lvert 1\rangle \end{pmatrix}$	$\begin{pmatrix} qagent_1 & qagent_2 \\ \nu_1=\lvert 1\rangle & \nu_2=\lvert 0\rangle \end{pmatrix}$	$\begin{pmatrix} qagent_1 & qagent_2 \\ \nu_1=\lvert 0\rangle & \nu_2=\lvert 1\rangle \end{pmatrix}$
$\begin{pmatrix} qagent_1 & qagent_2 \\ \nu_1=\lvert 0\rangle & \nu_2=\lvert 0\rangle \end{pmatrix}$	$\begin{pmatrix} qagent_1 & qagent_2 \\ \nu_1=\lvert 1\rangle & \nu_2=\lvert 0\rangle \end{pmatrix}$	$\begin{pmatrix} qagent_1 & qagent_2 \\ \nu_1=\lvert 1\rangle & \nu_2=\lvert 0\rangle \end{pmatrix}$	$\begin{pmatrix} qagent_1 & qagent_2 \\ \nu_1=\lvert 1\rangle & \nu_2=\lvert 0\rangle \end{pmatrix}$	$\begin{pmatrix} qagent_1 & qagent_2 \\ \nu_1=\lvert 0\rangle & \nu_2=\lvert 0\rangle \end{pmatrix}$
$\begin{pmatrix} qagent_1 & qagent_2 \\ \nu_1=\lvert 0\rangle & \nu_2=\lvert 0\rangle \end{pmatrix}$	$\begin{pmatrix} qagent_1 & qagent_2 \\ \nu_1=\lvert 0\rangle & \nu_2=\lvert 1\rangle \end{pmatrix}$	$\begin{pmatrix} qagent_1 & qagent_2 \\ \nu_1=\lvert 0\rangle & \nu_2=\lvert 1\rangle \end{pmatrix}$	$\begin{pmatrix} qagent_1 & qagent_2 \\ \nu_1=\lvert 0\rangle & \nu_2=\lvert 0\rangle \end{pmatrix}$	$\begin{pmatrix} qagent_1 & qagent_2 \\ \nu_1=\lvert 0\rangle & \nu_2=\lvert 1\rangle \end{pmatrix}$
$\begin{pmatrix} qagent_1 & qagent_2 \\ \nu_1=\lvert 0\rangle & \nu_2=\lvert 0\rangle \end{pmatrix}$	$\begin{pmatrix} qagent_1 & qagent_2 \\ \nu_1=\lvert 0\rangle & \nu_2=\lvert 0\rangle \end{pmatrix}$	$\begin{pmatrix} qagent_1 & qagent_2 \\ \nu_1=\lvert 0\rangle & \nu_2=\lvert 0\rangle \end{pmatrix}$	$\begin{pmatrix} qagent_1 & qagent_2 \\ \nu_1=\lvert 0\rangle & \nu_2=\lvert 0\rangle \end{pmatrix}$	$\begin{pmatrix} qagent_1 & qagent_2 \\ \nu_1=\lvert 0\rangle & \nu_2=\lvert 0\rangle \end{pmatrix}$

$$v_k \vee \eta_k \in \{(1,2),(5,6)\} \cap \{(1,2),(3,4)\} = \{(1,2)\}$$

where $S_1, \ldots, S_n$ are logic value true and false defined by logic operation

For example we have

$$v(p \bullet q) = \begin{bmatrix} qagent_1 & qagent_2 & \ldots & qagent_n \\ S_1 & S_2 & \ldots & S_n \end{bmatrix}$$

for which $p \bullet q = p \vee q$, that is represented by the subset $S = \{(1,2)\}$, $S_1, \ldots, S_n$ are different logic values of the operation union generated by the unitary matrix in quantum mechanics and the quantum measure. Now with the agent interpretation of the quantum mechanics, we can use the quantum superposition principle and obtain the aggregation of the qagents logic values.

$$v(p \vee q) = \begin{bmatrix} qagent_1 & qagent_2 & \ldots & qagent_n \\ S_1 & S_2 & \ldots & S_n \end{bmatrix}$$

For instance, for two conflicting qagents we have the superposition

$$\mu(p \bullet q) = \lvert p \bullet q\rangle = w_1\lvert S_1\rangle + w_2\lvert S_2\rangle + \ldots + w_n\lvert S_n\rangle$$

Thus, the AUT establishes a bridge between the quantum computer based on the Boolean logic with the many-valued logic and conflicts based on the quantum superposition phenomena. These processes are represented by fusion process in AUT. The example of many-valued logic in quantum computer is given in Table 1. In the table 1 any qagent can assume states $\mu(p) = \lvert p\rangle = w_1\lvert S_1\rangle + w_2\lvert S_2\rangle = w_1\lvert 1\rangle + w_2\lvert 0\rangle$ for a single proposition. For two different propositions p and q, a qagent can assume one of the following possible states:

$$\lvert 1\rangle \text{ or } \lvert 0\rangle$$

Also for the superposition we have Table 2. Now we can assign a fractional value to the true logic value as shown in Table 3.

Table 2. Quantum superposition as logic fusion

$p \wedge q$	$\begin{pmatrix} qagent_1 & qagent_2 \\ \nu_1=\lvert 0\rangle & \nu_2=\lvert 0\rangle \end{pmatrix}$	$\begin{pmatrix} qagent_1 & qagent_2 \\ \nu_1=\lvert 1\rangle & \nu_2=\lvert 1\rangle \end{pmatrix}$	$\begin{pmatrix} qagent_1 & qagent_2 \\ \nu_1=\lvert 1\rangle & \nu_2=\lvert 0\rangle \end{pmatrix}$	$\begin{pmatrix} qagent_1 & qagent_2 \\ \nu_1=\lvert 0\rangle & \nu_2=\lvert 1\rangle \end{pmatrix}$
$\begin{pmatrix} qagent_1 & qagent_2 \\ \nu_1=\lvert 0\rangle & \nu_2=\lvert 0\rangle \end{pmatrix}$	$\begin{pmatrix} qagent_1 & qagent_2 \\ \nu_1=\lvert 1\rangle & \nu_2=\lvert 1\rangle \end{pmatrix}$	$\alpha\lvert 1\rangle + \beta\lvert 1\rangle$	$\alpha\lvert 1\rangle + \beta\lvert 0\rangle$	$\alpha\lvert 0\rangle + \beta\lvert 1\rangle$
$\alpha\lvert 0\rangle + \beta\lvert 0\rangle$	$\begin{pmatrix} qagent_1 & qagent_2 \\ \nu_1=\lvert 1\rangle & \nu_2=\lvert 0\rangle \end{pmatrix}$	$\alpha\lvert 1\rangle + \beta\lvert 0\rangle$	$\alpha\lvert 1\rangle + \beta\lvert 0\rangle$	$\alpha\lvert 0\rangle + \beta\lvert 0\rangle$
$\alpha\lvert 0\rangle + \beta\lvert 0\rangle$	$\begin{pmatrix} qagent_1 & qagent_2 \\ \nu_1=\lvert 0\rangle & \nu_2=\lvert 1\rangle \end{pmatrix}$	$\alpha\lvert 0\rangle + \beta\lvert 1\rangle$	$\alpha\lvert 0\rangle + \beta\lvert 0\rangle$	$\alpha\lvert 0\rangle + \beta\lvert 1\rangle$
$\alpha\lvert 0\rangle + \beta\lvert 0\rangle$	$\begin{pmatrix} qagent_1 & qagent_2 \\ \nu_1=\lvert 0\rangle & \nu_2=\lvert 0\rangle \end{pmatrix}$	$\alpha\lvert 0\rangle + \beta\lvert 0\rangle$	$\alpha\lvert 0\rangle + \beta\lvert 0\rangle$	$\alpha\lvert 0\rangle + \beta\lvert 0\rangle$

Table 3. Many-valued logic in the quantum system

$p \wedge q$	$\alpha\lvert 0\rangle + \beta\lvert 0\rangle$	$\begin{pmatrix} qagent_1 & qagent_2 \\ \nu_1=\lvert 1\rangle & \nu_2=\lvert 1\rangle \end{pmatrix}$	$\begin{pmatrix} qagent_1 & qagent_2 \\ \nu_1=\lvert 1\rangle & \nu_2=\lvert 0\rangle \end{pmatrix}$	$\begin{pmatrix} qagent_1 & qagent_2 \\ \nu_1=\lvert 0\rangle & \nu_2=\lvert 1\rangle \end{pmatrix}$
$\begin{pmatrix} qagent_1 & qagent_2 \\ \nu_1=\lvert 0\rangle & \nu_2=\lvert 0\rangle \end{pmatrix}$	True	½ True	½ True	False
$\begin{pmatrix} qagent_1 & qagent_2 \\ \nu_1=\lvert 1\rangle & \nu_2=\lvert 1\rangle \end{pmatrix}$	½ True	½ True	False	False
$\begin{pmatrix} qagent_1 & qagent_2 \\ \nu_1=\lvert 1\rangle & \nu_2=\lvert 0\rangle \end{pmatrix}$	½ True	False	½ True	False
$\begin{pmatrix} qagent_1 & qagent_2 \\ \nu_1=\lvert 0\rangle & \nu_2=\lvert 1\rangle \end{pmatrix}$	False	False	False	False

Correlation (entanglement) in quantum mechanics and second order conflict. Consider two interacting particles as agents. These agents are interdependent. Their correlation is independent from any physical communication by any type of fields. We can say that this correlation is rather a characteristic of a logical state of the particles than dependence. We can view the quantum correlation as a conflicting state because we know from quantum mechanics that there is a correlation but when we try to measure the correlation, we cannot check the correlation itself. The measurement process destroys the correlation. Thus, if the spin of one electron is up and the spin of another

Figure 8. Many-valued logic in the AUT neural network

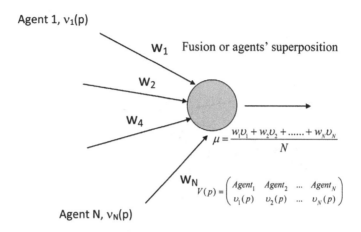

Figure 9. Many-valued logic operation AND in the AUT neural network

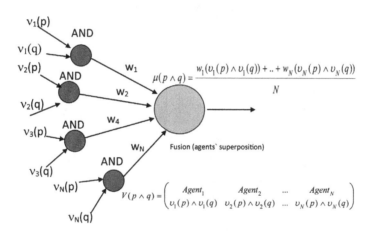

electron is down the first spin is changed when we have correlation or entanglement. It generates the change of the other spin instantaneously and this is manifested in a statistic correlation different from zero. For more explanation see D'Espagnat (1999).

NEURAL IMAGE OF AUT

In this section, we show the possibility for a new type of neural network based on the AUT. This type of the neural network is dedicated to com-

putation of many-valued logic operations used to model uncertainty process. The traditional neural networks model Boolean operations and classical logic. In a new fuzzy neural network, we combine two logic levels (classical and fuzzy) in the same neural network as presented in Figure 8.

Figures 9-11 show that at the first level we have the ordinary Boolean operations (AND, OR, and NOT). At the second level, the network fuses results of different Boolean operations. As a result, a many value logic value is generated as Figure 8 shows.

Figure 10. AUT Many-valued logic operation OR in the AUT neural network

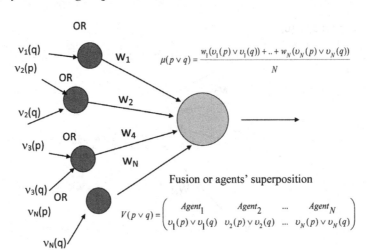

Figure 11.Many-valued logic operation NOT in the AUT neural network

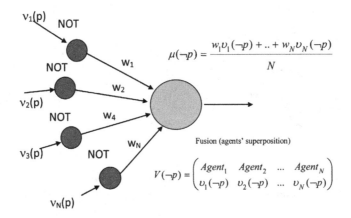

Now we show an example of many-valued logic operation AND with the agents and fusion in the neural network. Table 4 presents the individual AND operation for a single agent in the population of two agents.

$$\begin{pmatrix} qagent_1 & qagent_2 \\ \nu_1 = \begin{vmatrix} 0 \\ 0 \end{vmatrix} & \nu_2 = \begin{vmatrix} 0 \\ 0 \end{vmatrix} \end{pmatrix}$$

With the fusion process in AUT we have the many-valued logic in the neural network.

In Table 5 we use the aggregation rule that can generate a many-valued logic structure with the following three logic values:

$$\begin{pmatrix} nagent_1 & nagent_2 \\ false & false \end{pmatrix}$$ with equivalent notations

$$\Omega = \left\{ true, \frac{true}{2} = \frac{false}{2}, false \right\}$$

Now, having the commutative rule we derive,

$$\frac{false + false}{2} = false$$

The previous composition rule can be written in the simple form shown in Table 6 where different

Table 4. Agent Boolean logical rule

$p \wedge q$	$V(p\lor q)=\begin{pmatrix}Agent_1 & Agent_2 & ... & Agent_N \\ v_1(p)\lor v_1(q) & v_2(p)\lor v_2(q) & ... & v_N(p)\lor v_N(q)\end{pmatrix}$	$\begin{pmatrix} nagent_1 & nagent_2 \\ true & true \end{pmatrix}$	$\begin{pmatrix} nagent_1 & nagent_2 \\ true & false \end{pmatrix}$	$\begin{pmatrix} nagent_1 & nagent_2 \\ false & true \end{pmatrix}$
$\begin{pmatrix} nagent_1 & nagent_2 \\ false & false \end{pmatrix}$	$\begin{pmatrix} nagent_1 & nagent_2 \\ true & true \end{pmatrix}$	$\begin{pmatrix} nagent_1 & nagent_2 \\ true & true \end{pmatrix}$	$\begin{pmatrix} nagent_1 & nagent_2 \\ true & false \end{pmatrix}$	$\begin{pmatrix} nagent_1 & nagent_2 \\ false & true \end{pmatrix}$
$\begin{pmatrix} nagent_1 & nagent_2 \\ false & false \end{pmatrix}$	$\begin{pmatrix} nagent_1 & nagent_2 \\ true & false \end{pmatrix}$	$\begin{pmatrix} nagent_1 & nagent_2 \\ true & false \end{pmatrix}$	$\begin{pmatrix} nagent_1 & nagent_2 \\ true & false \end{pmatrix}$	$\begin{pmatrix} nagent_1 & nagent_2 \\ false & false \end{pmatrix}$
$\begin{pmatrix} nagent_1 & nagent_2 \\ false & false \end{pmatrix}$	$\begin{pmatrix} nagent_1 & nagent_2 \\ false & true \end{pmatrix}$	$\begin{pmatrix} nagent_1 & nagent_2 \\ false & true \end{pmatrix}$	$\begin{pmatrix} nagent_1 & nagent_2 \\ false & false \end{pmatrix}$	$\begin{pmatrix} nagent_1 & nagent_2 \\ false & true \end{pmatrix}$
$\begin{pmatrix} nagent_1 & nagent_2 \\ false & false \end{pmatrix}$	$\begin{pmatrix} nagent_1 & nagent_2 \\ false & false \end{pmatrix}$	$\begin{pmatrix} nagent_1 & nagent_2 \\ false & false \end{pmatrix}$	$\begin{pmatrix} nagent_1 & nagent_2 \\ false & false \end{pmatrix}$	$\begin{pmatrix} nagent_1 & nagent_2 \\ false & false \end{pmatrix}$

Table 5. Neuronal fusion process

$p \wedge q$	$\frac{true+false}{2}=\frac{false+true}{2}=\frac{true}{2}=\frac{false}{2}\;\; \frac{1}{2}true\;\frac{1}{2}false$	$\frac{true+true}{2}=true$	$\frac{true+false}{2}=\frac{true}{2}$	$\frac{false+true}{2}=\frac{true}{2}$
$\frac{false+false}{2}=false$	$\frac{true+true}{2}=true$	$\frac{true+true}{2}=true$	$\frac{true+false}{2}=\frac{true}{2}$	$\frac{false+true}{2}=\frac{true}{2}$
$\frac{false+false}{2}=false$	$\frac{true+false}{2}=\frac{true}{2}$	$\frac{true+false}{2}=\frac{true}{2}$	$\frac{true+false}{2}=\frac{true}{2}$	$\frac{false+false}{2}=false$
$\frac{false+false}{2}=false$	$\frac{false+true}{2}=\frac{true}{2}$	$\frac{false+true}{2}=\frac{true}{2}$	$\frac{false+false}{2}=false$	$\frac{false+true}{2}=\frac{true}{2}$
$\frac{false+false}{2}=false$	$\frac{false+false}{2}=false$	$\frac{false+false}{2}=false$	$\frac{false+false}{2}=false$	$\frac{false+false}{2}=false$

results for the same pair of elements are located in the same cell.

Table 6 contains two different results for the AND operation,

$$\frac{false+false}{2}=false$$

for p = ½ true and q = ½ true.

In this case we have no criteria to choose one or the other. Here the operation AND is not uniquely defined, we have two possible results one is false and the other is ½ true. The first one is shown in Table 7 and the second one is shown in Table 8.

The neuron image of the previous operation is shown in Figure 12.

Table 6. Simplification of the Table 5

$p \wedge q$	$\dfrac{true + false}{2} = \dfrac{false + true}{2} = \dfrac{true}{2}$	$\dfrac{true + true}{2} = true$	$\dfrac{true}{2}$
$\dfrac{false + false}{2} = false$	$\dfrac{true + true}{2} = true$	$\dfrac{true + true}{2} = true$	$\dfrac{true + false}{2} = \dfrac{true}{2}$
$\dfrac{false + false}{2} = false$	$\dfrac{true}{2}$	$\dfrac{false + true}{2} = \dfrac{true}{2}$ $\dfrac{false + false}{2} = false$	$\dfrac{false + true}{2} = \dfrac{true}{2}$
$\dfrac{false + false}{2} = false$	$\dfrac{false + false}{2} = false$	$\dfrac{false + false}{2} = false$	$\dfrac{false + false}{2} = false$

Table 7. First operation

$p \wedge q$	$p \wedge q = \dfrac{true}{2} \wedge \dfrac{true}{2} = \left\{ false, \dfrac{true}{2} \right\}$	$\dfrac{true + true}{2} = true$	$\dfrac{true}{2}$
$\dfrac{false + false}{2} = false$	$\dfrac{true + true}{2} = true$	$\dfrac{true + true}{2} = true$	$\dfrac{true + false}{2} = \dfrac{true}{2}$
$\dfrac{false + false}{2} = false$	$\dfrac{true}{2}$	$\dfrac{false + true}{2} = \dfrac{true}{2}$	$\dfrac{false + false}{2} = false$
$\dfrac{false + false}{2} = false$	$\dfrac{false + false}{2} = false$	$\dfrac{false + false}{2} = false$	$\dfrac{false + false}{2} = false$

Table 8. Second operation

$p \wedge q$	$\dfrac{false + false}{2} = false$	$\dfrac{true + true}{2} = true$	$\dfrac{true}{2}$
$\dfrac{false + false}{2} = false$	$\dfrac{true + true}{2} = true$	$\dfrac{true + true}{2} = true$	$\dfrac{true + false}{2} = \dfrac{true}{2}$
$\dfrac{false + false}{2} = false$	$\dfrac{true}{2}$	$\dfrac{false + true}{2} = \dfrac{true}{2}$	$\dfrac{true}{2}$
$\dfrac{false + false}{2} = false$	$\dfrac{false + false}{2} = false$	$\dfrac{false + false}{2} = false$	$\dfrac{false + false}{2} = false$

Figure 12. The neuron image of the operation presented in Table 6

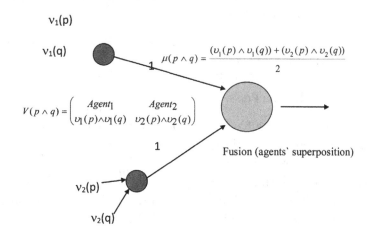

CONCLUSION

In this chapter, we have summarized the Agent-based Uncertainty Theory (AUT) and had shown how it can model the inconsistent logic values of the quantum computer and represent a new many-valued logic neuron.

We demonstrated that the AUT modeling of the inconsistent logic values of the quantum computer is a way to solve this old problem of inconsistency in quantum mechanics and quantum computer with mutually exclusive states. The general classical approach is based on states and differential equations without introduction of elaborated logic analysis. In the quantum computer, the unitary transformation has been introduced in the literature as media to represent classical logic operations in the form of Boolean calculus in the quantum phenomena. In the quantum computer, superposition gives us a new type of logic operation for which the mutual exclusion principle is not true. The same particle can be in two different positions in the same time. This chapter explained in a formal way how the quantum phenomena of the superposition can be formalized by the AUT

Many uncertainty theories emerged such as fuzzy set theory, rough set theory, and evidence theory. The AUT allows reformulating them in a similar way as we did for the quantum mechanics and quantum computer in this chapter. Now with the AUT, it is possible to generate neuron models and fuzzy neuron models that deal with intrinsic conflicting situations and inconsistency. The AUT opens the opportunities to rebuild previous models of uncertainty with the explicit and consistent introduction of the conflicting and inconsistent phenomena by the means of the many-valued logic.

REFERENCES

Abbott, A., Doering, C., Caves, C., Lidar, D., Brandt, H., & Hamilton, A. (2003). Dreams versus Reality: Plenary Debate Session on Quantum Computing. *Quantum Information Processing*, *2*(6), 449–472. doi:10.1023/B:QINP.0000042203.24782.9a

Atanassov, K. T. (1999). *Intuitionistic Fuzzy Sets, Physica Verlag*. Heidelberg: Springer.

Baki, B., Bouzid, M., Ligęza, A., & Mouaddib, A. (2006). A centralized planning technique with temporal constraints and uncertainty for multi-agent systems. *Journal of Experimental & Theoretical Artificial Intelligence*, *18*(3), 331–364. doi:10.1080/09528130600906340

Benenti, G. (2004). *Principles of Quantum Computation and Information (Vol. 1)*. New Jersey: World Scientific.

Carnap, R., & Jeffrey, R. (1971). *Studies in Inductive Logics and Probability (Vol. 1*, pp. 35–165). Berkeley, CA: University of California Press.

Chalkiadakis, G., & Boutilier, C. (2008). Sequential Decision Making in Repeated Coalition Formation under Uncertainty, In: Proc. of 7th Int. Conf. on Autonomous Agents and Multi-agent Systems (AA-MAS 2008), Padgham, Parkes, Müller and Parsons (eds.), May, 12-16, 2008, Estoril, Portugal, http://eprints.ecs.soton.ac.uk/15174/1/BayesRLCF08.pdf

Colyvan, M. (2004). The Philosophical Significance of Cox's Theorem. *International Journal of Approximate Reasoning, 37*(1), 71–85. doi:10.1016/j.ijar.2003.11.001

Colyvan, M. (2008). Is Probability the Only Coherent Approach to Uncertainty? *Risk Analysis, 28*, 645–652. doi:10.1111/j.1539-6924.2008.01058.x

D'Espagnat, B. (1999). *Conceptual Foundation of Quantum mechanics* (2nd ed.). Perseus Books.

DiVincenzo, D. (1995). Quantum Computation. *Science, 270*(5234), 255–261. doi:10.1126/science.270.5234.255

DiVincenzo, D. (2000). The Physical Implementation of Quantum Computation. *Experimental Proposals for Quantum Computation.* arXiv:quant-ph/0002077

Edmonds, B. (2002). Review of Reasoning about Rational Agents by Michael Wooldridge. *Journal of Artificial Societies and Social Simulation, 5*(1). Retrieved from http://jasss.soc.surrey.ac.uk/5/1/reviews/edmonds.html.

Fagin, R., & Halpern, J. (1994). Reasoning about Knowledge and Probability. *Journal of the ACM, 41*(2), 340–367. doi:10.1145/174652.174658

Ferber, J. (1999). *Multi Agent Systems*. Addison Wesley.

Feynman, R. (1982). Simulating physics with computers. *International Journal of Theoretical Physics, 21*, 467. doi:10.1007/BF02650179

Flament, C. (1963). *Applications of graphs theory to group structure*. London: Prentice Hall.

Gigerenzer, G., & Selten, R. (2002). *Bounded Rationality*. Cambridge: The MIT Press.

Halpern, J. (2005). *Reasoning about uncertainty*. MIT Press.

Harmanec, D., Resconi, G., Klir, G. J., & Pan, Y. (1995). On the computation of uncertainty measure in Dempster-Shafer theory. *International Journal of General Systems, 25*(2), 153–163. doi:10.1080/03081079608945140

Hiroshi, I., & Masahito, H. (2006). *Quantum Computation and Information*. Berlin: Springer.

Hisdal, E. (1998). *Logical Structures for Representation of Knowledge and Uncertainty*. Springer.

Jaeger, G. (2006). *Quantum Information: An Overview*. Berlin: Springer.

Kahneman, D. (2003). Maps of Bounded Rationality: Psychology for Behavioral Economics. *The American Economic Review, 93*(5), 1449–1475. doi:10.1257/000282803322655392

Kovalerchuk, B. (1990). Analysis of Gaines' logic of uncertainty, In I.B. Turksen (Ed.), *Proceedings of NAFIPS '90* (Vol. 2, pp. 293-295).

Kovalerchuk, B. (1996). Context spaces as necessary frames for correct approximate reasoning. *International Journal of General Systems, 25*(1), 61–80. doi:10.1080/03081079608945135

Kovalerchuk, B., & Vityaev, E. (2000). *Data mining in finance: advances in relational and hybrid methods*. Kluwer.

Montero, J., Gomez, D., & Bustine, H. (2007). On the relevance of some families of fuzzy sets. *Fuzzy Sets and Systems, 16*, 2429–2442. doi:10.1016/j.fss.2007.04.021

Nielsen, M., & Chuang, I. (2000). *Quantum Computation and Quantum Information*. Cambridge: Cambridge University Press.

Priest, G., & Tanaka, K. Paraconsistent Logic. (2004). Stanford Encyclopedia of Philosophy. http://plato.stanford.edu/entries/logic-paraconsistent.

Resconi, G., & Jain, L. (2004). *Intelligent agents*. Springer Verlag.

Resconi, G., Klir, G. J., Harmanec, D., & St. Clair, U. (1996). Interpretation of various uncertainty theories using models of modal logic: a summary. *Fuzzy Sets and Systems, 80*, 7–14. doi:10.1016/0165-0114(95)00262-6

Resconi, G., Klir, G. J., & St. Clair, U. (1992). Hierarchical uncertainty metatheory based upon modal logic. *International Journal of General Systems, 21*, 23–50. doi:10.1080/03081079208945051

Resconi, G., Klir, G.J., St. Clair, U., & Harmanec, D. (1993). The integration of uncertainty theories. *Intern. J. Uncertainty Fuzziness knowledge-Based Systems, 1*, 1-18.

Resconi, G., & Kovalerchuk, B. (2006). The Logic of Uncertainty with Irrational Agents In *Proc. of JCIS-2006 Advances in Intelligent Systems Research, Taiwan*. Atlantis Press

Resconi, G., Murai, T., & Shimbo, M. (2000). Field Theory and Modal Logic by Semantic field to make Uncertainty Emerge from Information. *International Journal of General Systems, 29*(5), 737–782. doi:10.1080/03081070008960971

Resconi, G., & Turksen, I. B. (2001). Canonical Forms of Fuzzy Truthoods by Meta-Theory Based Upon Modal Logic. *Information Sciences, 131*, 157–194. doi:10.1016/S0020-0255(00)00095-5

Ruspini, E. H. (1999). A new approach to clustering. *Information and Control, 15*, 22–32. doi:10.1016/S0019-9958(69)90591-9

Stolze, J., & Suter, D. (2004). *Quantum Computing*. Wiley-VCH. doi:10.1002/9783527617760

Sun, R., & Qi, D. (2001). Rationality Assumptions and Optimality of Co-learning, In *Design and Applications of Intelligent Agents* (LNCS 1881, pp. 61-75). Berlin/Heidelberg: Springer.

van Dinther, C. (2007). *Adaptive Bidding in Single-Sided Auctions under Uncertainty: An Agent-based Approach in Market Engineering (Whitestein Series in Software Agent Technologies and Autonomic Computing)*. Basel: Birkhäuser.

Vandersypen, L.M.K., Yannoni, C.S., & Chuang, I.L. (2000). *Liquid state NMR Quantum Computing*.

Von-Wun Soo. (2000). Agent Negotiation under Uncertainty and Risk In *Design and Applications of Intelligent Agents* (LNCS 1881, pp. 31-45). Berlin/Heidelberg: Springer.

Wooldridge, M. (2000). *Reasoning about Rational Agents*. Cambridge, MA: The MIT Press.

Wu, W., Ekaette, E., & Far, B. H. (2003). Uncertainty Management Framework for Multi-Agent System, Proceedings of ATS http://www.enel.ucalgary.ca/People/far/pub/papers/2003/ATS2003-06.pdf

KEY TERMS AND DEFINITIONS

Logic of Uncertainty: A field that deals with logic aspects of uncertainty modeling.

Conflicting Agents: Agents that have self-conflict or conflict with other agents in judgment of truth of specific statements.

Fuzzy Logic: A field that deals with modeling uncertainty based on Zadeh's fuzzy sets.

The Agent–Based Uncertainty Theory: (AUT): A theory that model uncertainty using the concept of conflicting agents.

Neural Network: Used in this chapter to denote any type of an artificial neural network.

Quantum Computer: Used in this chapter to denote any systems that provide computations based on qubits.

Fusion: Used in this chapter to denote a fusion of multi-dimensional conflicting judgments of agents.

Section 3
Bio-Inspired Agent-Based
Artificial Markets

Chapter 5

Bounded Rationality and Market Micro-Behaviors:
Case Studies Based on Agent-Based Double Auction Markets

Shu-Heng Chen
National Chengchi University, Taiwan

Ren-Jie Zeng
Taiwan Institute of Economic Research, Taiwan

Tina Yu
Memorial University of Newfoundland, Canada

Shu G. Wang
National Chengchi University, Taiwan

ABSTRACT

We investigate the dynamics of trader behaviors using an agent-based genetic programming system to simulate double-auction markets. The objective of this study is two-fold. First, we seek to evaluate how, if any, the difference in trader rationality/intelligence influences trading behavior. Second, besides rationality, we also analyze how, if any, the co-evolution between two learnable traders impacts their trading behaviors. We have found that traders with different degrees of rationality may exhibit different behavior depending on the type of market they are in. When the market has a profit zone to explore, the more intelligent trader demonstrates more intelligent behaviors. Also, when the market has two learnable buyers, their co-evolution produced more profitable transactions than when there was only one learnable buyer in the market. We have analyzed the trading strategies and found the learning behaviors are very similar to humans in decision-making. We plan to conduct human subject experiments to validate these results in the near future.

DOI: 10.4018/978-1-60566-898-7.ch005

INTRODUCTION

It is not from the benevolence of the butcher, the brewer, or the baker that we expect our dinner, but from their regard to their own interest. (Adam Smith, The Wealth of Nations, 1776)

In the classic *An Inquiry into the Natures and Causes of the Wealth of Nations*, the great economist Adam Smith demonstrated that an individual pursuing his own self-interest also promotes the good of his community as a whole, through a principle that he referred to as "invisible hand". Since then, the study of individual behaviors in a market economy has evolved into the field of *microeconomics*.

In a standard market, buyers and sellers interact to determine the price of a commodity or service. During the trading process, individuals maximize their own profits by adopting different strategies based on their experiences, familiarity with the commodity and the information they acquired. These differences in individual qualities in decision-making can also be explained by the concept of *bounded rationality* introduced by Herbert Simon (1997), who pointed out that perfectly rational decisions are often not feasible in practice due to the finite computational resources available for making them. As a result, humans employ heuristics to make decisions rather than a strict rigid rule of optimization. The difference in human qualities in decision-making is referred to as *the degree of rationality or intelligence*.

In a market that is composed of multiple self-interest traders, each of whom has a different degree of rationality, many unexpected behaviors may emerge. Our interest in studying the dynamics of these behaviors is motivated by the increasing popularity of Internet auction markets, such as *eBay* and *Amazon*. When designing an auction e-market, in addition to the maximization of macro market efficiency, the auction rules also have to consider the dynamics of auctioneers'

behaviors. In particular, would a rule create the opportunity for an auctioneer to engage in unfair bidding practices? If so, how can we prevent them from happening?

This type of preventive study is not new in the Internet auction market business. For example, to prevent "sniping" (the act of submitting a slightly higher bid than the current one at the last-minute), *eBay* has incorporated software agents in the Internet bidding process. There, each auctioneer is asked to provide his/her highest bid to an assigned agent, who then carries out the auction on his/her behalf. By contrast, *Amazon* adopts a different approach by extending the auction period for 10 more minutes if a sniper appears at the end of an auction (Roth & Ockenfels, 2002). This type of preventive study is important in order to design fair and successful auction markets.

In this chapter, we present our work using an agent-based genetic programming (GP) system (Chen & Tai, 2003) to analyze the behavior of traders with different degrees of rationality in an artificial double-auction (DA) market. This approach is different from that of experimental economics (Smith, 1976) in that instead of conducting experiments using human subjects, software agents are used to represent traders and to conduct market simulations under controlled settings. This paradigm of agent-based computational economics complements experimental economics to advance our knowledge of the dynamics of micro market behavior.

The rest of the chapter is organized as follows. Section 2 explains market efficiency and double-auction markets. Section 3 summarizes related work. In Section 4, the three types of DA market we studied are described. Section 5 presents the agent-based GP system used to conduct our experiments. The analysis of the dynamics of trading behaviors is given in Section 6. Finally, Section 7 concludes the chapter and outlines our future work.

BACKGROUND

In a standard market environment, the demand for and supply of a commodity (e.g., cotton, electricity) can be satisfied under different market prices. The demand curve gives the maximum price that consumers can accept for buying a given commodity, and the supply curve gives the minimum price at which the producers are willing to sell that commodity. For example, in Figure 1, the maximum price that buyers are willing to pay for the second unit is 26, and the minimum price that sellers are prepared to accept is 2.

Transactions only take place when the market price is below the demand curve and above the supply curve. The area between the demand and supply curves is the *surplus* region generated by the transactions. For example, for the second unit, the surplus is $26 - 2 = 24$. The distribution of the surplus depends on the transaction price. If the transaction price is 20, the surplus distributed to the consumer is $26 - 20 = 6$, and the rest 18 is distributed to the producer.

In an efficient market, a commodity's price is between the demand and supply curves so that all potential surpluses can be fully realized. When the market price is outside the curves, no transaction can occur. Consequently, the commodity stays with the producers leaving consumers unsatisfied. This market is not efficient. A double auction (DA) is one type of market structure that results in high market efficiency, and hence is very popular in the world. For example, the New York Stock Exchange (NYSE) and the Chicago Mercantile Exchange are organized as DA markets. In a DA market, both buyers and sellers can submit bids and asks. This is in contrasts with a market in which only buyers shout bids (as in an *English Auction*) or only sellers shout asks (as in a *Dutch Auction*). There are several variations of DA markets. One example is the clearinghouse DA of the Santa Fe Token Exchange (SFTE) (Rust, Miller, & Palmer, 1993) on which this work is based.

Figure 1. Demand and Supply Curves of a Market

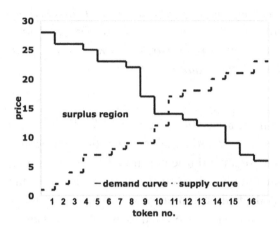

On the SFTE platform, time is discretized into alternating *bid/ask* (BA) and *buy/sell* (BS) steps. Initially, the DA market opens with a BA step in which all traders are allowed to simultaneously post bids and asks. After the clearinghouse informs the traders of each others' bids and asks, the holders of the *highest bid* and *lowest ask* are matched and enter into a BS step. During the BS step, the two matched traders perform the transaction using the *mid-point* between the *highest bid* and the *lowest ask* as the transaction price. Once the transaction is cleared, the market enters into a BA stage for the next auction round. The DA market operations are a series of alternating BA and BS steps.

RELATED WORK

Since the concept of *bounded rationality* (Simon, 1997) was introduced more than a decade ago, various studies on the impact of bounded rationality in DA markets have been reported. However, the focus of these works is on macro market efficiency instead of the dynamics of traders' behavior. For example, Gode and Sunder (1993) conducted experiments using traders with *"zero-intelligence"*, whose bids and offers were randomly generated within their budget constraints (i.e., traders were not permitted to sell below their costs or buy above their values). The DA market with only zero-intelligence traders was able to achieve almost 100% market efficiency. Based on the results, Gode and Sunder argued that the rationality of individual traders accounts for a relatively small fraction of the overall market efficiency.

To investigate the generality of Gode and Sunder's result, other researchers have conducted similar experiments using *zero-intelligence* traders in various types of DA markets. For example, Cliff and Burten (1997) studied a DA market with asymmetric supply and demand curves. They found that zero-intelligence traders gave rise to poor market efficiency. They then assigned the traders with the ability to use the closing price in the previous auction round to determine the current bid. Such traders, which they referred to as *zero-intelligence-plus*, performed better and improved market efficiency. Thus, individual traders' cognitive ability does impact overall market efficiency.

GP to Implement Bounded Rationality

Mapping an evolutionary process to model bounded rationality in human decision-making was first proposed in (Arthur, 1993). There, the author extended the key precept of bounded rationality – limits to information and cognition – by positing that *learning from experience* is important in explaining sub-optimality, the creation of heuristics, and limits to information. During the learning process, individuals improve their strategies through a Darwin process of learning-by-doing, that balance the path-dependent exploration of new strategies by extending current strategies versus simply exploiting existing strategies (Palmer, Arthur, Holland, LeBaron, & Tayler, 1994).

Genetic Programming (GP) (Koza, 1992) is one evolutionary system that has been used to implement bounded rationality. However, most

of the studies are not related to the co-evolution dynamics of individual behaviors. For example, (Manson, 2006) implemented bounded rationality using GP symbolic regression to study land change in the southern Yucatan peninsular region of Mexico. In that work, each household decision maker is represented as a GP symbolic regression. To best represent bounded rationality in his problem domain, the author investigated 5 different GP parameter settings: the fitness function, creation operator, selection operator, population size and the number of generations.

Previously, we have used GP to implement bounded rationality to study the co-evolution dynamics of traders' behaviors in an artificial DA market (Chen, Zeng, & Yu, 2009; Chen, Zeng, & Yu, 2009a). In that study, the market has two types of traders: GP traders who have the ability to learn and improve their trading strategies and naive (no-learning ability) truth-telling traders who always present the assigned prices during an auction. To distinguish the cognitive abilities of GP traders, different population sizes were assigned to these traders.

The rationale of this design decision is based on the learning from experience analogy of (Arthur, 1993). In a DA market, a trader's strategies are influenced by two factors: the trader's original ideas of how to bid/ask, and the experiences he/she learned during the auction process. In GP learning, the population is the brain that contains the possible strategies to be used for the next bid/ask. It is therefore reasonable to argue that a GP trader with a bigger population size has a larger reservoir to store and process new strategies, and hence is more intelligent.

We have designed two controlled settings to conduct the experiments. In the first setting, there was only one GP buyer among a group of truth-telling traders. The experimental results show that when assigned with a larger population size, the GP buyer was able to evolve a higher-profit strategy, which did not exist when the population size was smaller. Meanwhile, this higher-profit strategy has a more sophisticated structure, which combined two simpler strategies. These results suggest that when all other traders have no learning ability, more "intelligent" traders can make more profit.

In the second setting, the market has two GP buyers, who co-evolve their strategies to outdo each other and earn more profits. Various strategies have emerged from this competitive co-evolution environment. First, the GP buyer who was only able to evolve the higher-profit strategy under a large population size became able to evolve the same strategy with a smaller population size in the presence of another GP buyer in the market. In other words, the competitive environment led the GP buyer to learn more. Second, the strategy that was most used by the GP buyer in the first setting is replaced by a less-profitable strategy in this setting. In other words, the new GP buyer blocked the other GP buyer from learning a more profitable strategy to protect its own profit. Third, when both GP traders were given a larger population size (i.e., increased their intelligence), they learned to use more profitable strategies more often and both gained more profits.

These observed GP learning behaviors make intuitive sense. However can they be generalized to all market types? To answer this question, we have devised three less conventional market types to conduct our experiments. The following section describes these three markets.

THE DA MARKET ENVIRONMENT

The artificial DA market has 4 buyers and 4 sellers, each of whom is assigned 4 private token values for trading. For buyers, these are the 4 highest prices that they are willing to pay to purchase 4 tokens and, for sellers, they are the 4 lowest prices that they are prepared to accept to sell these tokens. All sellers in the market are truth-tellers, who always gave the assigned true token value during an auction. For buyers, however, two setups were made: one with one GP buyer and one

Figure 2. The 8 Traders' DA Market

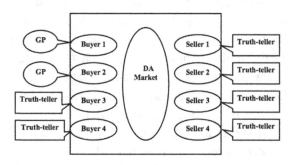

with two GP buyers. The first setting allows us to analyze GP buyers' learning behaviors under stable conditions and the second one is used to analyze the co-evolution dynamics of the two GP buyers. Figure 2 shows this market environment.

Three different markets defined by different supply and demand curves are investigated in this study. Market 1 has its supply and demand curves intersect multiple prices (see Figure 3). This market is unique in that the four buyers have 4 identical token values and the four sellers have the same 4 token values (see Table 1). When all traders are truth-tellers, only 12 of 16 tokens will be traded. The remaining 4 tokens have their supply price (cost) higher than the demand prices, and hence no transaction can take place. Among the 12 traded tokens, only 4 transactions generate a profit while the other 8 do not, because the 8 tokens have the same demand and supply prices. Also, each of the 4 profitable transactions generates profit of 4, which is allocated equally to the buyer (profit 2) and the seller (profit 2). Since each trader has only 1 profitable transaction, they all have the same daily profit of 2.

However what would happen if one or two buyers were equipped with GP learning ability? Are they able to devise strategies that generate a daily profit that is greater than 2? The answer to this question will be given in Section 6.1.

In Market II (Figure 4) and III (Figure 5), their supply and demand curves do not intersect. In

Figure 3. Market I Demand and Supply Curves

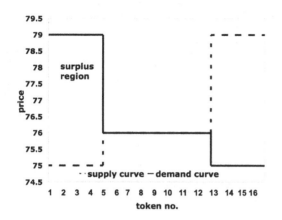

Figure 4. Market II Demand and Supply Curves

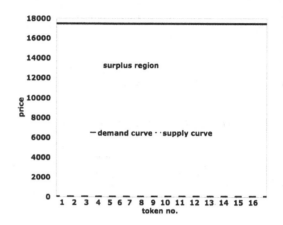

other words, there is no inequality between supply and demand. When all traders are truth-tellers, all buyers can purchase the tokens they want and all sellers will sell all the tokens they own. The token values for Market II are given in Table 2 and the token values for Market III are given in Table 3.

In Market II, the daily profit of each truth-telling trader is between 34, 859.5 and 34,863.5. In Market III, the daily profit is between 11,937.5 and 11,947.5. Would a GP buyer devise more profitable strategies in these types of markets? The answers will be given in Sections 6.2 and 6.3.

The following section presents the agent-based GP system (Chen & Tai, 2003) we used to conduct our experiments.

Figure 5. Market III Demand and Supply Curves

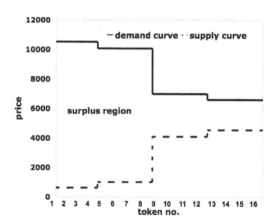

THE AGENT-BASED GP SYSTEM

In this agent-based GP system, each GP buyer evolves a population of strategies to use during an auction. The auction strategies are represented as rules. We provide 3 types of information for GP to construct these rules (see Table 4):

- Past experiences: terminals 1–9 and 16–17;
- Time information: terminals 10–11;

- Private information: terminals 12–14;

In this implementation, only transaction information on the previous day is provided for GP to compose bidding strategies. In our future work, we will investigate if more profitable strategies can be evolved when GP is provided with longer memory.

The three types of information are combined using logical and mathematical operators to decide the bidding prices. Table 5 lists these operators.

Table 1. Market I Token Value Table

Buyer1	Buyer2	Buyer3	Buyer4	Seller1	Seller2	Seller3	Seller4
79	79	79	79	75	75	75	75
76	76	76	76	76	76	76	76
76	76	76	76	76	76	76	76
75	75	75	75	79	79	79	79

Table 2. Market II Token Value Table

Buyer1	Buyer2	Buyer3	Buyer4	Seller1	Seller2	Seller3	Seller4
17473	17473	17473	17473	33	34	34	34
17471	17470	17470	17471	34	34	34	34
17465	17465	17465	17465	40	39	39	40
17464	17465	17465	17465	42	42	42	42

Table 3. Market III Token Value Table

Buyer1	Buyer2	Buyer3	Buyer4	Seller1	Seller2	Seller3	Seller4
10518	10519	10516	10521	622	622	618	619
10073	10072	10071	10071	1013	1010	1014	1016
6984	6981	6985	6987	4102	4101	4100	4100
6593	6593	6589	6590	4547	4548	4545	4550

Each DA market simulation is carried out with a fixed number of GP generations (g), where each generation lasts n (n= $2 \times pop\_size$) days. On each day, 4 new tokens are assigned to each of the buyers and sellers. The 8 traders then start the auction rounds to trade the 16 tokens. A buyer will start from the one with the highest price and then move to the lower priced ones while a seller will start from the one with the lowest price and then move to the higher priced ones. The day ends when either all 16 tokens have been successfully traded or the maximum number of 25 auction rounds is reached. Any un-traded tokens (due to no matching price) will be cleared at the end of each day. The following day will start with a new set of 16 tokens.

On each day, a GP buyer will randomly select one strategy from its population and use it the en-tire day to decide the bidding prices. The strategy might be to *pass* the round without giving a bid. By contrast, a truth-telling trader never passes an auction. A truth-telling buyer bids with the highest value of the tokens it owns while a truth-telling seller asks for the lowest value of the token it has. The same 8 strategies will play for the day's 25 auction rounds, during which a GP trader may give a different bidding price if the auction strategy uses information from the previous round/day. The truth-teller, however, will always present the same bid/ask through out the 25 rounds.

In each auction round, after all 8 traders have presented their prices, the highest bid and the lowest ask will be selected. If there are multiple buyers giving the same highest bid or multiple sellers giving the same lowest ask, one of them will be selected based on their order, i.e. buyer

Table 4. Terminal Set

Index	Terminal	Interpretation
1	PMax	The highest transaction price on the previous day
2	PMin	The lowest transaction price on the previous day
3	PAvg	The average transaction price on the previous day
4	PMaxBid	The highest bidding price on the previous day
5	PMinBid	The lowest bidding price on the previous day
6	PAvgBid	The average bidding price on the previous day
7	PMaxAsk	The highest asking price on the previous day
8	PMinAsk	The lowest asking price on the previous day
9	PAvgAsk	The average asking price on the previous day
10	Time1	The number of auction rounds left for today
11	Time2	The number of auction rounds that have no transaction
12	HTV	The highest token value
13	NTV	The second highest token value
14	LTV	The lowest token value
15	Pass	Pass the current auction round
16	CASK	The lowest asking price in the previous auction round
17	CBID	The highest bidding price in the previous auction round
18	Constant	Randomly generated constant number

Table 5. Function Set

+, -, *, %, min
>, exp, abs, log, max
sin, cos, if-then-else, if-bigger-then-else

(seller) 1 will be picked prior to buyer (seller) 2; buyer (seller) 2 will be picked before buyer (seller) 3 and so on. If the highest bid is equal to or more than the lowest ask, there is a match and the transaction takes place using the average of the bid and ask as the final price. The profit from the two strategies (the difference between the transaction and the given token values) is recorded. The fitness of the strategy F is the accumulated profit from the traded tokens during the day:

$$F = \sum_{i=1}^{m} \left| TokenValue_i - TransactionValue_i \right|$$

where m is the number of tokens traded using the strategy. Since one strategy is randomly selected each day to carry out the auction, after n= $2 \times pop\_size$ days, each strategy in the GP population will most likely be selected at least once and will have a fitness value at the end of each generation. This fitness value decides how each strategy will be selected and alternated to generate the next generation of new strategies. This sampling scheme, where each strategy is sampled either once or twice, might be too small to represent the GP learning process. We plan to carry out more studies in increasing the length of the auction period and evaluating the impact on GP learning behaviors.

When evaluating a GP-evolved trading strategy, it is possible that the output price is outside the token value range. That is, the GP buyer may buy above the value of a token. We might interpret this as a strategic way to win the auction. Since

Table 6. GP Parameters

Parameter	Value	Parameter	Value
tournament size	5	elitism size	1
initialization method	grow	max tree depth	5
population size (pop_size)	10, 50	no. of days	$2 \times$ pop_size
crossover rate	100%	subtree mutation	0.5%
no. of generation	200	point mutation	0.45%
no. of runs per setup	90	no. of GP trader	1, 2

the final price is the average of the winning bid and ask, the buyer might still make a profit from the transaction. However, such a risk-taking approach has been shown to make the market unstable and to reduce the market efficiency (Chen, S.-H., & Tai, C.-C., 2003). We therefore enforce the following rule on the price generated by a GP evolved strategy:

if Bid > 2 ×HTV then Bid=HTV

This rule protects the market from becoming too volatile and also allows GP to evolve rules that take on a small amount of risk to generate a profit.

Table 6 gives the GP parameter values used to perform simulation runs. With 2 different population sizes (110, 50) and 2 different ways to assign GP buyers, the total number of setups is 4. For each setup, we made 90 runs. The number of simulation runs made for each market is 360. With 3 market types, the total number of simulation runs is 1,080.

RESULTS AND ANALYSIS

For each market type, we analyze two scenarios: one GP buyer in the market and two GP buyers in the market. In the first case, the focus is on how GP population size influences the learning of strategies. In the second case, besides population size, we also investigate the co-evolution dynamics of two GP buyers and its impact on the learning of strategies.

To conduct our analysis, we collected all evolved strategies and their daily profit (F) generated during the last 10 generations of each run. We consider these strategies to be more "mature", and hence to better represent the GP buyers' trading patterns.

When the population size is 10, each generation is $2 \times 10 = 20$ days long. On each day, one strategy is picked randomly from the population to conduct the auction. The total number of strategies used during the last 10 generations is therefore $20 \times 10 = 200$. Since we made 90 runs for this setup, the number of strategies used to conduct our analysis is $200 \times 90 = 18,000$.

When the population size is 50, each generation is $2 \times 50 = 100$ days long. The total number of auction days (also the number of strategies picked to conduct the auction) during the last 10 generations for all 90 runs is $100 \times 10 \times 90 = 90,000$. The following subsections present our analysis of these GP evolved strategies under three different markets.

Market I

One GP Buyer in the Market

When there is one GP buyer (with population size 10) in this market, the daily profit (F) generated by the 18,000 strategies is between -41 and 3.5 (see Table 7). Among them, more than 95% of the strategies give a profit that is greater than 2, which is better than that produced by a naive

Table 7. Market I Evolved Strategies (Pop size 10)

Profit	-41	-12	-2.5	-2	0	0.5	1.5	2	2.5	3	3.5	Total
Count	6	8	37	72	378	92	15	181	475	126	16,610	18,000

Table 8. Market I Profit 3.5 Strategies (Pop size 10)

Strategy	Profit	Count	Ratio (Count/18,000)
Length ≥ 2	3.5	1,653	0.0918
NTV	3.5	14,957	0.8309
Total		16,610	0.9228

Table 9. Market I Profit 3.5 Strategies (Pop size 50)

Strategy	Profit	Count	Ratio (Count/90,000)
Length ≥ 2	3.5	5,958	0.0061
NTV	3.5	77,911	0.8657
Total		83,861	0.9318

truth-teller (see Section 4). This indicates that the GP buyer is more "intelligent" than the naive truth-telling buyers.

The strategies that generate a daily profit of 3.5 can be divided into two categories: NTV (the second highest token value) and those with length greater than or equal to 2. As shown in Table 8, NTV was used to conduct more than 83% of the auction, and we therefore decided to study how it generated the higher profit.

This strategy is actually quite smart: it bids with the second highest token value when all other truth-telling buyers bid the highest token price. During the first 3 auction rounds when at lease one truth-telling buyer bid the highest token value of 79, the GP buyer, who bid the second highest token value of 76, could not win the auction. However after the 3 truth-telling buyers purchased their first tokens and each earned a profit of 2, they moved to bid with the next highest token value of 76. Since buyer 1, the GP buyer, was preferred when there were multiple buyers giving the same highest bid (see Section

5), the GP buyer won the 4th auction round and performed the transaction using the average of the highest bid (76) and the lowest ask (75), which was 75.5. The token value that the GP buyer was purchasing was 79. So, the profit of this transaction for the GP buyer is $79 - 75.5 = 3.5$. In a market where all buyers have the same token values, this "waiting after all other buyers have purchased their tokens before winning the auction" is a more profitable strategy.

Did the more "intelligent" (population size 50) GP buyer devise a better strategy? We examined all 90,000 strategies but did not find one. Table 9 shows that the more "intelligent" GP buyer used profit -3.5 strategies slightly more often to conduct the auction (93% vs. 92%). Other than that, there was no significant difference between the behaviors of the GP buyers with population sizes of 10 and 50. This suggests that in a stable (all other traders are truth-tellers) market where all buyers have the same token values and all sellers have the same token values, a small degree of intelligence is sufficient to devise the optimal

Table 10. Market I: Strategies Used by 2 GP Buyers

Population size	Buyer	Strategy	Profit	Count	Ratio
P10	buyer 1	NTV	4	16,414	0.9119
	buyer 2	NTV	3	16,265	0.9036
P50	buyer 1	NTV	4	83,283	0.9254
	buyer 2	NTV	3	82,996	0.9222

strategy (the one that generates daily profit of 3.5 is the optimal one in this market). Any increase in the traders' intelligence/rationality has no significant impact on their behaviors. In other words, the relationship between intelligence and performance is not visible.

Two GP Buyers in the Market

When both buyers 1 and 2 are equipped with GP learning ability, the trading behaviors become more complicated. Table 10 gives information about the 2 most used strategies by the 2 GP buyers under population sizes of 10 and 50.

It appeared that both GP buyers learned the NTV strategy. When they used this strategy to bid against each other, GP buyer 1 earned a daily profit of 4 while GP buyer 2 earned a daily profit of 3. How did this happen?

We traced the market daily transactions and found that the bias in the market setup gives GP buyer 1 an advantage over GP buyer 2 who also has an advantage over buyers 3 & 4. During the first 2 auction rounds, each of the two truth-telling buyers (who bid 79) won one auction round and made a profit of 2 by carrying out the transaction using a price of $(79 + 75)/2 = 77$. In round 3, all buyers bid the second highest token value of 76. However, buyer 1, a GP buyer, is selected, based on the market setup, to carry out the transaction using the price of $(76 + 75)/2 = 75.5$. The profit earned by buyer 1 is therefore $79 - 75.5 = 3.5$. In the next auction round, all buyers bid 76 again and buyer 1 is again selected to carry out the transaction using the price of 75.5. Since GP buyer 1 is

purchasing the second token whose value is 76, the profit for this transaction is $76 - 75.5 = 0.5$. After that, GP buyer 1 did not make any profitable transaction and its total daily profit is 4.

The second buyer, who also has GP learning ability, only gets to win the auction in round 5 when the first GP buyer has purchased two tokens. In round 5, GP buyer 1 bids its next highest token value of 75 (see Table 1) and all other buyers bid 76. Buyer 2, a GP buyer, is selected over buyer 3 and 4 to carry out the transaction using the price $(76 + 76)/2 = 76$ (note that all 4 sellers are trading their second lowest token with a value of 76 as each has sold its 75 token during the first 4 auction rounds). Since GP buyer 2 is purchasing its first token with value 79, the profit gained in this transaction is $79 - 76 = 3$. After that, no market transactions are profitable due to the increase in seller token prices and the decrease in buyer token prices. The second GP buyer earned a total daily profit of 3.

When the population size of both GP buyers is increased to 50, Table 10 shows that there is no significant difference in their behaviors. This might also be due to the market type, as explained previously.

Market II

One GP Buyer in the Market

In Market II, the supply and demand curves are almost parallel with each other. A naive truth-telling strategy would trade all 16 tokens successfully. What kind of strategies would the GP

Table 11. Market II (Pop size 10)

Profit 69,705 Strategy	Count	Ratio
if bigger then else PMinAsk LT abs(sin PMin) Max CASK (- Pass Times2)	1	0.0001
42	22	0.0012
CASK	6,728	0.3738
PMaxAsk	3,428	0.1904
PMin	2,034	0.1130
PMinBid	56	0.0031
Total	12,269	0.6816

buyers evolve under this market environment? We examined all 18,000 strategies that were evolved by the GP buyer with population size 10 and found that they had 115 different daily profit values. Among them, the highest profit is 69,705, which is much higher than the profit earned by the truth-telling strategy (34,863.5). Since strategies with a profit of 69,705 were used the most (68%) during the auction, we decided to analyze how they earned the higher profit.

Table 11 gives the strategies that produce a daily profit of 69,705. Among them, the 3 mostly used strategies are:

- CASK: the lowest asking price in the previous auction;
- PMaxAsk: the highest asking price on the previous day;
- PMin: the lowest transaction price on the previous day;

One common feature of these 3 strategies is that they all used information from the previous transactions (either on the previous day or in the last auction) to decide the current bidding price. This is actually a very wise strategy in this type of market where the quantities of supply and demand are equal (16). Under such conditions, a buyer can bid any price to win 4 auction rounds as long as the price is above the sellers' token values. So, the closer the bidding price is to the sellers' token value, the higher the profit is for the buyer. However, where can a buyer find the sellers' token values? One source is the sellers' asking prices in the auction that took place on the previous day (PMaxAsk) or in the previous auction round (CASK). Another source is the transaction prices of the auction that took place on the previous day (PMin) or in the previous auction round. The GP buyer has learned that knowledge and has been able to use the acquired information to make the lowest possible bid. Consequently, its profit is way above the profit made using the truth-telling strategy.

Did the "smarter" GP buyer (with population size 50) devise a more profitable strategy for this type of market? We examined all 90,000 strategies but did not find one. Table 12 shows that the best strategies are still those that give a daily profit of 69,705. However, the "smarter" GP buyer has used this type of higher-profit strategy more frequently (86% vs. 68%) to conduct an auction. This indicates that in a stable (all other traders are truth-tellers) market and the supply and demand quantities are the same, more intelligent GP buyers exhibit more intelligent behavior by using the higher-profit strategies more frequently.

Two GP Buyers in the Market

When both buyers 1 and 2 are equipped with GP learning ability, the market becomes more competitive as both of them devise strategies to outdo each other to make a profit. Under such a

Table 12. Market II (Pop size 50)

Profit 69,705 strategies	Count	Ratio (Count/90,000)
Length ≥ 2	2,995	0.0333
42	11	0.0001
CASK	49,196	0.5466
PMaxAsk	10,652	0.1184
PMin	11,452	0.1272
PMinBid	35,58	0.0395
Total	77,864	0.8652

Table 13. Market II: Strategies Used by GP Buyers

Population size	Buyer	Profit	Count	Ratio	Total
P10	buyer 1	69,705	5,079	0.2822	
		69,710	3,903	0.2168	
		69,715	5,454	0.3030	0.8020
	buyer 2	69,705	9,885	0.5492	
		69,710	4,395	0.2442	0.7933
P50	buyer 1	69,705	26,480	0.2942	
		69,710	32,174	0.3575	
		69,715	16,281	0.1809	0.8326
	buyer 2	69705	51,756	0.5751	
		69,710	19,739	0.2193	0.7944

competitive environment, we found that both GP buyers evolved more profitable strategies than were evolved when the market only had one GP buyer. Table 13 gives the strategies evolved by the two GP buyers with population sizes 10 and 50.

Although both GP buyers evolved higher-profit strategies, GP buyer 1 evolved one group of strategies that GP buyer 2 did not evolve: the group that gives the highest profit of 69,715. Moreover, GP buyer 1 applied the two higher profit strategies (with profits 69,710 and 69,715) more frequently than GP buyer 2 did. This suggests that GP buyer 1 won the competition during the co-evolution of bidding strategies in this market. Was this really the case?

We examined all strategies and found that both GP buyers have learned to use past transac-

tion information, such as CASK, PMaxAsk and PMin, to make more profitable bids. Depending on which of these strategies were used against each other, GP buyers 1 and 2 each earned a different amount of profit. If both GP buyers used CASK, they would give the same bids. Under the market mechanism where buyer 1 was preferred over buyer 2, GP buyer 1 won the bid and earned the higher profit (69,715 vs. 69,705). However, if one GP buyer used PMaxAsk and the other used CASK, the one that used PMaxAsk would earn more profit (69,710 vs. 69,705). This is because PMaxAsk bid the highest token price of the sellers (42) while CASK bid the lowest asking price of the sellers in the previous auction round, which can be 33, 34, 39, 40 or 42 (see Table 2). Consequently, PMaxAsk won the auction and

Table 14. Market III: Most Used Strategies

Population size	Strategy	Profit	Count	Ratio	Total
P10	CASK	15,978	14,882	0.8268	0.8268
P50	Length ≥2	15,978	4331	0.0481	
	CASK	15,978	71091	0.7899	0.8380

earned the higher profit. Table 13 shows that GP buyer 1 used the profit 69,715 strategies most frequently. This indicates that GP buyer 1 won the competition due to the advantage it received from the market setup bias.

When the population size of the two GP buyers was increased to 50, the market setup bias no longer dominated the market dynamics. Instead, GP buyer 2 started to use PMaxAsk more often against GP buyer 1's CASK and earned the higher profit. As a result, the frequency with which GP buyer 1 earned profit of 69,715 was reduced (30% to 18%). Again, more intelligent GP buyers exhibited different behavior under the co-evolution setting in this market.

Market III

One GP Buyer in the Market

Market III is similar to Market II in that the quantities of supply and demand are equal (16). Did the GP buyer learn to use information from the previous auction to obtain the sellers' token price and make the lowest bid to earn the most profit? Table 14 gives the most used strategies by the GP buyer with population size 10 and 50. It is clear that the GP buyer has learned that knowledge. The most frequently used strategy is CASK, which earned a daily profit of 15,978. This profit is much higher than the profit earned by the truth-telling strategy (11,947.5).

The more intelligent GP buyer (who had a population size of 50) developed a similar style of strategies that gave a profit of 15,978. However, a small number of these strategy had more

complex structures, such as Max CASK CASK and If bigger then else CASK CASK CASK CASK. This is an understandable behavior change since a larger population size gives GP more room to maintain the same profit strategies with more diversified structures.

Two GP Buyers in the Market

Similar to the strategies in Market II, when both buyers 1 and 2 were equipped with GP learning ability, they both evolved strategies that earned more profit than that evolved when there was only 1 GP buyer in the market. Another similarity is that GP buyer 1 evolved the strategies that earned the highest profit of 17,765, which GP buyer 2 did not evolve. Table 15 gives the most used strategies by the 2 GP buyers with population sizes of 10 and 50.

The strategies also have similar dynamics to that in Market II. When both GP buyers used CASK, GP buyer 1 had an advantage and earned 17,765 while GP buyer 2 earned 15,975. When one GP buyer used CASK and the other used PMaxAsk, the one that used PMaxAsk earned a higher profit (16,863,5 vs. 15,978).

However, in this market, GP buyer 1 only used the strategies that earned 17,765 in 15% of the auction when both GP buyers had a population size of 10. This indicates that the market mechanism bias could not make buyer 1 win the competitive co-evolution in this market. Other factors, such as the supply and demand prices of the 16 tokens, also influenced the co-evolution dynamics.

Another type of GP buyers' behavior, which was different from that in Market II was that when

Table 15. Market III: Strategies of the 2 GP Buyers

Population size	Buyer	Profit	Count	Ratio	Total
P10	buyer 1	15,978	4,079	0.2266	
		16,866.5	2,050	0.1139	
		17,765	2,809	0.1561	0.4966
	buyer 2	15,975	7,754	0.4308	
		16,863.5	853	0.0474	0.4782
P50	buyer 1	15,978	32,764	0.3640	
		16,866.5	9,666	0.1074	
		17,765	26,300	0.2922	0.7637
	buyer 2	15,975	38,051	0.4228	
		16,863.5	18,945	0.2105	0.6333

the population size was increased to 50, both GP buyers increased the usage of the higher-profit strategies. In other words, more intelligent GP buyers learned to co-operate with each other, instead of competing with each other, which seems to be the behavior of the two GP buyers in Market II. More intelligent GP buyers also exhibited different behaviors in this market.

CONCLUDING REMARKS

In all three markets we have studied, the co-evolution of two self-interested GP buyers has produced more profitable transactions than when there was only one GP buyer in the market. This phenomenon was also observed in our previous work (Chen, Zeng, & Yu, 2009; Chen, Zeng, & Yu, 2009a): the overall buyer profit increases as the number of GP buyer increases in the market studied. In other words, an individual pursuing his own self-interest also promotes the good of his community as a whole. Such behavior is similar to that of humans in real markets as demonstrated by Adam Smith. Although we have only studied the case where only buyers have GP learning ability, this result suggests that to some degree, the GP trader agents have similar qualities to humans in decision-making. Meanwhile, the co-evolution

dynamics in the devised artificial DA market resembles the dynamics of real markets. We will continue to investigate the market dynamics when both buyers and sellers have GP learning ability.

Our analysis of the GP-evolved strategies shows that individual GP buyers with different degrees of rationality may exhibit different behavior depending on the type of market they are in. In Market I where all buyers have the same token values and all sellers have the same token values, the behavioral difference is not significant. However, in Markets II & III where the supply and demand prices have room to exploit a higher profit, more intelligent GP buyers exhibit more intelligent behavior, such as using higher-profit strategies more frequently or cooperating with each other to earn more profits. In (Chen, Zeng, & Yu, 2009; Chen, Zeng, & Yu, 2009a), a similar GP buyer behavioral difference in the market studied was reported. This suggests that the intelligent behavior of a GP trader becomes visible when the market has a profit zone to explore.

All of the observed individual traders' learning behaviors make intuitive sense. Under the devised artificial DA market platform, GP agents demonstrate human-like rationality in decision-making. We plan to conduct human subject experiments to validate these results in the near future.

REFERENCES

Arthur, W. B. (1993). On designing economic agents that behave like human agents. *Journal of Evolutionary Economics, 3,* 1–22. doi:10.1007/BF01199986

Chattoe, E. (1998). Just how (un)realistic are evolutionary algorithms as representations of social processes? *Journal of Artificial Societies and Social Simulation, 1.*

Chen, S.-H., & Tai, C.-C. (2003). Trading restrictions, price dynamics and allocative efficiency in double auction markets: an analysis based on agent-based modeling and simulations. *Advances in Complex Systems, 6*(3), 283–302. doi:10.1142/S021952590300089X

Chen, S.-H., Zeng, R.-J., & Yu, T. (2009). Co-evolving trading strategies to analyze bounded rationality in double auction markets . In Riolo, R., Soule, T., & Worzel, B. (Eds.), *Genetic Programming: Theory and Practice VI* (pp. 195–213). Springer. doi:10.1007/978-0-387-87623-8_13

Chen, S.-H., Zeng, R.-J., & Yu, T. (2009a). Analysis of Micro-Behavior and Bounded Rationality in Double Auction Markets Using Co-evolutionary GP . In *Proceedings of World Summit on Genetic and Evolutionary Computation.* ACM.

Cliff, D., & Bruten, J. (1997). *Zero is not enough: On the lower limit of agent intelligence for continuous double auction markets* (Technical Report HP-97-141). HP Technical Report.

Edmonds, B. (1998). Modelling socially intelligent agents. *Applied Artificial Intelligence, 12,* 677–699. doi:10.1080/088395198117587

Gode, D. K., & Sunder, S. (1993). Allocative efficiency of markets with zero-intelligence traders: markets as a partial substitute for individual rationality. *The Journal of Political Economy, 101,* 119–137. doi:10.1086/261868

Koza, J. R. (1992). *Genetic Programming: On the Programming of Computers by Means of Natural Selection.* MIT Press.

Manson, S. M. (2006). Bounded rationality in agent-based models: experiments with evolutionary programs. *International Journal of Geographical Information Science, 20*(9), 991–1012. doi:10.1080/13658810600830566

Palmer, R. G., Arthur, W. B., Holland, J. H., LeBaron, B., & Tayler, P. (1994). Artificial economic life: a simple model of a stock market. *Physica D. Nonlinear Phenomena, 75*(1-3), 264–274. doi:10.1016/0167-2789(94)90287-9

Roth, A. E., & Ockenfels, A. (2002). Last-minute bidding and the rules for ending second-price auction: evidence from Ebay and Amazon auctions on the Internet. *The American Economic Review, 92,* 1093–1103. doi:10.1257/00028280260344632

Rust, J., Miller, J., & Palmer, R. (1993). Behavior of trading automata in a computerized double auction market . In Friedmand, D., & Rust, J. (Eds.), *The Double Auction Market: Institutions, Theories and Evidence* (pp. 155–198). Addison-Wesley.

Simon, H. A. (1997). Behavioral economics and bounded rationality . In Simon, H. A. (Ed.), *Models of Bounded Rationality* (pp. 267–298). MIT Press.

Smith, V. (1976). Experimental economics: induced value theory. *The American Economic Review, 66*(2), 274–279.

KEY TERMS AND DEFINITIONS

Agent-based Modeling: A class of computational models for simulating the actions and interactions of autonomous agents (both individual or collective entities such as organizations or groups) with a view to assessing their effects on the system as a whole.

Genetic Programming (GP): An evolutionary algorithm-based methodology inspired by biological evolution to find computer programs that perform a user-defined task.

Bounded Rationality: A concept based on the fact that rationality of individuals is limited by the information they have, the cognitive limitations of their minds, and the finite amount of time they have to make decisions.

Co-evolution: "The change of a biological or artificial agents triggered by the change of a related agent"

Double Auction: A process of buying and selling goods when potential buyers submit their bids and potential sellers simultaneously submit their ask prices to an auctioneer, and then an auctioneer chooses some price p that clears the market: all the sellers who asked less than p sell and all buyers who bid more than p buy at this price p.

Chapter 6
Social Simulation with Both Human Agents and Software Agents:
An Investigation into the Impact of Cognitive Capacity on Their Learning Behavior

Shu-Heng Chen
National Chengchi University, Taiwan

Chung-Ching Tai
Tunghai University, Taiwan

Tzai-Der Wang
Cheng Shiu University, Taiwan

Shu G. Wang
Chengchi University, Taiwan

ABSTRACT

In this chapter, we will present agent-based simulations as well as human experiments in double auction markets. Our idea is to investigate the learning capabilities of human traders by studying learning agents constructed by Genetic Programming (GP), and the latter can further serve as a design platform in conducting human experiments. By manipulating the population size of GP traders, we attempt to characterize the innate heterogeneity in human being's intellectual abilities. We find that GP traders are efficient in the sense that they can beat other trading strategies even with very limited learning capacity. A series of human experiments and multi-agent simulations are conducted and compared for an examination at the end of this chapter.

DOI: 10.4018/978-1-60566-898-7.ch006

INTRODUCTION

The double auction is the core trading mechanism for many commodities, and therefore a series of human experiments and agent-based simulation studies have been devoted to studying the price formation processes or to looking for effective trading strategies in such markets. Among these studies, experiments on human-agent interactions such as Das, Hanson, Kephart, & Tesauro (2001), Taniguchi, Nakajima, & Hashimoto (2004), and Grossklags & Schmidt (2006), as well as computerized trading tournaments such as Rust, Miller, & Palmer (1993, 1994) have exhibited a general superiority of computerized trading strategies over learning agents, where the learning agents may stand for learning algorithms or human traders.

In Rust, Miller, & Palmer (1993, 1994)'s trading program tournaments, adaptive trading strategies did not exhibit human-like "intuitive leaps" that human traders seem to make in conjecturing good strategies based on limited trading experiences, although Rust, Miller, & Palmer (1993, 1994) also expected that most human traders will not be able to outperform software trading strategies because of their limited computational capability.

Rust, Miller, & Palmer (1993, 1994)'s conjecture can be evident when a series of human-agent interaction experiments is conducted. In Das, Hanson, Kephart, & Tesauro (2001), Taniguchi, Nakajima, & Hashimoto (2004), and Grossklags & Schmidt (2006)'s studies, in most of the situations, human traders cannot compete with their software counterparts. It seems that human traders could learn. However, due to some uncertain limitations, they just cannot win.

The ineffectiveness of learning behavior in double auction markets raises an interesting question: If learning is ineffective, then it implies that all human trading activities in double auction markets should have been replaced by trading programs since programs can perform better and more quickly. However, this is not the case in real

markets. So, what is the unique property of human learning behavior in double auction markets?

For the above question, Rust, Miller, & Palmer (1994) speculated that human traders do not outperform software strategies because they are constrained by their computational capacity, but they do have the advantage of being adaptive to a wide range of circumstances:

"The key distinction is adaptivity. Most of the programs are 'hardwired' to expect a certain range of trading environments: if we start to move out of this range, we would expect to see a serious degradation in their performance relative to humans. ... Anyone who has actually traded in one of these DA markets realizes that the flow of events is too fast to keep close track of individual opponents and do the detailed Bayesian updating suggested by game theory." (Rust, Miller, & Palmer, 1994, p.95)

Thus, learning ability constrained by computational capacity could be not only an explanation of how human traders are different from software strategies, but also the source of heterogeneity observed among human decision makers.

Obviously, unless the innate property of human learning behavior is captured and well characterized in software agents, we cannot build a sufficiently adequate model to describe what happens in real auction markets, let alone move on to evaluate alternative market institutions with agent-based systems populated by autonomous agents. As a result, agent-based computational economists have contributed much in discovering or inventing proper algorithms based on human experiments to describe human learning processes. However, besides Casari (2004), not much has been done to consider the impact of cognitive capacity on human traders' learning ability.

Therefore, in this chapter we initiate a series of experiments to test the possibility of constructing learning agents constrained by their cognitive capacity. We achieve this by modeling learning agents with Genetic Programming (GP) and then we manipulate their "cognitive capacity" or "computational capacity" by assigning GP traders with populations of different sizes.

Unlike common practices, we do not construct our agents based on the results of human experiments because human factors are so difficult to control and observe, and therefore it might not be easy to elicit definitive conclusions from human experiments. Instead, we adopted the Herbert Simon way of studying human behavior—understanding human decision processes by conducting computer simulations. Thus agent-based simulations in this chapter are not only research instruments used to test our conjecture, but they also serve as the design platform for human experiments.

We first model learning agents with Genetic Programming, and the population sizes of GP traders are regarded as their cognitive capacity. These GP traders are sent to the double auction markets to compete with other designed strategies. With the discovery of the capability of GP learning agents, we further conduct human experiments where human traders encounter the same opponents as GP agents did. By comparing the behavior and learning process of human traders with those of GP agents, we have a chance to form a better understanding of human learning processes.

This chapter is organized as follows: Section 2 will introduce related research to supply the background knowledge needed for this study. Section 3 depicts the experimental design, including the trading mechanism, software trading strategies, the design of GP learning agents, and experimental settings. The results, evaluations, and analysis of the experiments are presented in Section 4. Section 5 provides the concluding remarks.

LITERATURE REVIEW

In this section, we will present a series of related studies, which inspired the research behind this

chapter or provided the foundation of our research method. First, we will go through several double auction experiments to support our research question. Second, we will conduct a brief survey of how cognitive capacity is found to be decisive in human decision making. In the end, we will talk about how to model cognitive capacity in agent-based models.

Trading Tournaments in Double Auction Markets

The pioneering work in exploring individual characteristics of effective trading strategies in double auction markets consists of Rust, Miller, & Palmer (1993, 1994)'s tournaments held in the Santa Fe Institute. Rust, Miller, & Palmer (1993, 1994) collected 30 trading algorithms and categorized them according to whether they were simple or complex, adaptive or non-adaptive, predictive or non-predictive, stochastic or non-stochastic, and optimizing or non-optimizing.

Rust, Miller, & Palmer (1993, 1994) conducted the double auction tournament in a very systematic way. They proposed a random token generation process to produce the demand and supply schedules needed in their tournaments. A large amount of simulations which cover various kinds of market structures were performed, and an overall evaluation was made to distinguish effective strategies from poor ones.

The result was rather surprising: the winning strategy was simple, non-stochastic, non-predictive, non-optimizing, and most importantly non-adaptive. In spite of this, other strategies possessing the same characteristics still performed poorly. As a result, it remains an open question "whether other approaches from the literature on artificial intelligence might be sufficiently powerful to discover effective trading strategies." (Rust, Miller, & Palmer, 1994, pp. 94–95)

It is important to note that there are certain sophisticated strategies in Rust, Miller, & Palmer (1993, 1994)'s tournaments, and some of them even make use of an artificial intelligence algorithm as the learning scheme. Compared to simple strategies, such learning agents did not succeed in improving their performance within a reasonable period of time. Therefore Rust, Miller, & Palmer (1994) deemed that humans may perform better because they can generate good strategies based on very limited trading experiences.

The comparisons of learning agents versus designed strategies have assumed a different form in a series of human-agent interaction studies. Three projects will be introduced here, including Das, Hanson, Kephart, & Tesauro (2001), Taniguchi, Nakajima, & Hashimoto (2004), and Grossklags & Schmidt (2006).

Das, Hanson, Kephart, & Tesauro (2001) employed a continuous double auction market as the platform and had human traders compete with two software trading strategies—ZIP and GD. ZIP is an adaptive strategy proposed by Cliff & Bruten (1997), and a GD strategy is proposed by Gjerstad & Dickhaut (1998). In order to investigate the potential advantage of software strategies due to their speed, Das, Hanson, Kephart, & Tesauro (2001) distinguished fast agents from slow agents by letting slow agents 'sleep' for a longer time. Human traders encounter three kinds of opponents, namely GD Fast, ZIP Fast, and ZIP Slow opponents in Das, Hanson, Kephart, & Tesauro (2001)'s experiments.

The results show that regardless of whether the agents are fast or slow, they all surpass human traders and keep a very good lead. Although human traders seem to improve over time, they still cannot compete with software strategies at the end of the experiments.

The superiority of software strategies is further supported by Taniguchi, Nakajima, & Hashimoto (2004)'s futures market experiments. Taniguchi, Nakajima, & Hashimoto (2004) use the U-Mart futures market as the experimental platform where human traders and random bidding agents compete to buy or sell contracts at the same time. Before the experiments, Taniguchi, Nakajima, & Hashi-

moto (2004) trained their human subjects with knowledge about the futures and stock markets, related technical and fundamental trading strategies, and the operations of trading interface for 90 minutes. In addition to this arrangement, the trading mechanism of U-Mart, named "Itayose," is special in that the market matches the outstanding orders every 10 seconds. As a result, human traders have more time to contemplate and make their bids or offers. Both of the above designs enable human traders to have more advantages to compete with the randomly bidding software strategies.

However, the results show that human traders have poorer performance than software agents, although there is a human trader who learns to speculate and can defeat software strategies. In spite of their results, Taniguchi, Nakajima, & Hashimoto (2004)'s experiments still exhibit the possibility of defeating software strategies with human intelligence.

Unlike previous studies, Grossklags & Schmidt (2006)'s research question is more distinguishing: they want to know whether human traders will behave differently when they know there are software agents in the same market. In their futures markets, they devised a software agent called "Arbitrageur." Arbitrageur's trading strategy is simple: sell bundles of contracts when their prices are above the reasonable price, and buy bundles of contracts when their prices are below the reasonable price. This is a very simple strategy which human traders may also adopt. However, software agents make positive profits in 11 out of the total of 12 experiments. Because this is a zero-sum game, the software agents' positive performance means that losses are incurred by human traders, although the differences are not statistically significant.

Similar to Rust, Miller, & Palmer (1993, 1994)'s results, Grossklags & Schmidt (2006)'s results, together with Das, Hanson, Kephart, & Tesauro (2001)'s and Taniguchi, Nakajima, & Hashimoto (2004)'s findings, demonstrate a general picture in which it is difficult for human traders to compete with software agents even if the software agents are very simple. Learning agents (either software ones or humans) can hardly defeat designed strategies in a short period of time. Nevertheless, we can also observe from these experiments that learning agents (either software or human) may have the chance to defeat software strategies if they have enough time to learn. Then some questions naturally emerge: in what situations can learning agents outperform other software strategies? Is there any mechanism which has an influence on learning agents' learning behavior?

To answer the first question, we need to conduct experiments where learning agents can exert all their potential to win the game. By considering the cost of testing human traders in various situations with long time horizons, we adopt another approach: we conduct agent-based simulation with learning agents to examine their winning conditions first, and then we run human experiments to see how things develop when learning software traders are replaced by learning humans. In selecting an appropriate algorithm to model our learning agent, we choose Genetic Programming because as Chen, Zeng, & Yu (2008)'s research shows, GP traders can evolve and adapt very efficiently in a double auction market.

Cognitive Ability and Learning Behavior

The answer to the second question raised at the end of the last section is not so obvious. To find possible factors influencing people's learning behavior, we have to consult science disciplines which have paid much attention to this issue. Fortunately, this question has been investigated by psychologists and decision scientists for a long time.

To look into possible factors of learning, we have to realize that the reason why people have to learn lies in their bounded rationality. Because

humans are boundedly rational, they cannot find the optimal solutions at the beginning and have to improve their performance based on their experiences.

Information and cognitive capacity are two important sources of bounded rationality to human decision makers. While economists, either theorists or experimentalist, have mainly emphasized the importance of information, the significance of cognitive capacity has been temporarily mislaid but has started to regain its position in economics experiments in recent years.

Some of the earliest experimental ideas concerning cognitive capacity came from Herbert Simon, who was the initiator of bounded rationality and was awarded the Nobel Memorial Prize in Economics. In the "concept formation" experiment (Gregg & Simon, 1979) and the arithmetic problem (Simon, 1981), Simon pointed out that the problem is strenuous or even difficult to solve, not because human subjects did not know how to solve the problem, but mainly because without decision supports such as paper and pencil, such tasks can easily overload human subjects' *working memory capacity* and influence their performance (Simon, 1981, 1996).

In the realm of psychology, Payne, Bettman, & Johnson (1993)'s research can be a good foundation for our research. Payne, Bettman, & Johnson (1993) pointed out that humans have different strategies for solving a specific problem, and humans will choose the strategy by considering both accuracy and the cognitive effort they are going to make. In the end, it is dependent on the cognitive capacity of human decision makers.

In addition to the above conjectures and theories, more concrete evidence is observed in economic laboratories. Devetag & Warglien (2003) found a significant and positive correlation between subjects' short-term memory scores and conformity to standard game-theoretic prescriptions in the games. Benjamin, Brown, & Shapiro (2006) imposed a cognitive load manipulation by asking subjects to remember a seven-digit number while performing the task. The results showed that the cognitive load manipulation caused a statistically significant increase in one of two measures of small-stakes risk aversion. Devetag & Warglien (2008) pointed out that subjects construct representations of games of different relational complexity and will play the games according to these representations. Their experimental results showed that both the differences in the ability to correctly represent the games and the heterogeneity of the depth of iterated thinking in games appear to be correlated with short-term memory capacity.

Consequently, we choose to have working memory capacity as the representative of cognitive capacity in this study. In order to obtain this inherent variable, we will give several tests to our subjects to measure their working memory capacity, apart from manipulating it by imposing cognitive loading tasks.

Cognitive Ability in Agent-based Models

We have seen from the last section that cognitive capacity, or more specifically, working memory, could be an important factor that has an influence on learning capability. Nevertheless, it is rarely mentioned in agent-based economic models. To the best of our knowledge, the only exception is Casari (2004)'s model about adaptive learning agents with limited working memory. Casari (2004) used Genetic Algorithms (GA) as agents' learning algorithms, and made use of the size of each agent's strategy set. The results show that the model replicates most of the patterns found in common property resource experiments.

Being inspired by different psychological studies, we adopt a similar mechanism to model learning agents' cognitive capacity in this research. We employ GP as traders' learning algorithms so that they can construct new strategies or modify old ones based on past experiences. The limits of working memory are concretized as GP traders'

population sizes. The bigger the population a GP trader has, the more capable it is of handling various concepts and structures to form its trading strategies.

EXPERIMENTAL DESIGN

In this chapter, we will report two kinds of experimental results. One is from agent-based simulations, and the other is from human experiments.

The idea of agent-based simulation in this chapter is to understand human dynamics with the tool of GP agents. The purpose of such simulations is two-fold. First, we let GP agents compete with other software trading strategies to see the potential of learning agents, and observe the conditions when learning agents can defeat other designed strategies. Second, we can test the influences of cognitive capacity by imposing different population sizes on our GP learning agents. Such manipulation of cognitive capacity is almost impossible with human subjects, and thus it will be very informative if we can have simulated results before we eventually perform human experiments.

A human experiment is conducted after the simulations. The results of human experiments will be compared with simulation results to verify whether we have found an adequate way to model human learning processes. Parameters and the design of both experiments will be presented in this section.

Market Mechanism

Experiments in this chapter were conducted on a AIE-DA (Artificial Intelligence in Economics-Double Auction) platform which is an agent-based discrete double auction simulator with built-in software agents.

AIE-DA is inspired by the Santa Fe double auction tournament held in 1990, and in this study we adopted the same token generation process as in Rust, Miller, & Palmer (1993, 1994)'s design. Our experimental markets consist of four buyers and four sellers. Each of the traders can be assigned a specific strategy–either a designed trading strategy or a GP agent.

During the transactions, traders' identities are fixed so they cannot switch between buyers and sellers. Each trader has four units of commodities to buy or to sell, and can submit only once for one unit of commodity at each step in a trading day. Every simulation lasts 7,000 trading days, and each trading day consists of 25 trading steps. AIE-DA is a discrete double auction market and adopts AURORA trading rules such that at most one pair of traders is allowed to make a transaction at each trading step. The transaction price is set to be the average of the winning buyer's bid and the winning seller's ask.

At the beginning of each simulation, each trader will be randomly assigned a trading strategy or as a GP agent. Traders' tokens (reservation prices) are also randomly generated with random seed 6453. Therefore, each simulation starts with a new combination of traders and a new demand and supply schedule.

Software Strategies

In order to test the ability of GP agents, we programmed several trading strategies from the double auction literature as GP agents' competitors:

- **Truth Teller**: Truth-telling traders who simply use their reservation prices as their bids or asks.
- **Kaplan**, **Ringuette**, and **Skeleton** from Rust, Miller, & Palmer (1993, 1994)'s tournament: Skeleton is the strategy supplied to all entrants in their competition, and it makes safe bids/asks according to current bids or asks in the market. Kaplan and Ringuette were the best and second-best traders, respectively, their trading philosophy being to wait in the background

and let others negotiate, and then steal the deal when the bids and asks got close enough. In our simulations, we modified these three strategies so that they became more conservative in their bids and offers: when they are going to send their orders to the market, they will choose a number based on their next token values instead of current ones, which means their bids and offers are less competitive but are more profitable if they succeed in trading.

- **ZIC** (Zero-Intelligence Constrained) from Gode & Sunder (1993): ZIC traders send random bids or asks to the market in a range bounded by their reservation prices; hence they can avoid transactions which incur losses.
- **ZIP** (Zero-Intelligence Plus) from Cliff & Bruten (1997): A ZIP trader forms bids or asks by a chosen profit margin, and tries to choose a reasonable profit margin by inspecting its status, the latest shout price, and whether the shouted prices are accepted or not.
- **Markup** from Zhan & Friedman (2007): Markup traders setup certain markup rates and consequently determine their shouted prices. In this chapter, the markup rate was set to be 0.1. We choose 0.1 because Zhan and Friedman's simulations show that the market efficiency will be maximized when traders all have 0.1 markup rates.
- **Gjerstad-Dickhaut** (GD) from Gjerstad & Dickhaut (1998): A GD trader scrutinizes the market history and calculates the possibility of successfully making a transaction with a specific shouted price by counting frequencies of past events. After that, the trader simple chooses a price as her bid/ask if it maximizes her expected profits.
- **BGAN** (Bayesian Game Against Nature) from Friedman (1991): BGAN traders treat the double auction environment as a game against nature. They form beliefs in other

traders' bids or asks distribution and then compute the expected profit based on their own reservation prices. Hence their bids/asks simply equal their reservation prices minus/plus the expected profit. Finally, BGAN traders employ Bayesian updating procedures to update their prior beliefs.

- **Easley-Ledyard** (EL) from Easley & Ledyard (1993): EL traders balance the profit and the probability of successfully making transactions by placing aggressive bids or asks in the beginning, and then gradually decrease their profit margin when they observe that they might lose chances based on other traders' bidding and asking behavior.
- **Empirical** strategy is inspired by Chan, LeBaron, Lo, & Poggio, and it works in the same way as Friedman's BGAN but develops its belief by constructing histograms from opponents' past shouted prices.

Named by or after their original designers, these strategies were modified to accommodate our discrete double auction mechanism in various ways. They were modified according to their original design concepts as much as possible. As a result, they might not be 100% the same as they originally were.

Although most of the strategies were created for the purpose of studying price formation processes, we still sent them to the "battlefield" because they can represent, to a certain degree, various types of trading strategies which can be observed in financial market studies.

GP Trading Agents

GP agents in this study adopt only standard crossover and mutation operations, by which it is meant that no election, ADFs (Automatic Defined Functions), nor other mechanisms are implemented. We provide GP traders with simple but basic market information as their terminals, such as traders' own reservation prices, current market shouts, and average price in the last period, etc. We adopt the same design of genetic operations as well as terminal and function sets of GP traders as Chen, Chie, & Tai (2001) describes, apart from a different fitness calculation. The fitness value of GP traders is defined as the individual efficiency achieved, which will be explained later in this chapter.

We did not train our GP traders before they were sent to the double auction tournament. At the beginning of every trading day, each GP trader randomly picks a strategy from his/her population of strategies and uses it throughout the whole day. The performance of each selected strategy is recorded, and if a specific strategy is selected more than once, a weighted average will be taken to emphasize later experiences.

GP traders' strategies are updated–with selection, crossover, and mutation–every N days, where N is called the "select number." To avoid the flaw that a strategy is deserted simply because it was not selected, we set N as twice the size of the population so that theoretically each strategy has the chance to be selected twice. Tournament selection is implemented and the size of the tournament is 5, however big the size of the population is. We also preserve the elite for the next generation, and the size of the elite is 1.[1] The mutation rate is 5%, in which 90% of this operation consists of a tree mutation.[2]

Experimental Procedures

Since we have only eight traders (four buyers and four sellers) in the market while there are twelve trading strategies to be tested, we have to compare these strategies by randomly sampling (without replacement) eight strategies and inject them into the market one at a time. However, considering the vast amount of combinations and permutations of strategies, we did not try out all the possibilities. Instead, 300 random match-ups were created for each series of experiment. In each of these match-ups, any selected strategy will face strategies completely different from its own kind. That is, a certain type of strategy such as ZIC will never meet another ZIC trader in the same simulation. Thus, there is at most one GP trader in each simulated market, and this GP trader adjusts its bidding/asking behavior by learning from other kinds of strategies. There is no co-evolution among GP traders in our experiments.

In order to examine the validity of using population sizes as GP traders' intelligence, a series of experiments were conducted, and GP traders' population sizes were set at 5, 20, 30, 40, 50, 60, 70, 80, 90, and 100 respectively. As a result, we carry out 10 multi-agent experiments for different-size GP traders. In each experiment, there are 300 simulations due to random match-ups of strategies. On each simulation, the same market demand and supply is chosen and kept constant throughout 7,000 trading days. In each trading day, buyers' and sellers' tokens are replenished so that they can start over for another 25 trading steps.

Human Subject Experiments

The final stage of this research is to compare learning agents' behavior with human traders' learning dynamics. Therefore, we have to carry out corresponding experiments on agent simulations and human subject experiments.

In a way that is different from the 10 agent-based experiments described above, we manually choose 3 different market structures to test both our GP agents and human traders. The demand and supply schedules of these markets are shown in Figure 1. These markets are chosen because of their unique properties—market 1 (M1) is a symmetric market where buyers and sellers share the market surplus, and half of each trader's tokens are intramarginal; market 2 (M2) is asymmetric and competitive in that each trader has only one intramarginal token, but buyers can compete to grasp the chance to exchange with the lowest four units of commodities; market 3 (M3) is a

Figure 1. The Demand and the Supply Curves of Human Experiments. From the top to the bottom these are M1, M2, and M3, respectively

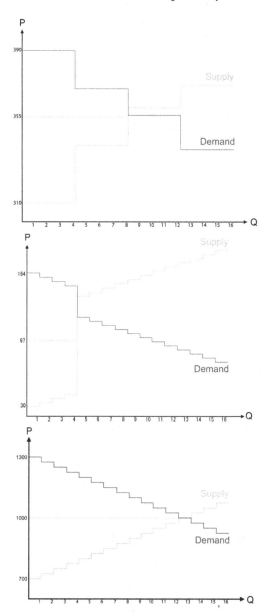

symmetric market similar to M1, but it is more profitable and there may be more space for strategic behavior.

In order to test the GP agents and human agents, we propose an environment where the opponents of GP traders and human traders are so simple that we can exclude many possible factors from interfering with the comparisons. To achieve this, we simply have truth tellers as GP traders' and human traders' opponents. As a result, each GP/ human trader will be facing seven truth tellers in his/her own market experiment.

Another important factor to test in human experiments is the effect of cognitive capacity on learning. We adopt working memory as the measure of cognitive capacity, and assess subjects' working memory capacity with a series of computerized psychological tests (Lewandowsky, Oberauer, Yang, & Ecker, 2009). The working memory tests employed here are:

- SS: Sentence Span test
- MU: Memory Updating test
- OS: Operation Span test
- SSTM: Spatial Short-term Memory test
- BDG: Backward Digit Span test

These tests are carried out after human subjects' double auction experiments so as to avoid any suggestive influence on their behavior in double auction markets.

Human subjects are randomly assigned to one of the market traders, and remain in that position throughout the three-market experiment so that they can learn from repeated experimentation.

ANALYSIS OF SIMULATIONS AND HUMAN EXPERIMENTS

In this section, we will report and analyze the results of multi-agent simulations as well as human experiments. Before proceeding with the actual analysis, we are going to acquaint the readers with the evaluation methods we used in this chapter.

In this research, we evaluate the traders' performances with a profit-variation point of view. Profitability is measured in terms of *individual efficiencies*.

Figure 2. Individual Surplus

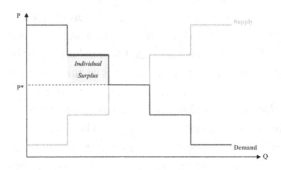

Considering the inequality in each agent's endowment due to random matching of strategies as well as random reservation prices, direct comparisons of raw profits might be biased since luck may play a very important role. To overcome this problem, a general index which can evaluate traders' relative performances in all circumstances is necessary. The idea of individual efficiency meets this requirement, and it can be illustrated by Figure 2. The individual surplus, which is the sum of the differences between one's intramarginal reservation prices and the market equilibrium price, measures the potential profit endowed by the specific position of a trader in the market. Individual efficiency is calculated as the ratio of one's actual profits to one's individual surplus, and thus measures the ability of a trader to explore its potential interests endowed in various circumstances. As demonstrated in Equation 1, individual efficiency is a ratio and can be easily compared across simulations without other manipulations such as normalization.

In addition to profits, a strategy's profit stability is also taken into account because in double auction markets, the variation in profits might be considered in human trading strategies, which are determined by the human's risk attitudes. Here we procure variations in strategies by calculating the standard deviation of each strategy's individual efficiencies.

Learning Capabilities of GP Agents

In investigating the GP traders' learning capability, we simply compare GP agents with designed strategies collected from the literature. We are interested in the following questions:

1. Can GP traders defeat other strategies?
2. How many resources are required for GP traders to defeat other strategies?

GP traders with population sizes of 5, 20, and 50 are sampled to answer these questions. Figure 3 is the result of this experiment. Here we represent GP traders of population sizes 5, 20, and 50 with P5, P20, and P50, respectively. We have the following observations from Figure 3:

*   No matter how big the population is, GP traders can gradually improve and defeat other strategies.
*   GP traders can still improve themselves even under the extreme condition of a population of only 5. The fact that the tournament size is also 5 means that strategies in the population might converge very quickly. Figure 4 shows the evolution of the average complexity of GP strategies. In the case of P5, the average complexity almost equals 1 at the end of the experiments, meaning that GP traders could still gain superior advantages by constantly updating their strategy pools composed of very simple heuristics. In contrast with P5, in the case of bigger populations, GP develops more complex strategies as time goes by.
*   What is worth noticing is that GP might need a period of time to evolve. The bigger the population, the fewer the generations that are needed to defeat other strategies. In any case, it takes hundreds to more than a thousand days to achieve good performances for GP traders.

*   Figure 3 also shows the results from a profit-variation viewpoint. Other things being equal, a strategy with higher profit and less variation is preferred. Therefore, one can draw a frontier connecting the most efficient trading strategies. Figure 3 shows that GP traders, although with more variation in profits in the end, always occupy the ends of the frontier.[3]

The result of this experiment shows that learning GP traders can outperform other (adaptive) strategies, even if those strategies may have a more sophisticated design.

Cognitive Capacity and Learning Speed

Psychologists tell us that the intelligence of human beings involves the ability to "learn quickly and learn from experiences" (Gottfredson, 1997). To investigate the influence of individual intelligence on learning speed, we think of a GP trader's population size as a proxy for his/her cognitive capacity. Is this parameter able to generate behavioral outcomes consistent with what psychological research tells us?

Figure 5 delineates GP traders' learning dynamics with a more complete sampling. Roughly speaking, we can see that the bigger the population size, the less time that GP traders need to perform well. In other words, GP traders with higher cognitive capacity tend to learn faster and consequently gain more wealth.

However, if we are careful enough, we may also notice that this trend is not as monotonic as we might think. It seems that there are three groups of learning dynamics in this figure. From P5 to P30, there exists a clearly positive relationship between "cognitive capacity" and performance. P40 and P50 form the second group: they are not very distinguishable, but both of them are better than traders with lower "cognitive capacity". The most unexplainable part is P60 to P100. Although

Figure 3. Comparison of GP Traders with Designed Strategies. From the top to the bottom rows are comparisons when GP traders' population sizes are 5, 20, and 50, respectively. (a) The left panels of each row are the time series of individual efficiencies. (b) The right panels of each row are the profit-variation evaluation on the final trading day. The horizontal axis stands for their profitability (individual efficiency, in percentage terms), and the vertical axis stands for the standard deviation of their profits.

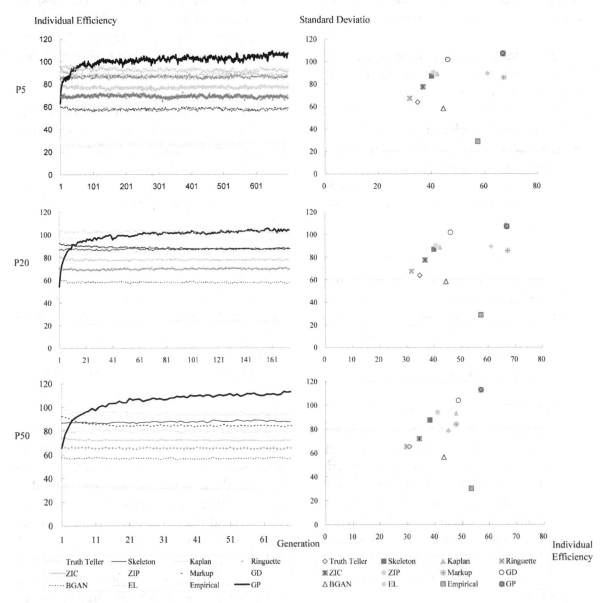

Figure 4. The Average Complexities of GP Strategies. The GP traders' population sizes are 5, 20, and 50, respectively (from the left panel to the right panel). The complexity is measured in terms of the number of terminal nodes and function nodes of GP traders' strategy parse trees.

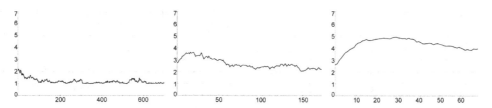

Figure 5. GP Traders' Performances at Different Levels of Cognitive Capacity. The horizontal axis denotes generations; the vertical axis consists of the individual efficiencies obtained by GP traders.

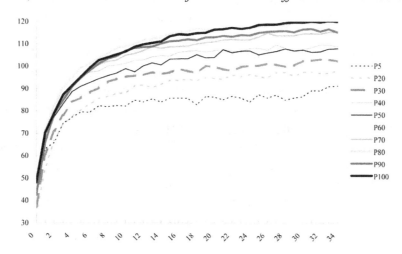

Table 1. Wilcoxon Rank Sum Tests for GP Traders' Performances on Individual Efficiencies

	P5	P20	P30	P40	P50	P60	P70	P80	P90	P100
P5	X									
P20	0.099*	X								
P30	0.010**	0.328	X							
P40	0.002**	0.103	0.488	X						
P50	0.000**	0.009**	0.129	0.506	X					
P60	0.000**	0.000**	0.003**	0.034**	0.130	X				
P70	0.000**	0.000**	0.015**	0.121	0.355	0.536	X			
P80	0.000**	0.000**	0.003**	0.036**	0.131	1.000	0.558	X		
P90	0.000**	0.000**	0.011**	0.079*	0.250	0.723	0.778	0.663	X	
P100	0.000**	0.000**	0.000**	0.002**	0.009**	0.284	0.093*	0.326	0.150	X

" * " denotes significant results under the 10% significance level; " ** " denotes significant results under the 5% significance level.

this group apparently outperforms traders with lower "cognitive capacity," the inner-group relationship between "cognitive capacity" and performance is quite obscure.

For a better understanding of this phenomenon, a series of nonparametric statistical tests were performed upon these simulation results. The outcomes of these tests are presented in Table 1. Pairwise Wilcoxon Rank Sum Tests show that when the "cognitive capacity" levels are low, small differences in cognitive capacity may result in significant differences in final performances. On the contrary, among those who have high cognitive capacity, differences in cognitive capacity do not seem to cause any significant discrepancy in performances. Therefore, there seems to be a decreasing marginal contribution in terms of performance.

This phenomenon can be an analogy of what the psychological literature has pointed out: high intelligence does not always contribute to high performance–the significance of intelligent performance is more salient when the problems are more complex. As to the decreasing marginal value of intelligence, please see Detterman & Daniel (1989) and Hunt (1995).

Human Subject Experiments

As mentioned in the section on experimental design, we conduct multi-agent simulations with GP traders for the three markets specified in Figure 1. In order to make it easier to observe and compare, we choose GP traders with population sizes based on a log scale: 5, 25, and 125. Figure 6 depicts the evolution of GP traders' performance over time.

As Figure 6 shows, GP traders learn very quickly, but they attain different levels of individual efficiencies in different markets. Does GP traders' cognitive capacity (population size) play any decision role in their performances? To have a more precise description, detailed test statistics are computed and the output is presented in Table 2.

It is shown in Table 2 that GP traders with different cognitive capacity do not have significant differences in their performances in market 1, while the differences in their cognitive capacity do bring about significant discrepancies in final performances in market 3—the bigger the population size, the better the results they can achieve. Market 2 is somewhere in between market 1 and market 3.

After a quick overview of GP traders' performance in these three markets, we now turn our attention to the results of human experiments. Unlike GP traders, it is impossible to know human traders' true cognitive capacity. Fortunately, we can have access to them via various tests which have been validated by psychologists. In our human experiments, we have twelve human subjects recruited from among graduate and undergraduate students. We measure their cognitive capacity with five working memory tests (see Table 3). In Table 3, we normalize subjects' working memory scores so that a negative number means their working memory capacity is below the average of the twelve subjects.

Each subject was facing seven truth-telling opponents in their own auction markets, and three markets (M1, M2, and M3, see Figure 1) were experienced in order by each trader. The dynamics of the human traders' performance in terms of individual efficiency is plotted in Figure 7.

We have several observations from Figure 7:

1.   Human traders have quite diverse learning patterns in market 1 and market 2, but the patterns appear to be more similar. This may be due to the idiosyncrasies of the markets, or it may be due to the learning effect of human traders so that they have come up with more efficient strategies market by market.

2.   Although there are some exceptions, human traders who have above-average working memory capacity seem to have better performance than those with below-average working memory capacity. We can see from

Figure 6. GP Traders' Performances over Time in Market 1, Market 2, and Market 3. The horizontal axis denotes generations; the vertical axis consists of the individual efficiencies (in percentage terms) obtained by GP traders.

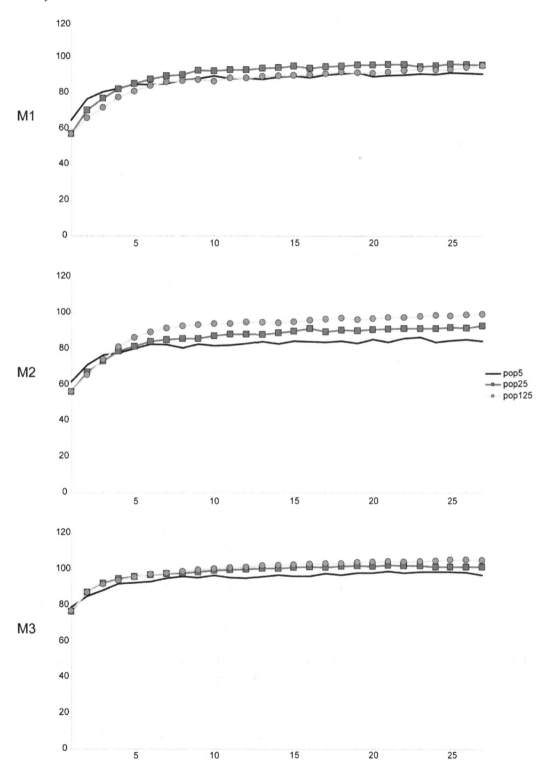

Table 2. Wilcoxon Rank Sum Tests for GP Traders' Performances in M1, M2, and M3

	P5	**P25**	**P125**
P5	X		
P25	0.2168 0.3690 0.004758**	X	
P125	0.1416 0.003733** 0.00000007873**	0.3660 0.1467 0.0004625**	X

The numbers in each cell are the p-values for the null hypothesis of no influence resulting from the difference in population sizes in M1, M2, and M3 respectively (from the top to the bottom). " * " denotes significant results under the 10% significance level; " ** " denotes significant results under the 5% significance level.

Table 3. Working Memory Capacity of Human Subjects

Subject	**SS**	**OS**	**MU**	**SSTM**	**BDG**	**WMC**	**Rank**
B	-0.36	-2.48	0.29	0.27	-1.13	-0.68	10
D	-1.79	-0.12	0.67	-0.28	1.06	-0.09	8
F	0.32	-0.59	-0.94	-1.15	-2.03	-0.88	11
G	0.65	-0.24	0.66	-0.05	0.68	0.34	4
H	-2.15	0.63	-1.42	-1.70	-0.48	-1.02	12
I	0.83	1.53	1.29	0.74	0.93	1.06	1
K	0.57	0.76	-0.17	0.35	0.55	0.41	3
L	0.06	0.35	-0.28	0.74	0.29	0.23	5
O	0.22	0.54	1.21	1.68	0.93	0.92	2
R	0.47	-0.35	0.90	0.19	-1.00	0.04	7
S	0.08	-0.55	-1.68	0.66	-0.48	-0.39	9
T	1.10	0.52	-0.53	-1.46	0.68	0.06	6

Tests scores of each test item are normalized, and the final scores (working memory capacity, WMC) are obtained by averaging these five scores.

the figure that the solid series tend to lie in a higher position than the dashed series.

3. On average, it takes human traders less than six trading periods to surpass 90%. GP traders' learning speed is about the same: it takes GP traders less than ten generations to achieve similar levels. However, we have to notice that the GP traders' generation consists of several trading periods—10 periods for P5, 50 periods for P50, and 250 periods for P125.

Thus there seems to be a big difference between their learning speeds. If we are going to have GP traders compete with human traders in the same market, we can obviously observe the difference in their learning speeds.

Although there is a difference in the GP traders' and human traders' learning speeds, suggesting that human traders may have different methods or techniques of updating and exploring their strategies, it is still possible to modify GP traders to catch up with human traders. However, what is important in this research is to delve into

Figure 7. Human Traders' Performances over Time in the Three Markets. Solid series are the learning dynamics of human traders whose working memory capacity is above average; dashed series are the learning dynamics of traders whose memory capacity is below average (the left panel). The right panels are the average performances of above- and below-average traders. The horizontal axis denotes generations; the vertical axis denotes the individual efficiencies (in percentage terms) obtained by the traders.

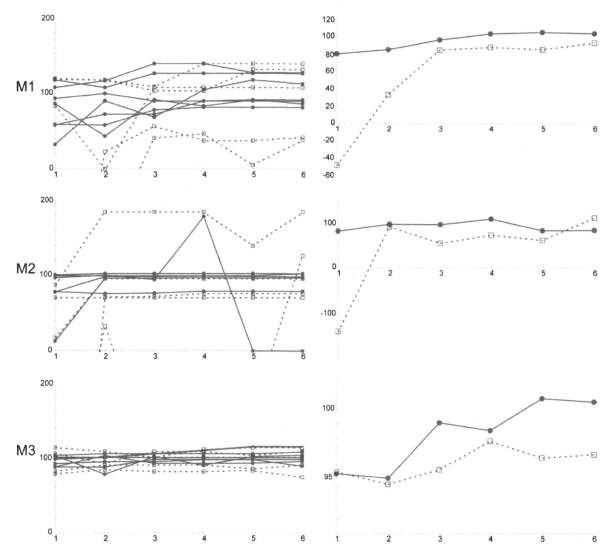

the relationship between cognitive capacity and learning behavior. Do our GP traders exhibit corresponding patterns as human traders?

Since we have seen that GP traders' cognitive capacity does not play a significant role in market 1 and market 2, but that it has a positive and significant influence on performance in market 3, will it be the same in our human experiment?

Because we cannot precisely categorize human traders according to their normalized scores, we choose to run linear regressions to see how working memory capacity contributes to human traders' performances. The results are rejected and the explanatory power of the regression model is very poor. However, we can go back to the raw data and see what might be neglected in our analysis.

Table 4. Results of Linear Regression of Working Memory Capacity on Human Traders' Performances

		Estimate	Standard Error	t statistic	p-value	Multiple R-squared Adjusted R-squared †
M1	intercept	91.444**	9.501	9.625	2.75e-05	0.3326
	wmc	31.673	16.960	1.868	0.104	0.2372 †
M2	intercept	83.44**	12.19	6.847	0.000243	0.0724
	wmc	16.08	21.76	0.739	0.483859	-0.06012 †
M3	intercept	96.556**	3.313	29.144	1.44e-08	0.3586
	wmc	11.701*	5.914	1.978	0.0884	0.267 †

"*" denotes significant results under the 10% significance level; "**" denotes significant results under the 5% significance level.

When we try to emphasize the potential influence of the cognitive capacity (here we mean the working memory capacity) on human traders' performances, we are suggesting that cognitive capacity may play a key role in the processing of information as well as the combination and construction of strategies. The assumption here is that people have to get acquainted with the problems and form their strategies from the beginning. However, this may be far from true because people come to the lab with different background knowledge and different experiences, and experimentalists can control this by excluding experienced subjects from participating in similar experiments. How can experienced subjects be excluded even if they did not participate in similar experiments before?

In this study, we can approach this problem by excluding subjects who have participated in markets which use double auctions as their trading mechanisms. From a survey after the experiments, we can identify three subjects who have experience in stock markets or futures markets.[4] Following this logic, we re-examine the relationship between the working memory capacity and human traders' performance, and the results are shown in Table 4.

As Table 4 shows, the working memory capacity only has a significant influence on traders' performances in market 3. We can compare this with the GP traders' results shown in Table 2. The results from the GP simulations tell us that the influences of cognitive capacity are significant in market 3, while they are insignificant in market 1, and only significant in market 2 when the difference in cognitive capacity is 25 times large. In brief, we have very similar patterns for the GP trader simulation and human subject experiments.

Does this prove anything related to our research goals? We have to realize that there is a limitation on our analysis so far, but the limits to our research may not necessarily work against our analytical results, but may suggest that we need more experiments and more evidence to clarify the entwined effects occurring during human traders' decision-making processes. We name several of them as follows:

1. The number of human subjects greatly limits the validity of our research. As a result, we have to conduct more human experiments to gain stronger support for our analytic results.
2. Does the significance of working memory capacity appear because of its influences in decision making, or it is because of a learning effect taking place when human subjects start from market 1 to market 3? We cannot identify the effects of various possible channels.
3. What does the pattern observed in GP simulations and human experiments mean, even if working memory capacity can really

influence economic decision making? Why does it not matter to have a larger working memory capacity in market 1 and market 2, and for it to become important in market 3?

Regarding the first point, it is quite striking that we can greatly increase the significance of our results simply by excluding three subjects. By recruiting more inexperienced subjects into our experiments, we may have more confidence in interpreting the results, either with positive proof or negative rejection.

As to the second point, there is no way but to collaborate more closely with psychologists. The effects and models of working memory have been studied by psychologists for such a long time that economists can get instant and clear help from their expertise in this. However, it will not be enough just to consult psychologists for their opinions, but tighter collaboration is a must from the design phase to the analytical stage because of the dissimilarities in the experimental logic and target problems, etc. between these two disciplines.

The final point, which is also the most concrete finding in the last part of our research, requires more contemplation. There are economic reasons for tackling this problem. Given the assumption of the effectiveness of the working memory capacity on economic performance, such results suggest that different market problems may bring decision makers different kinds of problems. It might be that it takes cognitive capacity to solve some of the problems, while this is not the case in other problems. Or, it may be that the potential effective strategies in market 3 require more cognitive capacity to be accomplished, while in other markets simple heuristics are already sufficiently profitable. If either of these explanations can be proved to be true, it will be very informative for economic experimentalists in the future in the sense that they can deploy agent simulations to understand the problem better before human experiments are launched.

CONCLUDING REMARKS

The significance of this chapter resides in two facets. First, it raises the issue of heterogeneity in individual cognitive capacity since most agent-based economic or financial models do not deal with it. Second, the research strategy we adopt in this chapter reveals the value of multi-directional and reciprocal relationships among agent-based computational modeling, experimental economics, and psychology.

In this chapter, we propose a method to model individual intelligence in agent-based double auction markets. We then run a series of experiments to validate our results according to what psychological studies have shown us.

Simulation results show that it is viable to use population size as a proxy of the cognitive capacity of GP traders. In general, the results are consistent with psychological findings—a positive relationship between intelligence and learning performance, and a decreasing marginal contribution of extra intelligence—and with our human experiments—the patterns of the influence of working memory capacity in agents' performances. Our study therefore shows that, by employing Genetic Programming as the learning algorithm, it is possible to model both the individual learning behavior and the innate heterogeneity of individuals at the same time.

The results of this study remind us of a possibility that there is another facet to connect human intelligence and artificial intelligence. Artificial intelligence not only can be used to model intellectual behavior individually, but it is also able to capture social heterogeneity through a proper parameterization.

The other contribution of this chapter is the relationship between human experiments, agent-based simulations, and psychology. It is already known that agent-based simulation is not just a complementary tool when it is too costly to conduct human experiments, for it can also help us test and verify economic theories without human experiments. However, even when being 'combined' together, human experiments are always the counselors of agent-based models, just as Duffy (2006) observes:

"with a few notable exceptions, researchers have not sought to understand findings from agent-based simulations with follow-up experiments involving human subjects. The reasons for this pattern are straightforward. ... As human subject experiments impose more constraints on what a researcher can do than do agent-based modeling simulations, it seems quite natural that agent-based models would be employed to understand laboratory findings and not the other way around." *(Duffy, 2006, p.951)*

Human experiments may be greatly constrained, but at the same time there are so many unobservable but intertwining factors functioning during human subjects' decision processes. On the other hand, agent-based models can be strictly controlled, and software agents are almost transparent. Thus, it would be an advantage to turn to agent-based simulations if researchers want to isolate the influence of a certain factor. In this regard, we think that agent-based simulations can also be a tool to discover unknown factors even before human experiments are conducted.

In this chapter, we actually follow this strategy by eliciting ideas from the psychological literature first, transplanting it in an economics environment in the form of agent-based simulations, and finally conducting corresponding human experiments after we have gained support from agent-based simulations. However, what we mean by multidirectional relationships among agent-based simulation, human experiments, and psychology has a deeper meaning. We believe that the knowledge from these three fields has a large space for collaboration, but it should be done not only by referring to the results of each other as a final source of reference. These three disciplines each possesses an experimental nature, and a cyclical

joint work including the design phase should be expected. We anticipate that researchers can acquire more precise findings by experimentation with the help from human subjects and software agents in a way delivered in this chapter.

ACKNOWLEDGMENT

The authors are grateful to an anonymous referee for very helpful suggestions. The authors are also thankful to Prof. Lee-Xieng Yang in the Research Center of Mind, Brain, and Learning of National Chengchi University for his professional support and test programs of working memory capacity. The research supports in the form of NSC grant no. NSC 95-2415-H-004-002-MY3, and NSC 96-2420-H-004-016-DR from the National Science Council, Taiwan are also gratefully acknowledged.

REFERENCES

Benjamin, D., Brown, S., & Shapiro, J. (2006). *Who is 'behavioral'? Cognitive ability and anomalous preferences.* Levine's Working Paper Archive 122247000000001334, UCLA Department of Economics.

Casari, M. (2004). Can genetic algorithms explain experimental anomalies? An application to common property resources. *Computational Economics, 24,* 257–275. doi:10.1007/s10614-004-4197-5

Chan, N. T., LeBaron, B., Lo, A. W., & Poggio, T. (2008). Agent-based models of financial markets: A comparison with experimental markets. MIT Artificial Markets Project, Paper No. 124, September. Retrieved January 1, 2008, from http://citeseer.ist.psu.edu/chan99agentbased.html.

Chen, S.-H., Chie, B.-T., & Tai, C.-C. (2001). Evolving bargaining strategies with genetic programming: An overview of AIE-DA Ver. 2, Part 2. In B. Verma & A. Ohuchi (Eds.), *Proceedings of Fourth International Conference on Computational Intelligence and Multimedia Applications (ICCIMA 2001)* (pp. 55–60). IEEE Computer Society Press.

Chen, S.-H., Zeng, R.-J., & Yu, T. (2008). Co-evolving trading strategies to analyze bounded rationality in double auction markets . In Riolo, R., Soule, T., & Worzel, B. (Eds.), *Genetic Programming Theory and Practice VI* (pp. 195–213). Springer.

Cliff, D., & Bruten, J. (1997). *Zero is not enough: On the lower limit of agent intelligence for continuous double auction markets* (Technical Report no. HPL-97-141). Hewlett-Packard Laboratories. Retrieved January 1, 2008, from http://citeseer.ist.psu.edu/cliff97zero.html

Das, R., Hanson, J. E., Kephart, J. O., & Tesauro, G. (2001). Agent-human interactions in the continuous double auction. In *Proceedings of the 17th International Joint Conference on Artificial Intelligence (IJCAI)*, San Francisco. CA: Morgan-Kaufmann.

Detterman, D. K., & Daniel, M. H. (1989). Correlations of mental tests with each other and with cognitive variables are highest for low-IQ groups. *Intelligence, 13,* 349–359. doi:10.1016/S0160-2896(89)80007-8

Devetag, G., & Warglien, M. (2003). Games and phone numbers: Do short-term memory bounds affect strategic behavior? *Journal of Economic Psychology, 24,* 189–202. doi:10.1016/S0167-4870(02)00202-7

Devetag, G., & Warglien, M. (2008). Playing the wrong game: An experimental analysis of relational complexity and strategic misrepresentation. *Games and Economic Behavior, 62,* 364–382. doi:10.1016/j.geb.2007.05.007

Duffy, J. (2006). Agent-based models and human subject experiments . In Tesfatsion, L., & Judd, K. (Eds.), *Handbook of Computational Economics* (*Vol. 2*). North Holland.

Easley, D., & Ledyard, J. (1993). Theories of price formation and exchange in double oral auction . In Friedman, D., & Rust, J. (Eds.), *The Double Auction Market-Institutions, Theories, and Evidence.* Addison-Wesley.

Friedman, D. (1991). A simple testable model of double auction markets. *Journal of Economic Behavior & Organization, 15,* 47–70. doi:10.1016/0167-2681(91)90004-H

Gjerstad, S., & Dickhaut, J. (1998). Price formation in double auctions. *Games and Economic Behavior, 22,* 1–29. doi:10.1006/game.1997.0576

Gode, D., & Sunder, S. (1993). Allocative efficiency of markets with zero-intelligence traders: Market as a partial substitute for individual rationality. *The Journal of Political Economy, 101,* 119–137. doi:10.1086/261868

Gottfredson, L. S. (1997). Mainstream science on intelligence: An editorial with 52 signatories, history, and bibliography. *Intelligence, 24*(1), 13–23. doi:10.1016/S0160-2896(97)90011-8

Gregg, L., & Simon, H. (1979). Process models and stochastic theories of simple concept formation. In H. Simon, *Models of Thought* (Vol. I). New Haven, CT: Yale Uniersity Press.

Grossklags, J., & Schmidt, C. (2006). Software agents and market (in)efficiency—a human trader experiment. *IEEE Transactions on System, Man, and Cybernetics: Part C . Special Issue on Game-theoretic Analysis & Simulation of Negotiation Agents*, *36*(1), 56–67.

Hunt, E. (1995). The role of intelligence in modern society. *American Scientist*, (July/August): 356–368.

Kagel, J. (1995). Auction: A survey of experimental research . In Kagel, J., & Roth, A. (Eds.), *The Handbook of Experimental Economics*. Princeton University Press.

Lewandowsky, S., Oberauer, K., Yang, L.-X., & Ecker, U. (2009). A working memory test battery for Matlab. under prepartion for being submitted to the *Journal of Behavioral Research Method*.

Payne, J., Bettman, J., & Johnson, E. (1993). *The Adaptive Decision Maker*. Cambridge University Press.

Rust, J., Miller, J., & Palmer, R. (1993). Behavior of trading automata in a computerized double auction market . In Friedman, D., & Rust, J. (Eds.), *Double Auction Markets: Theory, Institutions, and Laboratory Evidence*. Redwood City, CA: Addison Wesley.

Rust, J., Miller, J., & Palmer, R. (1994). Characterizing effective trading strategies: Insights from a computerized double auction tournament. *Journal of Economic Dynamics & Control*, *18*, 61–96. doi:10.1016/0165-1889(94)90069-8

Simon, H. (1981). Studying human intelligence by creating artificial intelligence. *American Scientist*, *69*, 300–309.

Simon, H. (1996). *The Sciences of the Artificial*. Cambridge, MA: MIT Press.

Taniguchi, K., Nakajima, Y., & Hashimoto, F. (2004). A report of U-Mart experiments by human agents . In Shiratori, R., Arai, K., & Kato, F. (Eds.), *Gaming, Simulations, and Society: Research Scope and Perspective* (pp. 49–57). Springer.

Zhan, W., & Friedman, D. (2007). Markups in double auction markets. *Journal of Economic Dynamics & Control*, *31*, 2984–3005. doi:10.1016/j.jedc.2006.10.004

KEY TERMS AND DEFINITIONS

Genetic Programming (GP): An automated method for creating a working computer program from a high-level problem statement of a problem. Genetic programming starts with a randomly created computer programs. This population of programs is progressively evolved over a series of generations. The evolutionary search uses the Darwinian principle of natural selection (survival of the fittest) and analogs of various naturally occurring operations, including crossover (sexual recombination), mutation, etc.

Cognitive Capacity: A general concept used in psychology to describe human's cognitive flexibility, verbal learning capacity, learning strategies, intellectual ability, etc. Although cognitive capacity is a very general concept and can be measured from different aspects with different tests, concrete concepts such as intelligence quotient (IQ) and working memory capacity are considered highly representative of this notion.

Double Auction: A system in which potential buyers submit their bids and potential sellers submit their ask prices (offers) simultaneously. The market is cleared when a certain price P is chosen so that all buyers who bid more than P and all sellers who ask less than P are matched to make transactions.

Working Memory: The mental resources used in the decision-making processes of humans and is highly related to general intelligence. It is generally assumed that working memory has a constrained capacity, hence this capacity plays an important role which determine people's performance in cognitive tasks, especially complex reasoning ones.

Boundedly Rational Agents: Experience limits in formulating and solving complex problems and in processing (receiving, storing, retrieving, transmitting) information, therefore, they solve problems by using certain heuristics instead of optimizing.

Individual Efficiency: A ratio used to evaluate agents' performance in the markets. In economic theory, once demand and supply determine the equilibrium price, agents' potential profits (individual surplus) can be measured as the differences between his/her reservation prices and the equilibrium price. Individual efficiency is calculated as the ratio of agents' actual profits over their potential profits.

ENDNOTES

[1] Elitism preserves the best strategy in current population to the next. While elitism helps preserve good strategies when there is no guarantee that every strategy will be sampled in our designed, we don't want it to be the main factor determining the compositions of the populations. Therefore, the number of elite is set to be 1.

[2] Generally speaking, the larger the mutation rate, the more diverse the genotypes of the strategies are. In most studies, the mutation rate ranges from 1% to 10%, therefore it is set to be 5% in this research.

[3] One may suspect that GP traders will perform very poorly from time to time since they also have the biggest variances in the profits. To evaluate how worse GP traders can be, we keep track of the rankings of their performances relative to other trading strategies. As a result, the average rankings of GP traders are the smallest among all the designed trading strategies. This means that although GP traders may use not-so-good strategies sometimes, their performances are still barely adequate as compared with other kinds of designed trading strategies.

[4] From the left panel of Figure 7, we can see that among the human traders with lower working memory capacity, there are about two traders who constantly perform quite well in every market. In fact, these traders are exactly those subjects with experience in stock markets or futures markets.

Chapter 7
Evolution of Agents in a Simple Artificial Market

Hiroshi Sato
National Defense Academy, Japan

Masao Kubo
National Defense Academy, Japan

Akira Namatame
National Defense Academy, Japan

ABSTRACT

In this chapter, we conduct a comparative study of various traders following different trading strategies. We design an agent-based artificial stock market consisting of two opposing types of traders: "rational traders" (or "fundamentalists") and "imitators" (or "chartists"). Rational traders trade by trying to optimize their short-term income. On the other hand, imitators trade by copying the majority behavior of rational traders. We obtain the wealth distribution for different fractions of rational traders and imitators. When rational traders are in the minority, they can come to dominate imitators in terms of accumulated wealth. On the other hand, when rational traders are in the majority and imitators are in the minority, imitators can come to dominate rational traders in terms of accumulated wealth. We show that survival in a finance market is a kind of minority game in behavioral types, rational traders and imitators. The coexistence of rational traders and imitators in different combinations may explain the market's complex behavior as well as the success or failure of various trading strategies. We also show that successful rational traders are clustered into two groups: In one group traders always buy and their wealth is accumulated in stocks; in the other group they always sell and their wealth is accumulated in cash. However, successful imitators buy and sell coherently and their wealth is accumulated only in cash.

INTRODUCTION

Economists have long asked whether traders who misperceive future prices can survive in a stock market. The classic answer, given by Friedman, is that they cannot. Friedman argued that mistaken investors buy high and sell low and as a result lose money to rational traders, eventually losing all their wealth.

DOI: 10.4018/978-1-60566-898-7.ch007

On the other hand, Shleifer and his colleagues questioned the presumption that traders who misperceive returns do not survive (De Long, 1991). Since noise traders who are on average bullish bear more risk than do investors holding rational expectations, as long as the market rewards risk-taking, noise traders can earn a higher expected return even though they buy high and sell low on average. Because Friedman's argument does not take into account the possibility that some patterns of noise traders' misperceptions might lead them to take on more risk, it cannot be correct as stated.

It is difficult to reconcile the regular functioning of financial markets with the coexistence of different populations of investors. If there is a consistently winning market strategy, then it is reasonable to assume that the losing population will disappear in the long run. It was Friedman who first advanced the hypothesis that in the long run irrational investors cannot survive because they tend to lose wealth and disappear. For agents prone to forecasting errors, the fact that different populations with different trading strategies can coexist still requires an explanation.

Recent economic and finance research reflects growing interest in marrying the two viewpoints, that is, in incorporating ideas from the social sciences to account for the fact that markets reflect the thoughts, emotions, and actions of real people as opposed to the idealized economic investors who underlie efficient markets (LeBaron, 2000). Assumptions about the frailty of human rationality and the acceptance of such drives as fear and greed underlie the recipes developed over the decades in so-called technical analysis. There is growing empirical evidence of the existence of herd or crowd behavior. Herd behavior is often said to occur when many people take the same action, because some mimic the actions of others (Sornette, 2003).

To adequately analyze whether noise traders are likely to persist in an asset market, we need to describe the long run distribution of wealth, not

just the level of expected returns. The question of whether there are winning and losing market strategies and how to characterize them has been discussed from a practical point of view in (Cinocotti, 2003). On the one hand, it seems obvious that different investors exhibit different investing behaviors that are responsible for the movement of market prices. On the other hand, it is difficult to reconcile the regular functioning of financial markets with the coexistence of heterogeneous investors with different trading strategies (Levy, 2000). If there exists a consistently winning market strategy, then it is reasonable to assume that the losing trading strategies will disappear in the long run through the force of natural selection.

In this chapter we take an agent-based model approach for a comparative study of different strategies. We examine how traders with various trading strategies affect prices and their success in the market measured by their accumulation of wealth. Specifically, we show that imitators may survive and come to dominate rational investors in wealth when the proportion of imitators is much less than that of rational traders.

The chapter is organized as follows: In Section 2 we survey the related literature. Section 3 describes the relationship between the Ising model and the Logit model. Sections 4 and 5 describe an artificial stock market as the main ingredient in our agent-based financial market. The simulation results and discussion are shown in Sections 6 and 7 respectively. Section 8 concludes the chapter.

RELATED LITERATURE

One can distinguish two competing hypotheses by their origins, one derived from the traditional Efficient Market Hypothesis (EMH) and a recent alternative that is sometimes called the Interacting Agent Hypothesis (IAH) (Tesfatsion, 2002). The EMH states that the price fully and instantaneously reflects any new information: The market is, therefore, efficient in aggregating available information with its invisible hand. The agents are assumed to be rational and homogeneous with respect to their access and their assessment of information; as a consequence, interactions among them can be neglected.

In recent literature, several papers try to explain the stylized facts as the macroscopic outcome of an ensemble of heterogeneous interacting agents (Cont, 2000; LeBaron, 2001). In this view, the market is populated by agents with different characteristics, such as differences in access to and interpretation of available information, different expectations or different trading strategies. The traders interact, for example, by exchanging information, or they trade by imitating the behavior of others. The market possesses, then, an endogenous dynamics, and the strict one-to-one relationship with the news arrival process does not hold any longer (although the market might still be efficient in the sense of a lack of predictability). The universality of the statistical regularities is seen as an emergent property of this internal dynamics, governed by the interactions among agents.

Boswijk et al. estimated an asset-pricing model using annual US stock price data from 1871 until 2003 (Boswijk, 2004). The estimation results support the existence of two expectation regimes. The first can be characterized as a fundamentalist regime because agents believe in mean reversion of stock prices toward the benchmark fundamental value. The second can be characterized as a chartist trend-following regime because agents expect deviations from the fundamental to trend. The fractions of agents using the fundamentalists forecasting rule and of agents using the trend-following forecasting rule show substantial time variation and switching between predictors.

They suggest that behavioral heterogeneity is significant and that there are two different regimes, a "mean reversion" regime and a "trend following" regime. To each regime, there corresponds a different investor type: fundamentalists and trend followers. These two investor types coexist and

their fractions show considerable fluctuation over time. The mean-reversion regime corresponds to the situation in which the market is dominated by fundamentalists who recognize overpricing or underpricing of the asset and who expect the stock price to move back towards its fundamental value. The trend-following regime represents a situation when the market is dominated by trend followers expecting continuation of, for example, good news in the (near) future and so expect positive stock returns.

They also allow the coexistence of different types of investors with heterogeneous expectations about future payoffs and evolutionary switching between different investment strategies. Disagreement in asset pricing models can arise because of two assumptions: differential information and differential interpretation. In the first case, there is an information asymmetry between one group of agents that observes a private signal and the rest of the population that has to learn the fundamental value from public information, such as prices. Asymmetric information causes heterogeneous expectations among agents.

Agents use different "models of the market" to update their subjective valuation based on the earnings news, and this might lead them to hold different beliefs. However, the heterogeneity of expectations might play a significant role in asset pricing. A large number of models have been proposed that incorporate this hypothesis. They assume that agents adopt a belief based on its past performance relative to the competing strategies. If a belief performed relatively well, as measured by realized profits, it attracts more investors while the fraction of agents using the "losing" strategies will decrease. Realized returns thus contribute more support to some of the belief strategies than others, which leads to time variation in the sentiment of the market.

The assumption of evolutionary switching among beliefs adds a dynamic aspect that is missing in most of the models with heterogeneous opinions mentioned above. In our model investors are boundedly rational because they learn from the past performance of the strategies which one is more likely to be successful in the near future. They do not use the same predictor in every period and make mistakes, but switch between beliefs in order to minimize their errors. Agents may coordinate expectations on trend-following behavior and mean reversion, leading to asset price fluctuations around a constant fundamental price.

(Alfarano, 2004) also estimated a heterogeneous agent model (HAM) to exchange rates with fundamentalists and chartists and found considerable fluctuation of the market impact of fundamentalists. All these empirical papers suggest that heterogeneity is important in explaining the data, but much more work is needed to investigate the robustness of this empirical finding. Our chapter may be seen as one of the first attempts to estimate a behavioral HAM on stock market data and investigate whether behavioral heterogeneity is significant.

INFERRING UTILITY FUNCTIONS OF SUCCESSES AND FAILURES

In this section, we try to infer the utility functions of traders by relating the so-called Ising model and the Logit model. We clarify the following fact: success calls success and failure calls failure.

Ising Model

Bornholdt and his colleagues analyzed profit margins and volatility by using the Ising model, which is a phase transition model in physics (Bornholdt, 2001; Kaizoji, 2001). The Ising model is a model of magnetic substances proposed by Ising in 1925 (Palmer, 1994). In the model, there are two modes of *spin*: upward ($S = +1$) and downward ($S = -1$). Investment attitude in the investor model plays the same role as spin plays in the Ising model. In the model, magnetic interactions seek to align

spins relative to one another. The character of the magnetic substance is determined by the interaction of the spins. In the investor model, the two spin states represent an agent's investment attitude. Each agent changes attitude according to the probability of the spin reversing.

The probability $P_i(t + 1)$ that agent i buys at time $t + 1$ is defined as

$$P_i(t+1) = \frac{1}{1 + \exp(-2\beta h_i(t))}, \qquad (3.1)$$

where

$$h_i(t) = \sum_j J_{ij} S_j(t) - \alpha S_i(t)\left|M(t)\right| \qquad (3.2)$$

$$M(t) = \sum_i S_i(t) / N \qquad (3.3)$$

In (3.1) $h_i(t)$, defined in (3.2), represents the *investment attitude* of agent i, the parameter β is a positive constant, and Jij represents the influence level of neighboring agent j. Therefore, the first term of (3.2) represents the influence of the neighborhood. The investment variables S_j, $j = 1, 2, \ldots, n$ take the value -1 when agent j sells and +1 when she buys. The second term of (3.2) represents the average investment attitude, with α a positive constant. If many agents buy, then the investment attitude decreases. The investment attitude represents the agent's conformity with neighboring agents.

The average investment attitude should rise at least so that prices may rise more than this time step. In other words, it is necessary that the number of agents who purchase be greater than this term. It is thought that the probability of the investment attitude changing rises as the absolute value of $M(t)$ approaches one. It can be said that the agent is "applying the brakes" to the action, where "action" refers to the opinion of the neighborhood.

Logit Model (Stochastic Utility Theory)

The Logit model is based on stochastic utility theory applied to individual decision-making (Durlauf 00). In stochastic utility theory, an agent is assumed to behave rationally by selecting the option that brings a high utility. But the individual's utility contains some random element. This uncertain factor is treated as a random variable in stochastic utility theory. The utilities associated with the choices of S_1 (buy) and S_2 (sell) are given as follows:

$U_1 = V_1 + \varepsilon_1$ the utility of choosing S_1,

$U_2 = V_2 + \varepsilon_2$ the utility of choosing S_2,

ε_i i=1, 2 random variables.

The probability of agent i buying is given by

$$p_i = \Pr(U_1 > U_2) = \Pr(V_1 + \varepsilon_1 > V_2 + \varepsilon_2)$$

$$= \Pr(V_1 - V_2 > \varepsilon_2 - \varepsilon_1). \tag{3.4}$$

By denoting the joint probability density function of the random variables ε_i, for $i = 1, 2$, by $f(\varepsilon_1, \varepsilon_2)$, we can derive

$$p_i = \int_{\varepsilon_1=-\infty}^{+\infty} \int_{\varepsilon_2=-\infty}^{V_1-V_2+\varepsilon_1} f(\varepsilon_1, \varepsilon_2) d\varepsilon_2 d\varepsilon_1. \tag{3.5}$$

Assume that random variables ε_i, for $i = 1, 2$, are independent and that they follow the Gumbel density function $F(x) = \exp\{-\exp(-x)\}$ (Levy, 2000). Then we can obtain the following expression by substitution and integration:

$$p_i = \frac{1}{1 + \exp\{-(V_1 - V_2)\}}. \tag{3.6}$$

The probability of agent i buying is given as a function of the difference between the utility of buying and the utility of selling.

Relating the Two Models

By equating (3.1) and (3.6), we can obtain the following relation:

$$\frac{1}{1 + \exp[-(V_1(t+1) - V_2(t+1))]} = \frac{1}{1 + \exp[-2\beta h_i(t)]}. \tag{3.7}$$

If we set $Jij = J$ for all i, j, then we have

$$V_1(t+1) - V_2(t+1) = 2\alpha\beta[\frac{J}{\alpha}\sum_j S_j(t) - S_i(t)|M(t)|]$$

$$= 2\alpha\beta[\frac{J}{\alpha}(n_1(t) - n_2(t)) - S_i(t)\frac{|N_1(t) - N_2(t)|}{N}]$$

$$= \frac{2\alpha\beta}{N}[\frac{JN}{\alpha}(n_1(t) - n_2(t)) - S_i(t)|N_1(t) - N_2(t)|]$$

$$= \frac{2\alpha\beta}{N}[\lambda(n_1(t) - n_2(t)) - S_i(t)|N_1(t) - N_2(t)|] \tag{3.8}$$

We also assume that $JN/\alpha = 1$ and $2\alpha\beta/N = 1$, and consider the following two cases:

(Case 1) $N_1(t) - N_2(t) \geq 0$: then we have

$$V_1(t+1) = \lambda n_1(t) - S_i(t)N_1(t)$$

$$V_2(t+1) = \lambda n_2(t) - S_i(t)N_2(t) \tag{3.9}$$

(Case 2) $N_1(t) - N_2(t) \leq 0$: then we have

$$V_1(t+1) = \lambda n_1(t) + S_i(t)N_1(t)$$

$$V_2(t+1) = \lambda n_2(t) + S_i(t)N_2(t) \tag{3.10}$$

If the sellers (or purchasers) are in the minority, the market is called a seller (purchaser) market. Profit is large for those on the minority side of the market. The act of selling (purchasing) in a seller (purchaser) market is called here "success", and the opposite action is called "failure". If the purchasers are in the majority ($N_1(t)$ - $N_2(t) > 0$), the market is a seller's market. On the other hand, if $N_1(t)$ - $N_2(t) < 0$, the market is called a buyer market since the sellers are in the majority.

We can now classify the utility functions of a success and a failure as follows:

(1)   The utility functions of a success are

$$V_1(t+1) = \lambda n_1 + N_1(t)$$

$$V_2(t+1) = \lambda n_2 + N_2(t) \qquad (3.11)$$

(2)   The utility functions of a failure are

$$V_1(t+1) = \lambda n_1 - N_1(t)$$

$$V_2(t+1) = \lambda n_2 - N_2(t) \qquad (3.12)$$

In the market, the number of failures is more than the number of successes. Therefore, the investment attitudes of an agent who succeeds and an agent who fails will become opposite. Furthermore, the successful agent will again be in the minority in the next period, and likewise the failures will again be in the majority in the next period. Therefore the successful agents will continue to win, and the failing agents will continue to fail.

AN AGENT-BASED FINANCIAL MARKET MODEL

We consider a dynamic asset pricing model consisting of heterogeneous agents. We assume that the fundamentals of the stock are constant and it is publicly available to all agents, but they have different beliefs about the persistence of deviations of stock prices from the fundamental benchmark.

The most important design question faced in market building comes in the representation and structure of the actual trading agents. Agents can vary from simple budget-constrained zero-intelligence agents to those used in sophisticated genetic programming models. This variation in design is due to the fact that trading agents must solve a poorly defined task. Given that there are many ways to process past data, there must be as many ways to construct trading agents (Johnson, 2003).

The simplest and most direct route is to model agents by well-defined dynamic trading rules modeled more or less from strategies used in the real world. This method can lead to very tractable precise results that give insight into the interactions between trading rules. Many markets of this type assume that the trading strategies will continue without modification, although the wealth levels under their control may be diminishing to zero. This leaves some open questions about coevolutionary dynamics with only a limited amount of new speciation. A second critique is that agents in these markets do not operate with any well-defined objective function. There is some usefulness to having well-defined objective functions for the agents. There may be an important tradeoff problem which only a simulation model can answer.

The second most important part of agent-based markets is the actual mechanism that governs the trading of assets. Once one leaves the relatively simple world of equilibrium modeling, it is necessary to think about the actual details of trading. This can be both a curse and a blessing to market designers. On the negative side, it opens up another poorly understood set of design questions. However, on the positive side, it may allow one to study the impact of different trading mechanisms, all of which would be inconsequential in an equilibrium world.

Most agent-based markets have solved this problem in one of three ways: by assuming a simple price response to excess demand, by building the market in such a way that a kind of local equilibrium price can be found easily, or by explicitly modeling the dynamics of trading to look like the continuous trading in an actual market. Most of the earliest agent-based markets used the first method to model price movements. Most markets of this type poll traders for their current demand, sum up the market demand and if there is excess demand, increase the price. If there is an excess supply, they decrease the price.

This has been interpreted as evidence that as a forecaster ages, evaluators develop tighter prior beliefs about the forecaster's ability, and hence the forecaster has less incentive to herd with the group. On the other hand, the incentive for a second-mover to discard his private information and instead mimic the market leader increases with his initial reputation, as he strives to protect his current status and level of pay. In a practical implementation of a trading strategy, it is not sufficient to know or guess the overall direction of the market. There are additional subtleties governing how the trader is going to enter (buy or sell in) the market. For instance, a trader will want to be slightly ahead of the herd to buy at a better price, before the price is pushed up by the bullish consensus. Symmetrically, she will want to exit the market a bit before the crowd, that is, before a trend reversal. In other words, she would like to be somewhat of a contrarian by buying when the majority is still selling and by selling when the majority is still buying, slightly before a change of opinion of the majority of her "neighbors". This means that she will not always want to follow the herd, at least at finer time scales. At this level, she cannot rely on the polling of her

"neighbors" because she knows that they, as well as the rest of the crowd, will have similar ideas about trying to outguess each other on when to enter the market. More generally, ideally she likes to be in the minority when entering the market, in the majority while holding her position and again in the minority when closing her position.

HYPOTHETICAL VALIDATION USING AN AGENT-BASED MODEL

In this section we introduce three population-types that have been already described in the literature and that represent more realistic trading behaviors. The aim is twofold: First, we want to study the behavior of these stylized populations in a realistic environment characterized by limited resources and a market clearing mechanism. Second, we want to address the important issue of whether or not winning strategies exist. The fractions of agents using the fundamental and trend-following forecasting rules show substantial time variation and switching between predictors.

Market Mechanism and Performance Measures

One of the most important parts of agent-based markets is the actual mechanism that governs the trading of assets. Most agent-based markets assume a simple price response to excess demand and the market is built so that finding a local equilibrium price is not difficult. If supply exceeds demand, then the price decreases. Agents maintain stock and capital, and stock is bought or sold in exchange for capital. The model generates fluctuations in the value of the stock by limiting transactions to one unit of stock.

Price model. The basic model assumes that the stock price reflects the level of excess demand, which is governed by

$$P(t) = P(t-1) + \chi[N_1(t) - N_2(t)], \qquad (5.1)$$

where $P(t)$ is the stock price at time t, $N_1(t)$ and $N_2(t)$ are the corresponding number of agents buying and selling respectively, and χ is a constant. This expression implies that the stock price is a function of the excess demand. That is, the price rises when there are more agents buying, and it descends when more agents are selling.

Price volatility.

$$v(t) = [P(t) - P(t-1)] / P(t-1). \qquad (5.2)$$

Individual wealth. We introduce the *notional wealth $W_i(t)$* of agent i into the model as follows:

$$W_i(t) = P(t){*}\Phi_i(t) + C_i(t), \qquad (5.3)$$

where Φ_i is the amount of assets (stock) held and C_i is the amount of cash (capital) held by agent i. It is clear from the equation that an exchange of cash for assets at any price does not in any way affect the agent's notional wealth. However, the important point is that the wealth $W_i(t)$ is only notional and not real in any sense. The only real measure of wealth is $C_i(t)$, the amount of capital the agent has available to spend. Thus, it is evident that an agent has to do a "round trip" (buy (sell) a stock and then sell (buy) it back) to discover whether a real profit has been made.

Trader Types

For modeling purposes, we use representative agents (rational agents) who make rational decisions in the following stylized terms: If they expect the price to go up then they buy, and if they expect the price to go down then they sell immediately. But this then leads to the problem, what happens if every trader behaves in this same way?

Here, some endogenous disturbances need to be introduced. Given this disturbances, the individual is modeled to behave differently. One

body of research has sought to explain the data with aggregate models in which a representative agent solves this optimization problem. If the goal is simply to fit the data, it is not unreasonable to attribute to agents the capacity to explicitly formulate and solve dynamic programming problems. However, there is strong empirical evidence that humans do not perform well on problems whose solution involves backward induction. For this reason, these models fail to provide a realistic account of the phenomenon. The model we describe will not invoke a representative agent, but will posit a heterogeneous population of individuals. Some of these will behave "as if" they were fully informed optimizers, while others will not. Social networks and social interactions–clearly absent from the prevailing literature–will play an explicit central role.

Heterogeneity turns up repeatedly as a crucial factor in many evolving systems and organizations. But the situation is not always as simple as saying that heterogeneity is desirable and homogeneity is undesirable. This remains a basic question in many fields: What is the right balance between heterogeneity and homogeneity? When heterogeneity is significant, we need to be able to show the gains associated with it. However, analysis of a collection of heterogeneous agents is difficult, often intractable.

The notion of type facilitates the analysis of heterogeneity. A *type* is a category of agents within the larger population who share some characteristics. We distinguish types by some aspects of the agents' unobservable internal model that characterize their observable behaviors. One can imagine how such learning models might evolve over time towards equilibria. In principle, this evolutionary element can be folded into a metalearning that includes both the short-term learning and long-term evolution.

Interaction between agents is a key feature of agent-based systems. Traditional market models do not deny that agents interact but assume that they only do so through the price system. Yet agents do, in fact, communicate with each other and learn from each other. The investor who enters the market forecasts stock prices by various techniques. For example, the investor makes a linear forecast of past price data, forecasts based on information from news media, and so forth. The types of typical investors are usually described based on differences in their forecast methods (or methods of deciding their investment attitude). Three typical investor types are as follows:

- *Fundamentalist*: investor with a fundamentalist investment attitude based on various economic indicators.
- *Chartist*: investor who uses analysis techniques for finding present value from charting past price movement.
- *Noise trader*: investor who behaves according to a strategy not based on fundamental analysis.

In this chapter, traders are segmented into two basic types depending on their respective trading behavior: rational traders and imitators. Rational traders are further classified into two types: momentum and contrarian traders.

1. **Rational traders.** If we assume the fundamental value is constant, their investment strategy is based on their expectation of the trend continuing or reversing.
    Momentum trader: These traders are trend followers who make decisions based on the trend of past prices. A momentum trader speculates that if prices are rising, they will keep rising, and if prices are falling, they will keep falling.
    Contrarian trader: These traders differ in trading behavior. Contrarian traders speculate that, if the price is rising, it will stop rising soon and will decrease, so it is better to sell near the maximum. Conversely, if the price is falling, it will stop falling soon and will rise. This trading behavior is present

among actual traders, albeit it is probably less popular than trend-following strategies.

2. **Imitators.** Traders may have incorrect expectations about price movements. If there are such misperceptions, imitators who do not affect prices may earn higher payoffs than strategic traders. Each imitator has a unique social network with strategic traders. Within this individual network, if the majority of strategic traders buy then she also buys, and if the majority of strategic traders sell then she also sells. It is now widely held that mimetic responses result in herd behavior and, crucially, that the properties of herding arise in financial markets.

Trading Rules of Trader Types

Agents are categorized by their strategy space. Since the space of all strategies is complex, this categorization is not trivial. Therefore, we might, for example, constrain the agents to be finite automata with a bounded number of states. Even after making this kind of limitation, we might still be left with too large a space to reason about, but there are further disciplined approaches to winnowing down the space. An example of a more commonly used approach, is to assume that the opponent is a "rational learner" and to place restrictions on the opponent's prior about our strategies. In this section we describe a trading rule for each type of trader discussed in the previous section.

1. **Rational traders (fundamentalists):** Rational traders observe the trend of the market and trade so that their short-term payoff will be improved. Therefore if the trend of the market is "buy", this agent's attitude is "sell". On the other hand, if the trend of the market is "sell", this agent's attitude is "buy". As has been explained, trading according to the minority decision creates wealth for the agent on performing the necessary trade, whereas trading according to

the majority decision loses wealth. However, if the agent has held the asset for a length of time between buying it and selling it back, his wealth will also depend on the rise and fall of the asset price over the holding period. On the other hand, the amount of stock that the purchaser (seller) can put in a single deal and buy (sell) is one unit. Therefore, when the numbers of purchasers and sellers are different, there exists an agent who cannot make her desired transaction:

- ○ **When sellers are in the majority:** There is an agent who cannot sell even if she is selected to sell exists. Because the price still falls in a buyer's market, it is an agent agents who sell are maintaining a large amount of properties. The agents who maintain the most property are the ones able to sell.
- ○ **When buyers are in the majority:** There is an agent who cannot buy even if she is selected to buy. Because the price rises, The agent still able to buy is the one maintaining a large amount of capital. The agents who maintain the most property are the ones able to buy.

The above trading behavior is formulated as follows. We use the following terminology:

$N_1(t)$: Number of agents who buy at time t.
N: Number of agents who participate in the market.
$R(t) = N_1(t) / N$: The rate of agents buying at time t.

We also denote the estimated rate of buying of agent i at time t as

$$R_F(t) = R(t - 1) + \mu_i \qquad (5.5)$$

where ε_i ($-0.5 < \varepsilon_i < 0.5$) is the rate of bullishness and timidity of the agent and differs depending on the agent.

Trading rule for rational traders:

If $R_F(t) > 0.5$ then sell
If $R_F(t) < 0.5$ then buy (5.6)
If ε_i is large, agent i has a tendency to "buy", and if it is small, agent i has a tendency to "sell".

2. **Imitators (chartists):** These agents watch the behavior of the rational traders. If the majority of rational traders "buy" then the imitators "buy", and if the majority of rational traders "sell" then the imitators "sell".

We can formulate the imitators' behavior as follows:

$R_S(t)$: The fraction of rational traders buying at time t
$P_1(t)$: The value of $R_S(t)$ estimated by imitator j

$$P_I(t) = R_F(t - 1) + \mu_j \qquad (5.7)$$

where ε_j ($-0.5 < \varepsilon_j < 0.5$) is the rate of bullishness and timidity of imitator j and in this experiments, ε_j is normally distributed.

Trading rule for imitators:

If $R_F(t) > 0.5$ then buy
If $R_F(t) < 0.5$ then sell (5.8)

SIMULATION RESULTS

We consider an artificial stock market consisting of 2,500 traders in total. In Figure 1, we show market prices over time for varying fractions of rational traders and imitators.

Figure 1. Market prices over time for varying fractions of rational traders and imitators

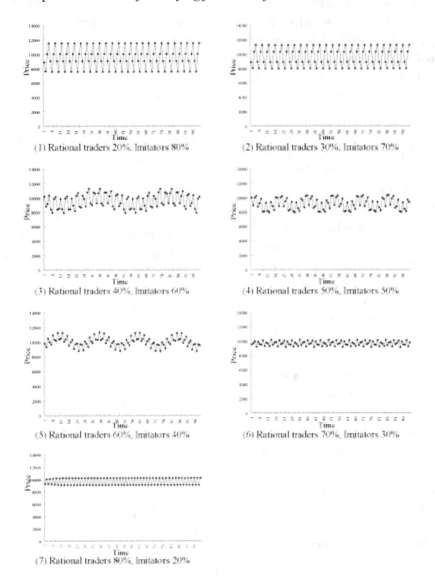

(1) Rational traders 20%, Imitators 80%

(2) Rational traders 30%, Imitators 70%

(3) Rational traders 40%, Imitators 60%

(4) Rational traders 50%, Imitators 50%

(5) Rational traders 60%, Imitators 40%

(6) Rational traders 70%, Imitators 30%

(7) Rational traders 80%, Imitators 20%

Stock Prices Over Time

Imitators mimic the movement of a small number of rational traders. If the rational traders start to raise the stock price, the imitators also act to raise the stock price. If the rational traders start to lower the stock price, the imitators lower the stock price further. Therefore, the actions of a large number of imitators amplify the price movement caused by the rational traders, increasing the fluctuation in the value of the stock.

Increasing the fraction of rational traders stabilizes the market. Maximum stability is achieved when the fraction of rational traders in the population is 70% and that of the imitators 30%. On the other hand, increasing the number of the rational traders further induces more fluctuation, and the price will cycle up and down if the fraction of rational traders is increased to 80%. Rational traders always trade in whatever direction places them in a minority position. In this situation, their actions do not induce fluctuations in the market

Figure 2. Movement of price over time when the fraction of rational traders increases gradually from 20% to 80%

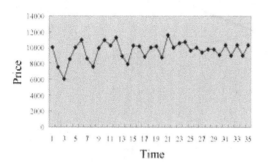

Figure 3. Movement of price over time when the fraction of rational traders moves randomly between 20% and 80%

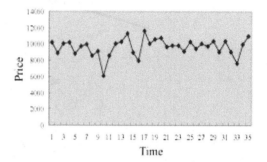

price. However, when rational traders are in the majority, their movements give rise to large market price fluctuations.

In Figure 2 we show the price movement when the fraction of rational traders is increased gradually from 20% to 80%. Figure 3 shows the price movement when the fraction of rational traders moves randomly between 20% and 80%.

Comparison of Wealth

We also show the average wealth of the rational traders and imitators over time for varying fractions of rational traders and imitators.

We now conduct a comparative study of rational traders and imitators. Imitators only mimic the actions of rational traders. On the other hand, rational traders deliberately consider the direction of movement of the stock price. Our question is which type is better off in terms of their accumulated wealth.

When rational traders are not in the majority (their fraction is less than 50%), their average wealth increases over time and that of the imitators decreases. Therefore, if the rational traders are in the minority, they are better off and their successful accumulation of wealth is due to losses by the majority, the imitators.

In the region where the number of the rational traders is almost the same as the number of imi-

tators, no trader is a winner or a loser and none accumulates wealth.

On the other hand, when the rational traders are in the majority, and the imitators are in the minority, the average wealth of the imitators increases over time and that of the rational traders decreases. Therefore, when the imitators are in the minority, they are better off and their successful accumulation of wealth is due to losses by the majority, the rational traders.

Evolution of the Population

We then change the composition of the traders using an evolutionary technique. Eventually, poor traders learn from other, wealthy traders. Figure 5 shows the two typical cases of evolution.

Domination occurs when traders evolve according to total assets because it takes some time to reverse the disparity in total assets between winners and losers. On the other hand, rational agents and imitators coexist when they evolve according to their gain in assets. An important point is that coexistence is not a normal situation. Various conditions are necessary for both types to coexist, including an appropriate updating scheme.

Figure 4. Changes in average wealth over time for different fractions of rational traders and imitators

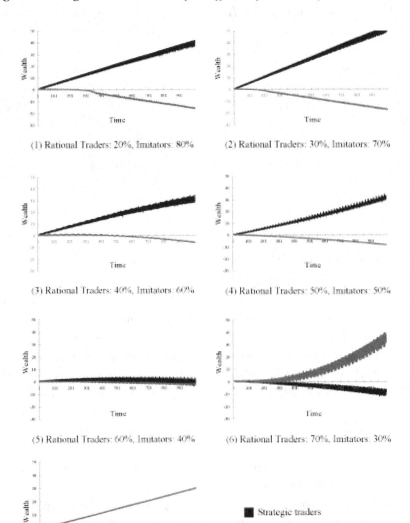

IMPLICATION OF SIMULATION RESULTS

The computational experiments performed using the agent-based model show a number of important results. First, they demonstrate that the average price level and the trends are set by the amount of cash present and eventually injected into the market. In a market with a fixed amount of stocks, a cash injection creates an inflationary pressure on prices. The other important finding of this work is that different populations of traders characterized by simple but fixed trading strategies cannot coexist in the long run. One population prevails and the other progressively loses weight and disappears. Which population will prevail and which will lose cannot be decided on the basis of their strategies alone.

Figure 5. Time path of the composition of traders' types (a) Evolution by wealth (sum of cash and stocks), (b) Evolution by gain in wealth

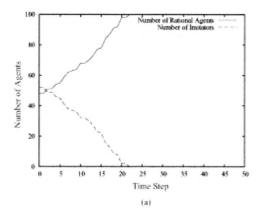

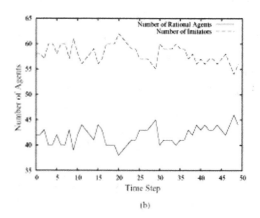

Trading strategies yield different results under different market conditions. In real life, different populations of traders with different trading strategies do coexist. These strategies are boundedly rational and thus one cannot really invoke rational expectations in any operational sense. Though market price processes in the absence of arbitrage can always be described as the rational activity of utility maximizing agents, the behavior of these agents cannot be operationally defined. This work shows that the coexistence of different trading strategies is not a trivial fact but requires explanation.

One could randomize strategies, imposing that traders statistically shift from one strategy to another. It is however difficult to explain why a trader embracing a winning strategy should switch to a losing strategy. Perhaps the market changes continuously and makes trading strategies randomly more or less successful. More experimental work is necessary to gain an understanding of the conditions that allow the coexistence of different trading populations. As noted earlier, there are two broad types of agents and we designate them "strategic traders" ("rational agents") and "imitators". The agents in our model fall into two categories. Members of one group (strategic traders) adopt the optimal decision rules. If they expect the price to go up, then they will buy, and if they

expect the price to go down, then they will sell immediately. In order to introduce heterogeneity among strategic agents we also introduce some randomness in the behavioral rules. The other group consists of imitators, who mimic the strategic traders of their social networks. The model we describe does not invoke a representative agent, but posits a heterogeneous population of agents. Some of these behave as if they are fully informed optimizers, while others do not.

SUMMARY AND FUTURE WORK

Experimental economics and psychology have now produced strong empirical support for the view that framing effects as well as contextual and other psychological factors put a large gap between homo-sapiens and individuals with bounded rationality. The question we pose in this chapter is as follows: Does that matter and how does it matter? To answer these questions, we developed a model in which imitation in social networks can ultimately yield high aggregate levels of optimal behavior. It should be noted that the fraction of agents who are rational in such an imitative system will definitely affect the stock market. But the eventual (asymptotic) attainment per se of such a state need not depend on the extent to which rationality is bounded. Perhaps the main issue then is not how much rationality there is at the micro level, but how little is enough to generate macro-level patterns in which most agents are behaving "as if" they were rational, and how various social networks affect the dynamics of such patterns.

We conclude by describing our plan for further research. An evolutionary selection mechanism based on relative past profits will govern the dynamics of the fractions and the switching of agents between different beliefs or forecasting strategies. A strategy attracts more agents if it performed relatively well in the recent past compared to other strategies. There are two related theoretical issues. One is the connection between individual rationality and aggregate efficiency, that is, between optimization by individuals and optimality in the aggregate. The second is the role of social interactions and social networks in individual decision-making and in determining macroscopic outcomes and dynamics. Regarding the first, much of mathematical social science assumes that aggregate efficiency requires individual optimization. Perhaps this is why bounded rationality is disturbing to most economists: They implicitly believe that if the individual is not sufficiently rational it must follow that decentralized behavior is doomed to produce inefficiency.

REFERENCES

Alfarano, S., Wagner, F., & Lux,T. (2004). *Estimation of Agent-Based Models: the case of an asymmetric herding model*.

Bornholdt, S. (2001). Expectation bubbles in a spin model of markets. *International Journal of Modern Physics C*, *12*(5), 667–674. doi:10.1142/S0129183101001845

Boswijk H. P., Hommes C. H, & Manzan, S. (2004). *Behavioral Heterogeneity in Stock Prices*.

Cincotti, S., Focardi, S., Marchesi, M., & Raberto, M. (2003). Who wins? Study of long-run trader survival in an artificial stock market. *Physica A*, *324*, 227–233. doi:10.1016/S0378-4371(02)01902-7

Cont, R., & Bouchaud, J.-P. (2000). Herd behavior and aggregate fluctuations in financial markets. *Macro-economics Dynamics*, *4*, 170–196.

De Long, J. B., Shleifer, A. L., Summers, H., & Waldmann, R. J. (1991). The survival of noise traders in financial markets. *The Journal of Business*, *64*(1), 1–19. doi:10.1086/296523

Durlauf, S. N., & Young, H. P. (2001). *Social Dynamics*. Brookings Institution Press.

Johnson, N., Jeffries, P., & Hui, P. M. (2003). *Financial Market Complexity*. Oxford.

Kaizoji. T, Bornholdt, S. & Fujiwara.Y. (2002). Dynamics of price and trading volume in a spin model of stock markets with heterogeneous agent. *Physica A*.

Le Baron, B. (2001). A builder's guide to agent-based financial markets. *Quantitative Finance*, *1*(2), 254–261. doi:10.1088/1469-7688/1/2/307

LeBaron, B. (2000). Agent based computational finance: suggested readings and early research. *Journal of Economic Dynamics & Control*, *24*, 679–702. doi:10.1016/S0165-1889(99)00022-6

Levy, M. Levy, H., & Solomon, S. (2000). *Microscopic Simulation of Financial Markets: From Investor Behavior to Market Phenomena*. San Diego: Academic Press.

Lux, T., & Marchesi, M. (1999). Scaling and criticality in a stochastic multi-agent model of a financial market. *Nature*, *397*, 498–500. doi:10.1038/17290

Palmer, R. G., Arthur, W. B., Holland, J. H., LeBaron, B., & Tayler, P. (1994). Artificial economic life: A simple model of a stock market. *Physica D. Nonlinear Phenomena, 75,* 264–274. doi:10.1016/0167-2789(94)90287-9

Raberto, M., Cincotti, S., Focardi, M., & Marchesi, M. (2001). Agent-based simulation of a financial market. *Physica A, 299*(1-2), 320–328. doi:10.1016/S0378-4371(01)00312-0

Sornette, D. (2003). *Why stock markets crash.* Princeton University Press.

Tesfatsion, L. (2002). Agent-based computational economics: Growing economies from the bottom up. *Artificial Life, 8,* 55–82. doi:10.1162/106454602753694765

KEY TERMS AND DEFINITIONS

Artificial Market: a research approach of the market by creating market artificially.

Agent-Based Model: a class of computational models for simulating the actions and interactions of autonomous agents

Rational Trader: a type of trader whose decisions of buy, sell, or hold are based on fundamental analysis

Noise Trader: a type of trader whose decisions of buy, sell, or hold are not based on fundamental analysis

Ising Model: a mathematical model of ferromagnetism in statistical mechanics.

Logit Model: a mathematical model of human decision in statistics.

Chapter 8
Agent–Based Modeling Bridges Theory of Behavioral Finance and Financial Markets

Hiroshi Takahashi[1]
Keio University, Japan

Takao Terano[2]
Tokyo Institute of Technology, Japan

ABSTRACT

This chapter describes advances of agent-based models to financial market analyses based on our recent research. We have developed several agent-based models to analyze microscopic and macroscopic links between investor behaviors and price fluctuations in a financial market. The models are characterized by the methodology that analyzes the relations among micro-level decision making rules of the agents and macro-level social behaviors via computer simulations. In this chapter, we report the outline of recent results of our analysis. From the extensive analyses, we have found that (1) investors' overconfidence behaviors plays various roles in a financial market, (2) overconfident investors emerge in a bottom-up fashion in the market, (3) they contribute to the efficient trades in the market, which adequately reflects fundamental values, (4) the passive investment strategy is valid in a realistic efficient market, however, it could have bad influences such as instability of market and inadequate asset pricing deviations, and (5) under certain assumptions, the passive investment strategy and active investment strategy could coexist in a financial market.

INTRODUCTION

Financial Economics researches have become active since 1950's and many prominent theories regarding asset pricing and corporate finance have been proposed (Markowitz, 1952; Modigliani, Miller, 1958; Sharpe, 1964; Shleifer, 2000). The

DOI: 10.4018/978-1-60566-898-7.ch008

assumption of the efficiency of financial markets plays an important role in the literature in traditional financial theory and many research have been conducted based on the assumption (Friedman, 1953; Fama, 1970). For example, CAPM (Capital Asset Pricing Model), one of the most popular asset pricing theory in the traditional financial literature, is derived based on the assumptions of the efficient market and rational investors. CAPM indicates that the optimal investment strategy is to hold market portfolio (Sharpe, 1964).

However, conventional finance theory meets severe critiques about the validities of the assumptions on the markets, or the capabilities to explain real world phenomena. For example, the worldwide financial crisis in 2008 was said to be the one, which would occur per ten decades. Recently, N. N. Taleb describes the role of accidental effects in a financial markets and human cognitions about the effects (Taleb, 2001). Also, researchers in behavioral finance have raised some doubts about the efficient market assumption, by arguing that an irrational trader could have influences on asset prices (Shiller, 2000; Shleifer, 2000; Kahneman, Tversky, 1979; Kahneman, Tversky, 1992).

To address the problems, we employ agent-based model (Arthur, 1997; Axelrod, 1997) in order to analyze the relation between micro-rules and macro-behavior (Axtell, 2000; Russell, 1995). In the literature, they have frequently reported that a variety of macro-behavior emerges bottom-up from local micro-rules (Epstein, 1996; Levy, 2000; Terano, 2001; Terano, 2003; Arthur, 1997; Tesfatsion, 2002). We have developed an artificial financial market model with decision making agents. So far, we have reported on micro-macro links among agents and markets, investors' behaviors with various mental models, and risk management strategies of the firms (Takahashi, 2003; Takahashi, 2004; Takahashi, 2006; Takahashi, 2007; Takahashi, 2010). In this chapter, based on our recent research, we will describe the basic principles and architecture of our simulator and explain our main findings. The objective of the research is to investigate (1) the influences of micro- and macro-level of investment strategies, (2) roles of the evaluation method, and (3) financial behaviors, when there are so many investors with different strategies.

The next section of this chapter describes the model utilized for this analysis, then analysis results are discussed in sections 3 and 4. Section 5 contains summary and conclusion.

DESCRIPTION OF AN AGENT-BASED FINANCIAL MARKET MODEL

Basic Framework and Architecture of Models of a Financial Market

In our research, first, we have observed the macro level phenomena of a real financial market, then, second, we have modeled the phenomena in an artificial market in a computer. To model the market, third, we have introduced micro level decision making strategies of human investors based on the recent research on behavioral financial theory and cognitive science (Shleifer, 2000). Forth, we have designed micro-macro level interactions in the artificial market, which are not able to be examined in the real world. Therefore, our method is a constructive approach to bridge the state-of-the art financial theory and real behaviors in a market through agent-based models. The framework is summarized in Figure 1.

Based on the framework, we have implemented a common artificial market model depicted in Figure 2. The market model is characterized as follows: (1) benefit and/or loss of a firm is randomly determined, (2) the information is observed by investor agents to make their investment decisions, (3) based on the decisions, agents trade the financial assets in the artificial market, and the market prices are determined, and (4) the determined prices of the market again give the effects of decision making of the agents. The detailed descriptions of the model are given below.

A agent-based simulator of the financial market involving 1,000 investors is used as the model for this research. Several types of investors exist in the market, each of them undertakes

Figure 1. Framework of agent-based financial research

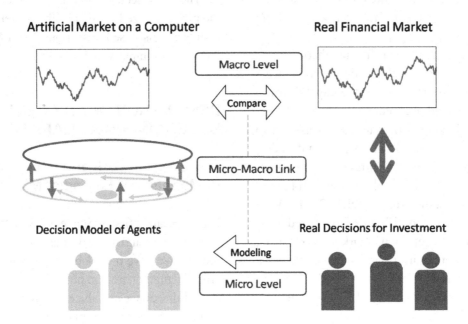

Figure 2. Outline of a common artificial market

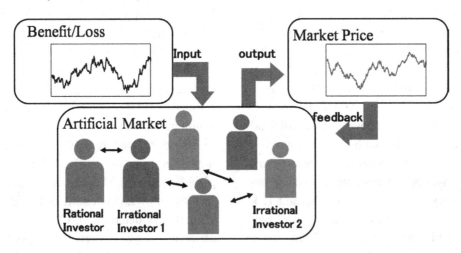

transactions based on their own stock calculations. They share and risk-free assets with the two possible transaction methods. The execution of the simulator consists of the three major steps: (1) generation of corporate earnings, (2) formation of investor forecasts, and (3) setting transaction prices. The market conditions will change through these steps. About the details of parameters of the simulator, please refer to the appendix 1.

Assets Traded in the Market

The market consists of both risk-free and risky assets. About the risky assets, all profits gained during each term are distributed to the shareholders. Corporate earnings (y_t) are expressed as $(y_t = y_{t-1} \cdot (1 + \varepsilon_t))$. They are generated according to the process $\varepsilon_t \sim N(0, \sigma_y^2)$. Risky assets are traded just after the public announcement of profit for the term. Each investor is given common asset holdings at the start of the term with no limit placed on debit and credit transactions.

Modeling Passive Investors

Passive investors of the simulation model invest their assets with the same ratio of the market benchmarks. This means that (1) each passive investors keeps one volume stock during the investment periods, (2) the investment ratio to the stocks is automatically determined, and (3) the trade strategy follows buy-and-hold of initial interests.

Modeling Active Investors

Active investors make decisions based on expected utility maximization method described in (Black, Litterman, 1992). Contrary to passive investors, active investors forecast the stock price. In the following section, we will explain the forecasting models of active investors.

Forecasting Models of Investors

(a)   Fundamentalists

We will refer to the investors who make investment decisions based on fundamental values as 'fundamentalists'. We adopt the dividend discount model, which is the most basic model to determine the fundamental value of stocks. The fundamentalists are assumed to know that the profit accrues according to Brownian motion. They forecast the stock price P_{t+1}^f and the profit y_{t+1}^f from the profit of current period (y_t) and the discount rate of the stock (δ) as $P_{t+1}^f = y_t/\delta$ and $y_{t+1}^f = y_t$, respectively.

(b)   Trend Predictors

We formulate a model of the investor who finds out the trends from randomly fluctuate stock prices. This type of investor predicts the stock price $\left(P_{t+1}^f\right)$ of the next period by extrapolating the latest stock trends (10 days). The trend predictors forecast the stock price $\left(P_{t+1}^f\right)$ and the profit $\left(y_{t+1}^f\right)$ from the trend at period t-1 as $P_{t+1}^f = P_{t-1} \cdot \left(1 + a_{t-1}\right)^2$ and $y_{t+1}^f = y_t \cdot \left(1 + a_{t-1}\right)$, where $a_{t-1} = \left(1/10\right) \cdot \sum_{i=1}^{10}\left(P_{t-i}/P_{t-i-1} - 1\right)$. Predicted price $\left(P_{t+1}^f\right)$ and profit $\left(y_{t+1}^f\right)$ are different when trend measurement period is different.

(c)   Loss over-estimation investors

We formulate a model in which the investor doubles the loss estimates from the reference stock price. In the model, the reference stock price is the one of the 10 periods beforehand. When the most recent price $\left(P_{t-1}\right)$ is lower than the price at the reference point $\left(P_t^{ref}\right)$, the "Loss over-estimation investors" forecast the stock price $\left(P_{t+1}^f\right)$ by converting the original predicted price $\left(P_{t+1}^{bef\ f}\right)$ using the formula $P_{t+1}^f = 2.25 \cdot P_{t+1}^{bef\ f} - 1.25 \cdot P_t^{ref}$. As for the original predicted price $\left(P_{t+1}^{bef\ f}\right)$, we use the dividend discount model.

(d)   Overconfident Investors

Bazerman reported that human beings tend to be overconfident in his/her own ability (Bazerman, 1998). In the area of Behavioral Finance, Kyle analyzed the influence of the overconfident investment behaviors on the markets with the analytical method (Kyle, 1997). Also in a real market, we often find that each investor talks about different future prospects with confidence. It seems that all investors tend to have overconfidence in varying degrees.

We formulate the model of investors who are overconfident in their own predictions by assuming that they underestimate the risk of the stock. The risk of the stock estimated by an overconfident investor (σ^s) is calculated from the historical volatility (σ^h) and the adjustment factor to determine the degree of overconfidence constant value $k (k = 0.6)$ as $(\sigma^s)^2 = k(\sigma^h)^2$.

Calculation of Expected Return Rate of the Stock

The investors in the market predict the stock price (P^f_{t+1}) and the corporate profit (y^f_{t+1}) at the term $t+1$ based on the corporate profit (y_t) at the term t and the stock prices at and before the term t-1 $(P_{t-1}, P_{t-2}, P_{t-3} \cdots)$. In the following, we represent the predicted values of the stock price and the corporate profit by the investor $i (i = 1, 2, 3 \cdots)$ as $P^{y,i}_{t+1}$ and $y^{f,i}_{t+1}$, respectively. The expected rate of return on the stock for the investor $i \left(r^{\text{int},i}_{t+1}\right)$ is calculated as follows:

$$r^{\text{int},i}_{t+1} = \left[r^{im}_t \cdot c^{-1} \cdot \left(\sigma^s_{t-1}\right)^{-2} + r^{f,i}_{t+1} \cdot \left(\sigma^s_{t-1}\right)^{-2}\right] \cdot \left[c^{-1} \cdot \left(\sigma^s_{t-1}\right)^{-2} + \left(\sigma^s_{t-1}\right)^{-2}\right]^{-1},$$

where $r^{f,i}_{t+1} = \left(\left(P^{f,i}_{t+1} + y^{f,i}_{t+1}\right)/P_t - 1\right) \cdot \left(1 + \varepsilon^i_t\right)$ and

$$r^{im}_t = 2\lambda \left(\sigma^s_{t-1}\right)^2 W_{t-1} + r_f$$

(Black, Litterman, 1992).

Determination of Trading Prices

The traded price of the stock is determined at the price the demand meets the supply (Arthur, 1997). Both of the investment ratio $\left(w^i_t\right)$ and the number

of the stock held by investors $\left(\sum^M_{i=1} \left(F^i_t \cdot w^i_{t_i}\right)/P_t\right)$ are the decreasing function of the stock price, and the total number of the stock issued in the market (N) is constant. We derive the traded price as $\sum^M_{i=1} \left(F^i_t \cdot w^i_{t_i}\right)/P_t = N$ by calculating the price (P_t), where the total amount of the stock retained by investors $\left(\left(F^i_t \cdot w^i_{t_i}\right)/P_t\right)$ meets the total market value of the stock.

Natural Selection Rules

Natural selection rules are equipped in the market to represent the changes of cumulative excess return for the most recent 5 terms (Takahashi, Terano, 2003). The rules are divided into the two steps: (1) appointment of investors who alter their investment strategy, and (2) alteration of investment strategy. With the alteration of investment strategy, existence of investment strategy alteration is decided based upon the most recent performance of each 5 term period after 25 terms have passed since the beginning of market transactions. In addition, for the simplicity, investors are assumed to decide to alter investment strategy with the higher cumulative excess return over at most recent 5 terms.

Using the models, we have investigated the roles of over-confidence agents and passive and active strategies in the following two sections.

Over-Confidence Agents will Survive in a Market

First, we have analyzed the initial forecasting model ratio where there was (1) a higher ratio of fundamental forecasting, and (2) a higher ratio of trend forecasting. As the results of this analysis have suggested that these higher ratio strength the degree of overconfidence in both cases, then, we have analyzed the random distribution of the initial ratio of each forecasting model to determine whether the same result could be obtained under different conditions. The results of this analysis are explained in detail below.

Searching for Investment Strategies

(a) When there is a High Ratio of Fundamental Forecasting

As fundamentalists enforce a strong influence on the market value under these conditions, the market value is almost in concord with the fundamental value (Figure 3). We have confirmed that the number of fundamentalists is on the increase due to the rules of natural selection in regard to the transition of investor numbers (Figure 4). Looking at transition in the degree of overconfidence, the degree of overconfidence strongly remains as the simulation steps go forward (Figure 5).

(b) When there is a High Ratio of Trend Forecasting

When there is a high ratio of investors using trend forecasting, the market value deviated greatly from the fundamental value. The number of investors using trend forecasting also increases as such investors enforce a strong influence on the market value. There is the investment environment, in which different forecasting methods were applied to obtain excess return. On the other hand, investors with a strong degree of overconfidence survive in the market even under these conditions.

(c) When the Initial Ratio is Applied Randomly

We have analyzed the case in which the initial ratio of investors is randomly applied. Although the case shown in Figure 6 indicates that a number of investors employ the strategy based on the fundamental values, the forecasting model employed by market investors is dependent on the ratio of each type of investors, changing along with circumstances such as trend forecasting and

the average value of past equity prices. In contrast, overconfident investors survive in the market even when a random initial value is applied for the degree of overconfidence (Figures 6 and 7).

This interesting analysis result suggests the possibility of universality when survival trends of overconfident investors are compared with the forecasting model.

Exploring Market Conditions

This subsection explores the conditions in which transaction prices reach the fundamental values.

(a) Inverse Simulation Method

Inverse Simulation Analysis consists of the following four steps: (1) Carry out 100 times simulations with an investment period of 100 terms; (2) Calculate the deviation between transaction prices and the fundamental value for each simulation; (3) Select the simulation cases in which the deviation value is small, and (4) Using recombination operations of Genetic Algorithms, set the calculated index better ones, (5) repeat the process 100 simulation generations. The index (q) of deviation between transaction prices and the fundamental value expresses the deviation ratio with the fundamental value and is specifically calculated as $q = E[x]^2 + Var[x]$. P_t^0 represents the fundamental value $x_t = (P_t - P_t^0)/P_t^0$ for time t.

(b) Experimental Results

The experimental results are summarized in Figures 8 and 9. In Figure 8, the transaction prices tend to coincide with the corresponding fundamental values, where there is a high percentage of fundamentalist investors. Also, Figure 9 shows that the number of fundamentalist investors are superior to the other kinds of investors. In addition,

Figure 3. Price transition

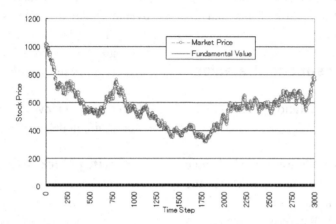

Figure 4. Transition of number of investors

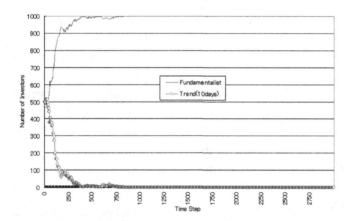

Figure 5. Transition of average degree of overconfidence (Fundamentalist: Trend=500:500)

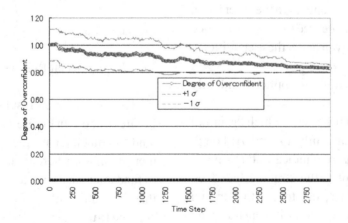

Figure 6. Transition of average number of overconfidence (Random)

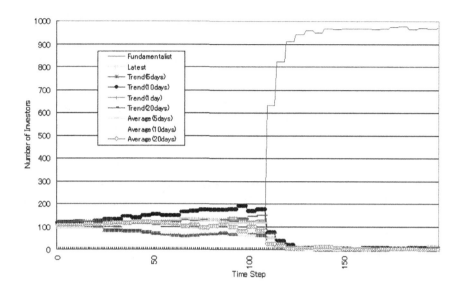

Figure 7. Transition of average degree of investors (Random)

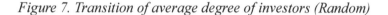
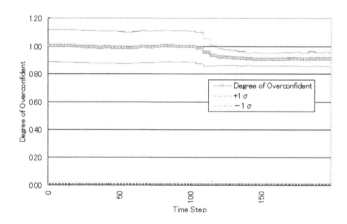

transaction prices almost match the fundamental value in this case.

Traditional finance argues that market survival is possible for those investors able to swiftly and accurately estimate both the risk and rate of return on stock, achieving market efficiency. However, analysis results obtained here regarding the influence irrational investors have on prices suggests a different situation, pointing to the difficulty of market modeling which takes real conditions into account.

HOW PASSIVE AND ACTIVE INVESTORS BEHAVE

The series of experiments on the behaviors of passive and active investors are divided into the two parts: First, we have fundamentalist agents and passive-investment agents in the market to investigate the influences of the two strategies. Next, in order to analyze the effects, we introduce the other kinds of investors, such as trend chasers,

Figure 8. Transition of average number of investors (Inverse Simulation)

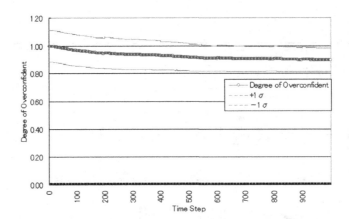

Figure 9. Transition of average degree of overconfidence (Inverse Simulation)

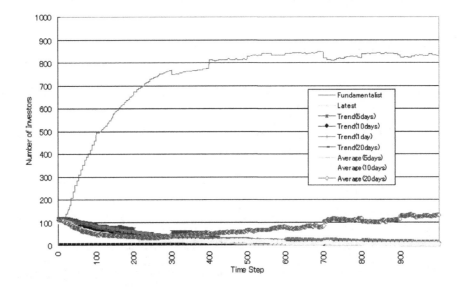

loss over estimation investors, or overconfidence investors.

Trading with Fundamentalist and Passive Investors

Figures 10 and 11 illustrate the case where there exist the same 500 numbers of the two kinds of investors (Case 0). Figure 10 shows the histories of stock prices. The solid line in Figure 10 represents the traded price and the line with x

mark represents the fundamental value. Figure 11 depicts the histories of cumulative excess returns of each investor. This graph shows that the fluctuation of the traded price agrees with the one of the fundamental value. The line with mark x in Figure 11 shows the performance of passive investment strategy and the dotted line shows the ones of fundamentalists. The performances of fundamentalists are slightly different among them, because each of fundamentalists respectively has a predicting error. As the traditional asset pricing

Figure 10. Price transition(Case 0)

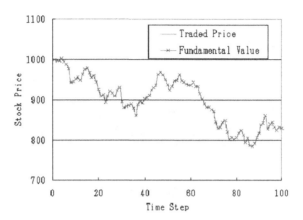

Figure 11. Cumulative excess returns(Case 0)

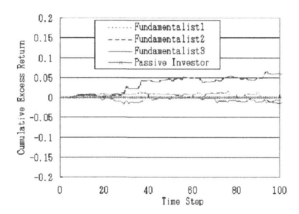

theory suggests, the trading prices are coincide with the fundamental values and fundamentalist and passive investors can get the same profit in average.

Next, using natural selection principles of Genetic Algorithms (see the appendix 2 for detail), let the investor agents change their strategies when (1) the excess returns are under the target (e.g., over 10%), (2) the excess returns are under 0%, and (3) the excess returns are too bad (e.g., under 10%).

The results of Case 1 are shown in Figures 12, 13, 14, and 15. The results of Case 2 are shown in Figures 16, 17, 18, and 19. Figures 12, 13, 14,

and 15 are obtained by 100 experiments, each of which consists of 3,000 simulation steps.

In Case 1, traded price changes in accordance with fundamental value and both of investors coexist in the market. On the other hand, in Case 2, traded price doesn't reflect the fundamental value and only passive investors can survive in the markets after around 1,600 time step. This result is quite different from the ones in Case 1. In Case 3, we have obtained the results similar to the ones in Case 2. These differences among each experiment are brought about by the difference of evaluation methods. In this sense, evaluation methods have a great influence on the financial markets.

Figure 12. Price transition

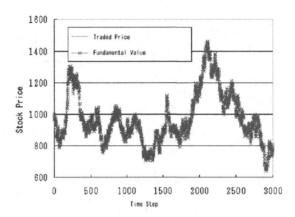

Figure 13. Transition of number of investors

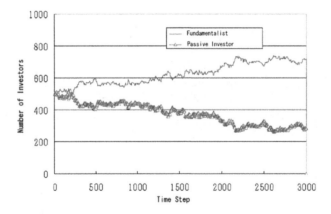

Figure 14. Distribution of fundamentalists

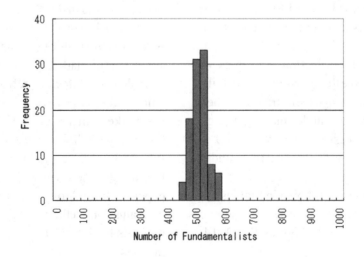

Figure 15. Distribution of fundamentalists

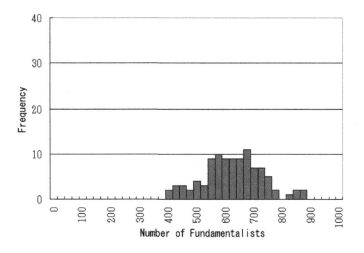

Figure 16. Price transition

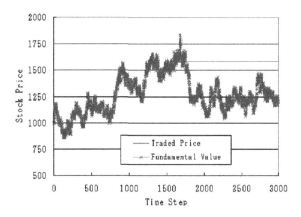

Figure 17. Transition of number of investors

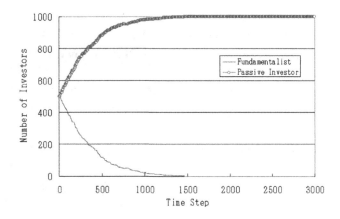

Figure 18. Distribution of fundamentalists

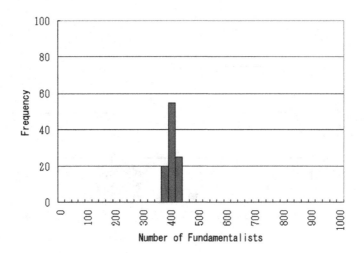

Figure 19. Distribution of fundamentalists

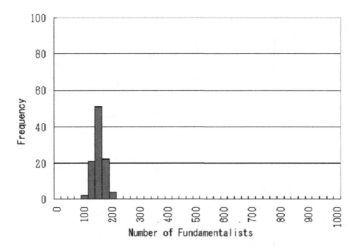

Throughout the experiments shown above, we have confirmed the effectiveness of passive investment strategy. Among them, the result in Figure 19 has indicated the superiority of passive strategy in more actual situation. However as is shown in Case 2 and Figure 18, we have also confirmed the unfavorable influence of passive investment on the market.

TRADING WITH FUNDAMENTALISTS, PASSIVE INVESTORS, AND OVER CONFIDENT INVESTORS, AND INVESTORS WITH PROSPECT THEORY

This section describes the experimental results with the five different investor agents: Fundamentalists, Passive Investors, Trend Chaser, Investors with Prospect Theory, and Over Con-

Figure 20. Price transition (Case 4)

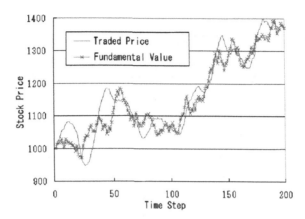

Figure 21. Cumulative excess returns (Case 4)

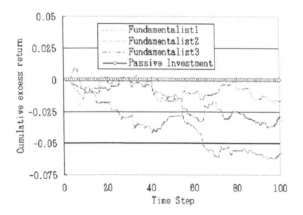

fident Investors. First, the results of Case 4 with 400 Fundamentalists, 400 trend chasers, and 200 passive investors are shown in Figures 20 and 21. Second, the results of Case 5 with 400 Fundamentalists, 400 Over Confident Investors, and 200 passive investors are shown in Figures 22 and 23. Third, the results of Case 6 with 400 Fundamentalists, 400 Investors with Prospect Theory, and 200 passive investors are shown in Figures 24 and 25.

In all cases, we have observed that passive investors keep their moderate positions positive, even when stock prices largely deviate from the fundamental value. In other words, passive investment strategy is the most effective way if investors

want not to get the worst result in any cases, even if they have failed to get the best result. In asset management business, some investors adopt the passive investment to avoid getting the worst performance.

Figures 26 and 27 show the results where the agents are able to change their strategy when the excess returns are less than 0. In this experiment, we have slightly modified the natural selection rule as is described in previous section. In the following experiments, investors change their strategy depending on their recent performance as is the same way in previous section and after that, investors change their strategy randomly in a small possibility (0.01%) which is correspond to

Figure 22. Price transition (Case 4)

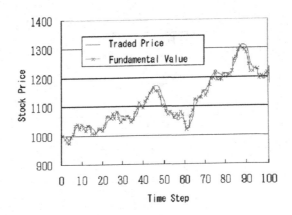

Figure 23. Cumulative excess returns(Case 4)

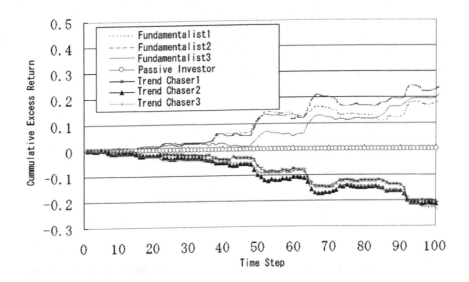

Figure 24. Price transition(Case 5)

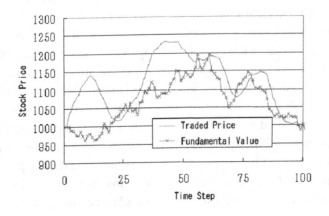

Figure 25. Cumulative excess returns(Case 5)

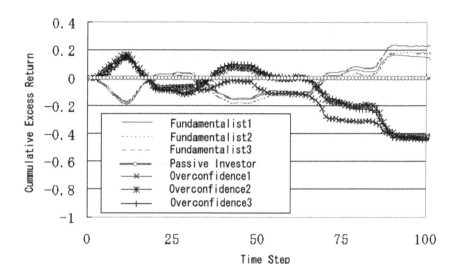

Figure 26. Price transition(Case 6)

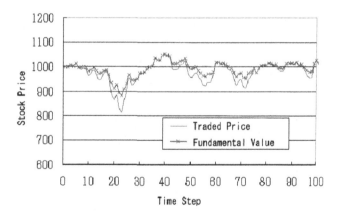

mutation in genetic algorithm. The result shown in Figure 27 suggests that there remain both fundamentalist and passive investors in the market and that they keep the market stable. The results shown in Figures 28 and 29 are quite different from the ones in Case 2. These results suggest that even slight differences in the market conditions and investors behavior could cause large changes in the markets. In this sense, these results are thought-provoking.

SUMMARY AND CONCLUSION

This chapter has described our recent results on Agent-Based Model to analyze both microscopic and macroscopic associations in the financial market. we have found that (1) investors' overconfidence plays various roles in a financial market, (2) overconfident investors emerge in a bottom-up fashion in a market, (3) they contribute to the achievement of a market which adequately reflects fundamental values, (4) the passive investment

Figure 27. Cumulative Excess Returns(Case 6)

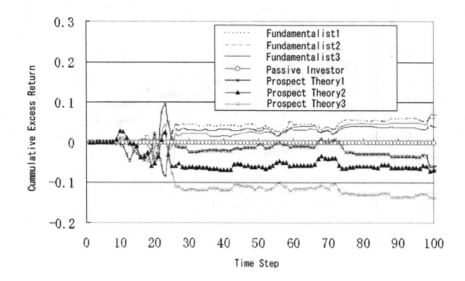

Figure 28. Price Transition

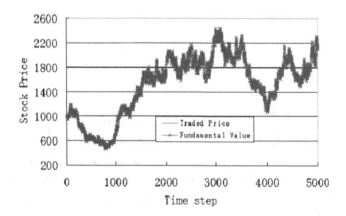

Figure 29. Transition of Number of Investors

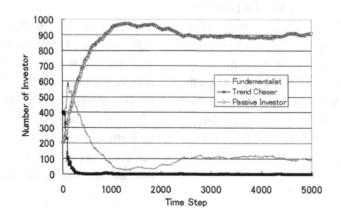

strategy is valid in a realistic efficient market, however, it could have bad influences such as instability of market and inadequate asset pricing deviations, and (5) under certain assumptions, the passive investment strategy and active investment strategy could coexist in a financial market. These results have been described in more detail elsewhere (e.g., Takahashi, Terano 2003, 2004, 2006a, 2006b, Takahashi, Takahashi, Terano 2007). Using a common simple framework presented in the chapter, we have found various interesting results, which may or may not coincide with both financial theory and real world phenomena. We believe that the agent-based approach would be fruitful for the future research on social systems including financial problems, if we would continue the effort to convince the effectiveness (Terano, 2007a, 2007b; Takahashi, 2010).

REFERENCES

Arthur, W. B., Holland, J. H., LeBaron, B., Palmer, R. G., & Taylor, P. (1997). Asset Pricing under Endogenous Expectations in an Artificial Stock Market. [Addison-Wesley.]. *The Economy as an Evolving Complex System, II*, 15–44.

Axelrod, R. (1997). *The Complexity of Cooperation -Agent-Based Model of Competition and Collaboration*. Princeton University Press.

Axtell, R. (2000). Why Agents? On the Varied Motivation For Agent Computing In the Social Sciences. *The Brookings Institution Center on Social and Economic Dynamics Working Paper*, November, No.17.

Bazerman, M. (1998). *Judgment in Managerial Decision Making*. John Wiley & Sons.

Black, F., & Litterman, R. (1992, Sept/Oct). Global Portfolio Optimization. *Financial Analysts Journal*, 28–43. doi:10.2469/faj.v48.n5.28

Brunnermeier, M. K. (2001). *Asset Pricing under Asymmetric Information*. Oxford University Press. doi:10.1093/0198296983.001.0001

Epstein, J. M., & Axtell, R. (1996). *Growing Artificial Societies Social Science From the The Bottom Up*. MIT Press.

Fama, E. (1970). Efficient Capital Markets: A Review of Theory and Empirical Work. *The Journal of Finance*, *25*, 383–417. doi:10.2307/2325486

Friedman, M. (1953). *Essays in Positive Economics*. University of Chicago Press.

Goldberg, D. (1989). *Genetic Algorithms in Search, Optimization, and Machine Learning*. Addison-Wesley.

Kahneman, D., & Tversky, A. (1979). Prospect Theory of Decisions under Risk. *Econometrica*, *47*, 263–291. doi:10.2307/1914185

Kahneman, D., & Tversky, A. (1992). Advances in. prospect Theory: Cumulative representation of Uncertainty. *Journal of Risk and Uncertainty*, *5*.

Kyle, A. S., & Wang, A. (1997). Speculation Duopoly with Agreement to Disagree: Can Overconfidence Survive the Market Test? *The Journal of Finance*, *52*, 2073–2090. doi:10.2307/2329474

Levy, M., Levy, H., & Solomon, S. (2000). *Microscopic Simulation of Financial Markets*. Academic Press.

Markowitz, H. (1952). Portfolio Selection. *The Journal of Finance*, *7*, 77–91. doi:10.2307/2975974

Modigliani, F., & Miller, M. H. (1958). The Cost of Capital, Corporation Finance and the Theory of Investment. *The American Economic Review*, *48*(3), 261–297.

Russell, S., & Norvig, P. (1995). *Artificial Intelligence*. Prentice-Hall.

Sharpe, W. F. (1964). Capital Asset Prices: A Theory of Market Equilibrium under condition of Risk. *The Journal of Finance, 19*, 425–442. doi:10.2307/2977928

Shiller, R. J. (2000). *Irrational Exuberance.* Princeton University Press.

Shleifer, A. (2000). *Inefficient Markets.* Oxford University Press. doi:10.1093/0198292279.001.0001

Takahashi, H. (2010), "An Analysis of the Influence of Fundamental Values' Estimation Accuracy on Financial Markets, " *Journal of Probability and Statistics*, 2010.

Takahashi, H., Takahashi, S., & Terano, T. (2007). Analyzing the Influences of Passive Investment Strategies on Financial Markets via Agent-Based Modeling . In Edmonds, B., Hernandez, C., & Troutzsch, K. G. (Eds.), *Social Simulation- Technologies, Advances, and New Discoveries* (pp. 224–238). Hershey, PA: Information Science Reference.

Takahashi, H., & Terano, T. (2003). Agent-Based Approach to Investors' Behavior and Asset Price Fluctuation in Financial Markets. *Journal of Artificial Societies and Social Simulation, 6*(3).

Takahashi, H., & Terano, T. (2004). Analysis of Micro-Macro Structure of Financial Markets via Agent-Based Model: Risk Management and Dynamics of Asset Pricing. *Electronics and Communications in Japan, 87*(7), 38–48.

Takahashi, H., & Terano, T. (2006a). Emergence of Overconfidence Investor in Financial markets. *5th International Conference on Computational Intelligence in Economics and Finance*.

Takahashi, H., & Terano, T. (2006b). Exploring Risks of Financial Markets through Agent-Based Modeling. In *Proc. SICE/ICASS 2006* (pp. 939-942).

Terano, T. (2007a). Exploring the Vast Parameter Space of Multi-Agent Based Simulation. In L. Antunes & K. Takadama (Eds.), *Proc. MABS 2006* (LNAI 4442, pp. 1-14).

Terano, T. (2007b). KAIZEN for Agent-Based Modeling. In S. Takahashi, D. Sallach, & J. Rouchier (Eds.), *Advancing Social Simulation -The First Congress-* (pp. 1-6). Springer Verlag.

Terano, T., Deguchi, H., & Takadama, K. (Eds.). (2003), *Meeting the Challenge of Social Problems via Agent-Based Simulation: Post Proceedings of The Second International Workshop on Agent-Based Approaches in Economic and Social Complex Systems*. Springer Verlag.

Terano, T., Nishida, T., Namatame, A., Tsumoto, S., Ohsawa, Y., & Washio, T. (Eds.). (2001). *New Frontiers in Artificial Intelligence*. Springer Verlag. doi:10.1007/3-540-45548-5

Tesfatsion, L. (2002). Agent-Based Computational Economics. *Economics Working Paper*, No.1, Iowa Sate University.

ENDNOTES

[1]   Graduate School of Business Administration, Keio University, 4-1-1 Hiyoshi, Yokohama, 223-8526, Japan, E-mail: htaka@kbs.keio.ac.jp

[2]   Department of Computational Intelligence and Systems Science, Tokyo Institute of Technology, 4259-J2-52 Nagatsuta-cho, Midori-ku, Yokohama, 226-8502, Japan, E-mail: terano@dis.titech.ac.jp

APPENDICES

1. List of Parameters of the Proposed Model

The parameters used in the proposed model are summarized as follows:

M: the number of investors (1,000)

N: the number of issued stocks (1,000)

F_t^i :the total amount of assets of the investor i at the term t (F_0^i =2,000:common)

W_t :the stock ratio in the market at the term t (W_0 =0.5)

w_t^i :the investment ratio of the stock of the investor i at the term t (w_0^i =0.5:constant)

σ_y :the standard deviation of the profit fluctuation($0.2 / \sqrt{200}$:constant)

δ :the discount rate of the stock (0.1/200:constant)

λ :the degree of risk aversion of the investor(1.25:common,constant)

c: the adjustment coefficient for variance(0.01)

σ_t^h :the historical volatility of the stock (for the recent 100 terms)

σ_n :the standard deviation of the dispersion of the short term expected rate of return on the stock (0.01:common)

k: the adjustment coefficient for confidence (0.6)

2. Rules of Natural Selection Principle

This section explains the rules of natural selection principle. The principle used in this chapter is composed of two steps: (1) selection of investors who change their investment strategies and (2) selection of new strategy. Each step is described in the following sections:

Selection of Investors Who Change Their Investment Strategies

After 25 terms pass since the market has started, each investor makes decision at regular interval (every five terms) whether he/she changes the strategy. The decision is made depending on the cumulative excess return during the recent five terms and the investors who obtain smaller return changes the strategy at higher probability. To be more precise, the investors who obtain negative cumulative excess return changes the strategy at the following probability:

$$p_i = \max\left(0.3 - a \cdot e^{r_i^{cum}}, 0\right),$$

p_i: probability at which investor i changes own strategy,

r_i^{cum} : cumulative return of investor i during recent 5 terms,

a: the coefficient for the evaluation criteria(0.2,0.3,0.4).

Selection of New Strategy

We apply the method of genetic algorithm (Goldberg (1989)) to the selection rule of new strategy. The investors who change the strategy tend to select the strategy that has brought positive cumulative excess return. The probability to select s_i as new strategy is given as: $p_i = e^{r_i^{cum}} \Big/ \sum_{j=1}^{M} e^{r_j^{cum}}$, where r_i^{cum} is the cumulative excess return of each investor.

Section 4
Multi–Agent Robotics

Chapter 9

Autonomous Specialization in a Multi–Robot System using Evolving Neural Networks

Masanori Goka
Hyogo Prefectural Institute of Technology, Japan

Kazuhiro Ohkura
Hiroshima University, Japan

ABSTRACT

Artificial evolution has been considered as a promising approach for coordinating the controller of an autonomous mobile robot. However, it is not yet established whether artificial evolution is also effective in generating collective behaviour in a multi-robot system (MRS). In this study, two types of evolving artificial neural networks are utilized in an MRS. The first is the evolving continuous time recurrent neural network, which is used in the most conventional method, and the second is the topology and weight evolving artificial neural networks, which is used in the noble method. Several computer simulations are conducted in order to examine how the artificial evolution can be used to coordinate the collective behaviour in an MRS.

INTRODUCTION

Artificial evolution is one of the emerging approach in the design of controllers for autonomous mobile robots. In general, a robot controller, which is represented as an evolving artificial neural network (EANN), is evolved in a simulated or a physical environment such that it exhibits the behaviour required to perform a certain task. The field of research on autonomous robots with EANNs is called evolutionary robotics (ER) (Cliff, et al., 1993) (Harvey, et al., 2005).

There has been a great deal of interest in EANNs. A good summary of EANNs carried out until 1999 can be found in a study of Yao (1999). Traditionally, EANNs have been classified into the following three categories on the basis of their network structure:

- the network structure is fixed and the connection weights evolve.
- the network structure evolves and the connection weights are trained by learning.

DOI: 10.4018/978-1-60566-898-7.ch009

- the network structure and the connection weights evolve simultaneously

The first type of network structure corresponds to the most conventional approach in the field of EANNs.

A typical application is presented in a study of Mondada and Floreano (1995). Recently, continuous time recurrent neural networks (CTRNNs) have been frequently used for evolving autonomous robots (Beer, 1996) (Blynel and Floreano, 2003).

In this chapter, we also consider the network structure of the third type, which are called topology and weight evolving artificial neural networks (TWEANNs) because the evolvability in the corresponding approach is the largest among those of the tree approaches (Stanley and Miikulainen, 2002). Thus far, many TWEANN approaches such as GNARL (Angeline et al., 1994), EPnet (Liu and Yao, 1996), ESP (Gomez and Miikkulainen, 1999) and NEAT (Stanley and Miikulainen, 2002) have been proposed. We are motivated by their work and also developed a robust and efficient approach to EANN design, called MBEANN (Ohkura et al., 2007). In the remainder of this chapter, we deal with the following two approaches: evolving CTRNN (eCTRNN) and MBEANN.

The evolutionary control of multi-robot systems (MRS) has not been well discussed. Only a few papers have been published thus far; however, some basic trail results such as those presented in a study of Triani (2007) or Acerbi (2007) have recently been published. This implies that although artificial evolution is a promising approach in behaviour coordination for a single robot, it is not very easy to evolve an MRS for developing a type of adaptively specialized team play.

Here, we mention two epochal papers. Baldassarre et al. (2003) evolved four simulated Khepera robots to develop a flocking formation and then discussed the variation in the formation patterns with changes in the fitness function. They used

Figure 1. Cooperative package pushing problem

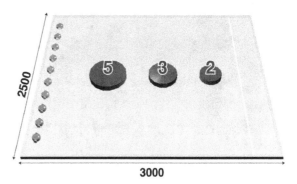

simple two-layered feed-forward EANNs as robot controllers and a binary Genetic Algorithms (GA) in which a real value is encoded with eight bits with their original reproduction procedure.

Quinn et al. (2001) evolved a physical robot group by adopting the so-called Sussex approach (Harvey and Husbands, 1997). Specifically, they first evolved the robot controllers in a simulated environment, and then, after achieving sufficiently good performance, they conducted physical experiments on the evolved controllers. They used their original recurrent EANNs as robot controllers and a steady-state real-coded GA.

In this chapter, we solve the task as in Figure 1: ten autonomous mobile robots have to push three packages to the goal line, which is drawn on the right side of the floor. The packages are assumed to be too heavy for a single robot to move. In order to move a package, the robot has to cooperate with each other to gather at a certain side of a package and push it in the same direction, as illustrated in Figure 2.

This chapter is organized as follows. In the next section, the task and our objectives are introduced and described in detail. In Section 3, the computer simulations performed for eCTRNN and MBEANN are described. The achieved behaviour is discussed from the viewpoint of autonomous specialization. Then, the appear is concluded in the last section.

Figure 2. Package being pushed by robots

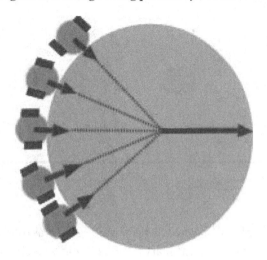

COOPERATIVE PACKAGE PUSHING PROBLEM

Figure 1 shows the cooperative package pushing problem involving ten autonomous mobile robots. In the initial state, ten robots are placed at equal interval on the left side of the floor. The objective is to push all the three packages, which are arranged as shown, to the right side of the floor within 2,000 time steps.

It is assumed that the three packages have weights such that they cannot be moved by a single robot. To move the heaviest package, which is the largest of the package shown in Figure 1, five robots are required.

Further, it is obvious that the five robots have to push the package in the same direction to move it. Similarly, it is also assumed that to move the other two packages, three and two robots, respectively, are required.

The ten autonomous robots have the same specifications, as shown in Figure 2. A robot has six IR sensors on its front, two IR sensors in its back, an omnidirectional vision camera at the center of its body, and a compass in order to detect its global direction. A robot also has two wheels for movement. The informat-ion obtained by these sensors serves as the inputs for the ANN

Table 1. Profit table

Contact between robot and package	+100
Package's final position	+x
Successful transport	+1000
Steps required	*-step*
Collision	*-P*

controller. The output layer has two neurons, which corresponds to the left and right motors. The details on the neural controller are shown below.

All the robots are assumed to have identical ANN controller. The behaviour of each robot will be different because the robots are placed at different positions. A group of robots collects points as shown in Table 1. A group is awarded gets 100 points when any of the robots touches one of the packages. The group is awarded points for moving packages according to the distances between the initial positions and the final positions in the direction of the goal line. The group is awarded 1,000 points each time the group successfully pushes a package to the goal line.

On the other hands, a group loses the points as the number of time steps required increases. It can also lose points when robots collide with each other. In the following computer simulations, the collision penalty is set as $P = 0$ and $P = 5.0 \times 10^{-6} = \Delta P$ because it was found in preliminary experiments that the behaviour of robot groups seems to be very sensitive to P, i.e., the groups show very different results for slight change in P.

COMPUTER SIMULATIONS

eCTRNN

The eCTRNN robot controller has an input layer consisting of 16 neurons: eight neurons correspond to eight IR sensors, three neurons are dedicated to the package nearest to the robot in terms of the polar representation, three neurons are dedicated

Figure 3. Specifications of a robot

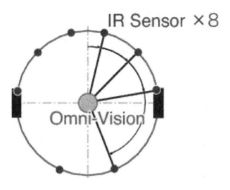

to the robot nearest in terms of the polar representation, and two neurons are used for representing the global direction, as shown in Figure 3. The output layer consists of two neurons, each of which connects to the left and right motors, respectively, as shown in the same figure. In this study, the hidden layer is assumed to have four neurons, each of which has a synaptic connection with all the neurons in the layer. Consequently, each neuron has a recurrent connection to itself.

Experimental Results

The eCTRNN was evolved by using (10, 70)-ES, i.e., by using the standard evolution strategies (Bäck, 1996) in which 10 individuals generate 70 offspring with an equal probability and only the top ten offspring are allowed to survive in the next generation. Ten independent runs were conducted.

The results of the computer simulations are summarized in Table 2. The success rates clearly indicate that the task is more difficult when $P = \Delta P$ than when $P = 0$.

Figure 5 and Figure 6 show the collective behaviour of the robot group in the last generation of a successful run when $P = 0$ or $P = \Delta P$, respectively. The robots showed almost the same collective behaviour for the two value of P: they pushed packages towards the lower right region, presumably to avoid collisions between packages.

Table 2. Success rates in eCTRNN

Penalty (-P/collision)	Success rate [%]
0.0	90
$5.0 \times 10^{-6} = \Delta P$	30

Figure 7 shows the transitions of artificial evolutions observed from the fitness values, the number of required steps, and finished packages. In the figure, (a) and (b) show the transitions of fitness values for all the runs. In the case where $P = 0$, the robot group succeeded in transporting all the three packages in nine runs. On the other hand, in the case where $P = \Delta P$, the robot group succeeded only in three runs. The gray lines in both the graphs show the average fitness transitions calculated using only the number of successful runs. Figure 7(c) and (d) show the transitions of the number of required time steps in the successful runs. The light gray lines show the averages of nine and three successful runs, respectively.

We soon found that when $P = 0$, the robot group showed stable performance in all the successful runs after about the 400th generation. Similarly, when $P = \Delta P$, stable performance was observed in two of three runs at around the 250th generation; further, in one run did not perform stably until the last generations. In Figure 7(e) and (f), indicate whether the three packages were transported to the goal line within the time-step limit. We also soon found that the robot group pushed the three packages to the goal line after around 100 and 150 generation when $P = 0$ and when $P = \Delta P$, respectively.

Next, let us investigate a typical run to determine the strategy that the robot group developed by artificial evolution. Figure 8 shows graphs for ten robots at the 500th generation in the cases when $P = 0$ and $P = \Delta P$. Both the graphs have ten lines, each of which corresponds to a robot. The horizontal axes represent time steps, which begin at 0 and

Figure 4. The eCTRNN robot controller

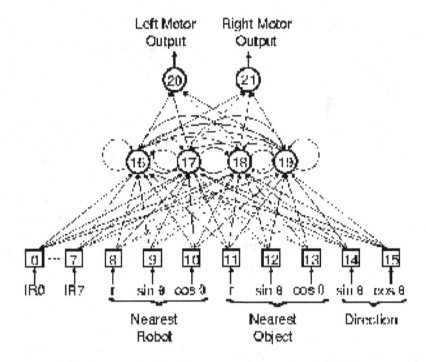

end at 2,000. A black dot is marked when a robot is in contact with the largest package. Similarly, a gray dot or a light gray dot is marked when a robot is in contact with a midsize package or the smallest package, respectively. As observed in Figure 8(a), the robots showed highly sophisticated team play to successfully and pushed almost simultaneously push all the packages to the goal line. Robots No.1 and No.2 pushed the smallest package, Robots No.3, 4, and 5 pushed the midsize package, and the others pushed the largest package to the goal line. The robots not only pushed the three packages but also coordinated their behaviour so that the packages did not collide with each other. Figure 8(b) illustrates similar results that were obtained in the case where $P = \Delta P$. In the last generation, the robots achieved almost the same team play as in the case where $P = 0$.

MBEANN

Recently, our research group proposed a novel EANN for TWEANNs; this EANN is called MBEANN (Ohkura et al., 2007).

Here, the MBEANN is briefly introduced for an understanding of some of its important characteristics.

The MBEANN is a type of TWEANN in which no crossover is used. Instead, the MBEANN adopts two types of structural mutations that work neutrally or almost neutrally in term of the fitness value. In order to utilize these special structural mutations, an individual is represented as a set of subnetworks, i.e., as a set of modules called operons. Figure 9 shows the concept of the genotype. As shown in the figure, the node information consists of the node type and the node identification number. The link information consists of the input node, the output node, the weight value, and the link identification number. The two identification numbers should be unique only to each indi-

Figure 5. The achieved collective behavior when P = 0 for eCTRNN

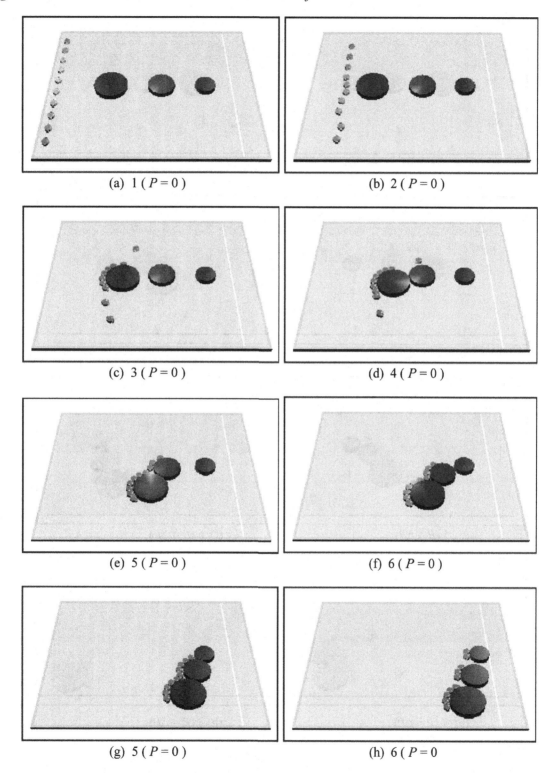

(a) 1 ($P = 0$)

(b) 2 ($P = 0$)

(c) 3 ($P = 0$)

(d) 4 ($P = 0$)

(e) 5 ($P = 0$)

(f) 6 ($P = 0$)

(g) 5 ($P = 0$)

(h) 6 ($P = 0$

Figure 6. The achieved collective behavior when P = ΔP for eCTRNN

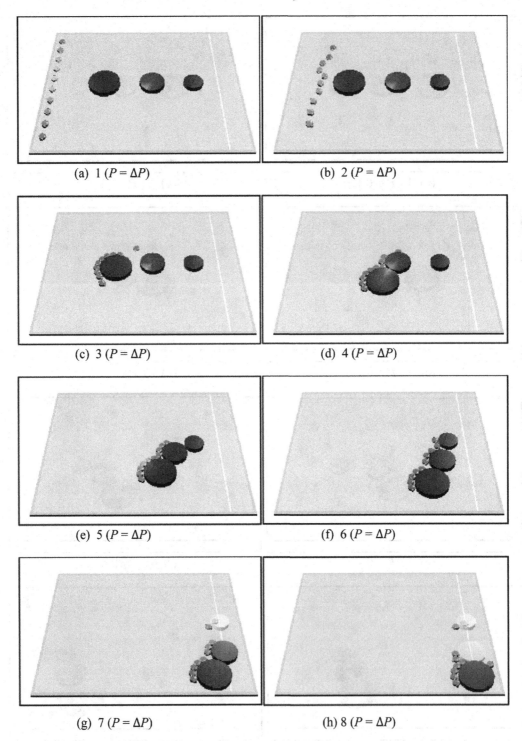

(a) 1 $(P = \Delta P)$

(b) 2 $(P = \Delta P)$

(c) 3 $(P = \Delta P)$

(d) 4 $(P = \Delta P)$

(e) 5 $(P = \Delta P)$

(f) 6 $(P = \Delta P)$

(g) 7 $(P = \Delta P)$

(h) 8 $(P = \Delta P)$

Figure 7. The experimental results when P = 0 (the left three graphs) and P = ΔP (the right three graphs) for eCTRNN

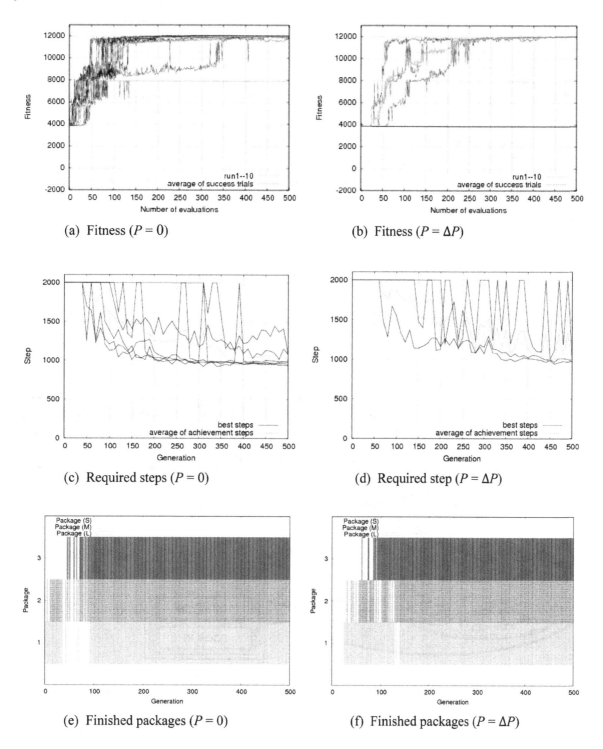

(a) Fitness (P = 0)

(b) Fitness (P = ΔP)

(c) Required steps (P = 0)

(d) Required step (P = ΔP)

(e) Finished packages (P = 0)

(f) Finished packages (P = ΔP)

Figure 8. The achieved collective behavior by eCTRNN at the last generation

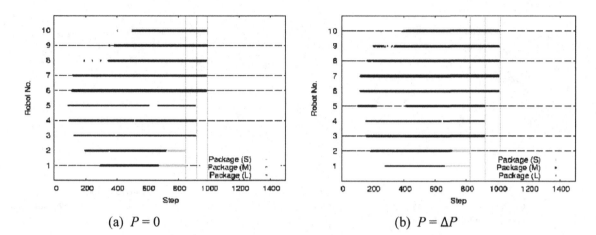

(a) $P = 0$ (b) $P = \Delta P$

Figure 9. Conceptual representation of genotype for MBEANN

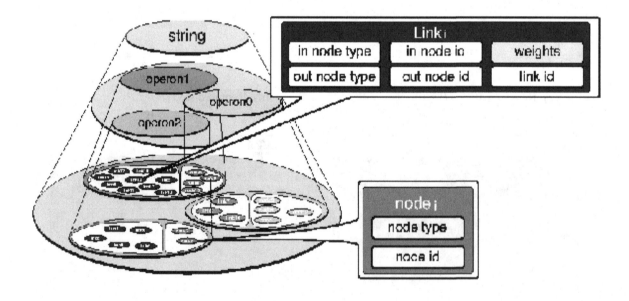

Figure 10. Initial controller for MBEANN

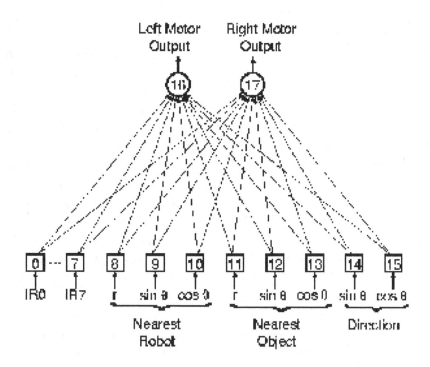

vidual. Thus, if I is the maximum number of operons, a genotype *string* is formulated as follows:

$$string = \{operon_0, operon_1, \ldots, operon_I\}$$

$$operon_i = \{\{node_j \mid j \in O_{N_i}\}, \{link_k \mid k \in O_{L_i}\}\}$$

$$O_{N_i} \quad O_{L_i}$$

As for an initial population, since MBEANN starts with the population consisting of the initial individuals having only one operon, *operon₀*, i.e., the minimal structure in which there is no hidden node, as shown in Figure 10. This is identical to the case of NEAT (Stanley and Miikkulainen, 2002).

Table 3. Parameters for MBEANN

add node rate	0.01
add connection rate	0.1
weight mutation rate	1.0
variance for weight mutation	0.05
tournament size	3
α in sigmoid function	0.5
β in sigmoid function	5.0
weight value in add node	1.0
max generation	500

Experimental Results

Similar to the evolution of the eCTRNN in the experiments, the MBEANN was evolved using (10, 70)-ES. The conditions in the computer simulations were the same as those in the eCTRNN experiments. Ten independent runs were conducted for the two cases where $P = 0.0$ and $P = \Delta P$. The results are summarized in Table 4. We found that at

Table 4. Success rate in MBEANN

Penalty (-P × step)	Success rate [%]
0.0	80
ΔP	30

the last generation, the MBEANN showed almost the same success rate as the eCTRNN .

Figure 11 and Figure 12 show the collective behaviour of the robot group at the final generation in

a successful run when $P = 0$ or $P = \Delta P$, respectively. In the case where $P = 0$, the robot group showed almost the same collective behaviour as observed in the cases of eCTRNN. The robots pushed the packages toward the lower right region. On the other hand, in the case where $P = \Delta P$, it was observed that the robots used completely different strategy. First, they found a partner to move around. After making five pairs, one of them moved toward the smallest package in an almost straight line, even though they could not see the smallest package, because the neural controller of the robot received the sensory information only for the nearest package from the omnivision camera. In addition, two of the robots started moving toward midsized package. During the movement toward the package, they contacted with the largest package; one of them remained at this location. The other three robots went on to push the midsized package toward the goal line. One robot and the remaining two pairs of robots started to pushing the largest package toward the goal line. This highly sophisticated collective behavior is the results of ``autonomous specialization''.

Figure 13 shows the transitions of artificial evolutions observed from the fitness values, required steps, and finished packages. In the figure, (a) and (b) show the transitions of the fitness values in all the runs.

In the case where $P = 0$, the robot group succeeded in transporting all the three packages at eight runs.

In the other case ($P = \Delta P$), the robot group succeeded only in three runs. The black lines in both the graphs show the average fitness transitions calculated using only the successful runs. Figure 13 (c) and (d) show the transitions of the required time steps in the successful runs. The gray lines show the averages of eight and three successful runs, respectively. We soon found that generally, the performance of the robot was poorer than in the case of eCTRNN, particularly in the early generations. When $P = 0$, the robot group did not show stable performance in all the successful runs until the last few generations. On the contrary, when $P = \Delta P$, very stable performance was observed in the three successful runs after around the 270th generation. Figure 13(e) and (f) indicate whether the three packages were transported to the goal line within the time-steps limit. We also soon found that the robot group stably pushed the three packages to the goal line after around the 420 and 270 generations when $P = 0$ and when $P = \Delta P$, respectively.

We suppose that the initial topological structure was too simple for evolutionary learning to solve this complex task, and thus, the robot group needed around 200 generations to obtain a sufficient topological complexity, i.e., the ability to solve this task. This might be validated by Figure 15 or Figure 16, which shows the structural evolution of the robot's neural controller in a typical run. As indicated by figure (a) and (b), which show the controller at the 100th generation, or (c) and (d), which show the controller at the 300 generation, in comparison to the case of the eCTRNN robot controller, the neural network seems too simple to solve the task. The topological structures at the last generation are shown in (e) and (f) in the same figure. The controllers might be sufficiently complex.

Next, similar to the cases of eCTRNN, let us investigate a typical run to determine the strategy

Figure 11. The achieved collective behavior when P = 0 for MBEANN

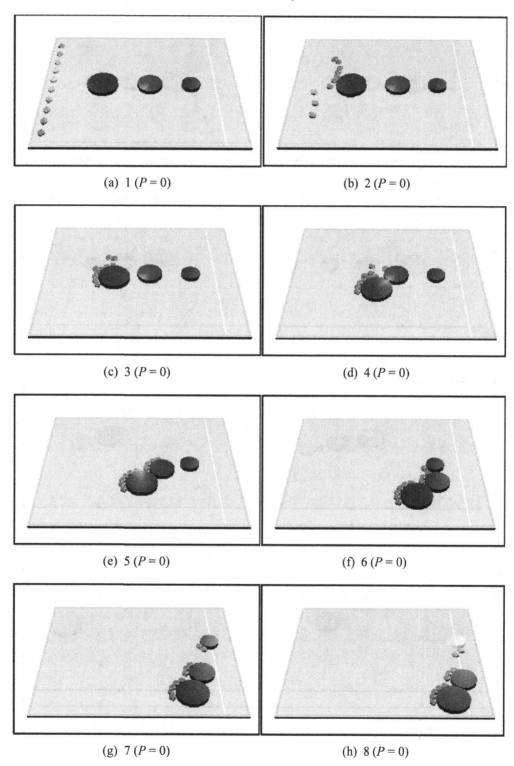

(a) 1 (*P* = 0)   (b) 2 (*P* = 0)

(c) 3 (*P* = 0)   (d) 4 (*P* = 0)

(e) 5 (*P* = 0)   (f) 6 (*P* = 0)

(g) 7 (*P* = 0)   (h) 8 (*P* = 0)

Figure 12. The achieved collective behavior when $P = \Delta P$ for MBEANN

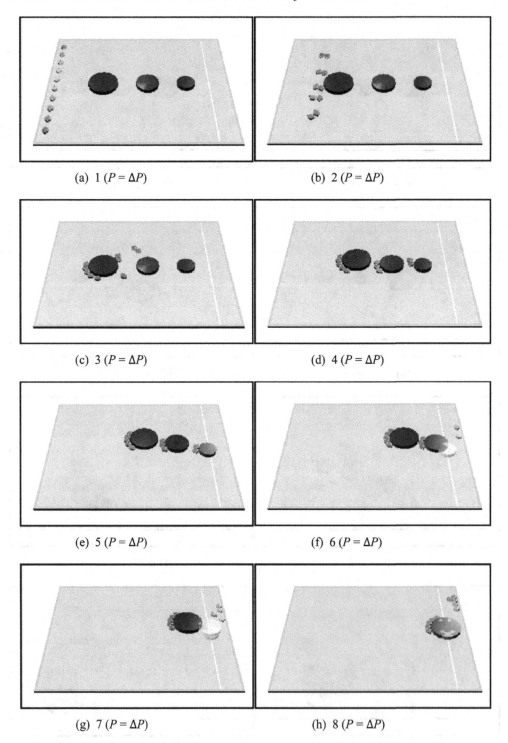

(a) 1 $(P = \Delta P)$ (b) 2 $(P = \Delta P)$

(c) 3 $(P = \Delta P)$ (d) 4 $(P = \Delta P)$

(e) 5 $(P = \Delta P)$ (f) 6 $(P = \Delta P)$

(g) 7 $(P = \Delta P)$ (h) 8 $(P = \Delta P)$

Figure 13. The experimental results when P = 0 (the left three graphs) and P= ΔP (the right three graphs) for MBEANN

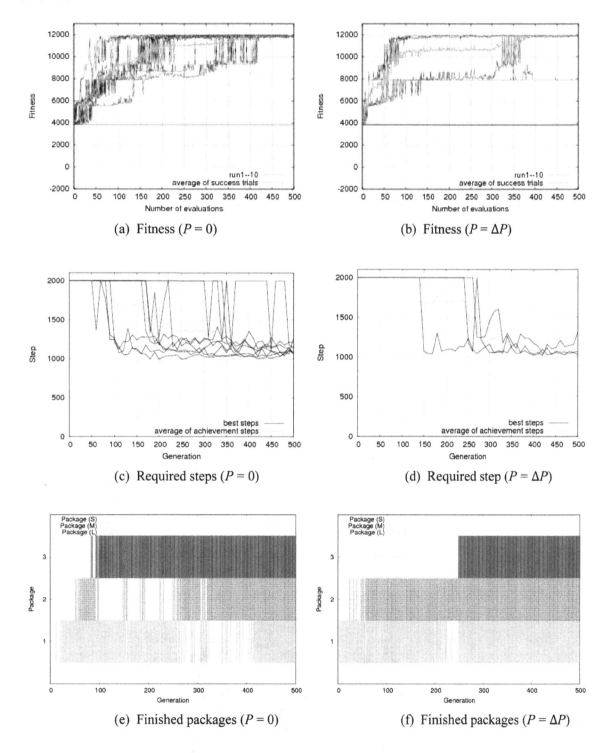

(a) Fitness ($P = 0$)

(b) Fitness ($P = \Delta P$)

(c) Required steps ($P = 0$)

(d) Required step ($P = \Delta P$)

(e) Finished packages ($P = 0$)

(f) Finished packages ($P = \Delta P$)

Figure 14. The achieved collective behavior by MBEANN at the last generation

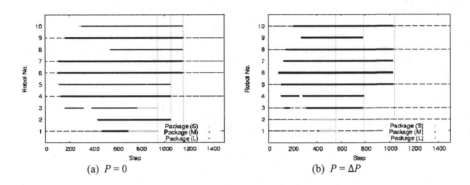

(a) $P = 0$ (b) $P = \Delta P$

Figure 15. A typical result of the structural of the neural network controller for MBEANN

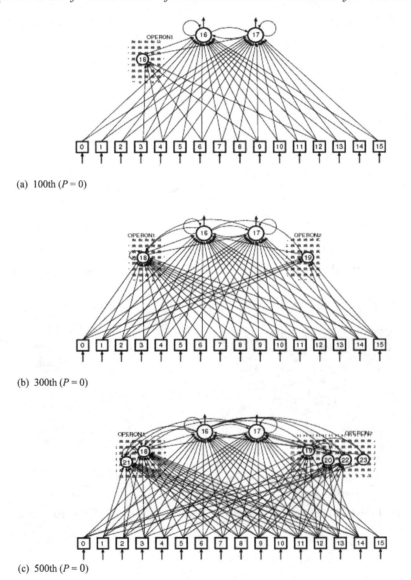

(a) 100th $(P = 0)$

(b) 300th $(P = 0)$

(c) 500th $(P = 0)$

Figure 16. A typical result of the structural of the neural network controller for MBEANN

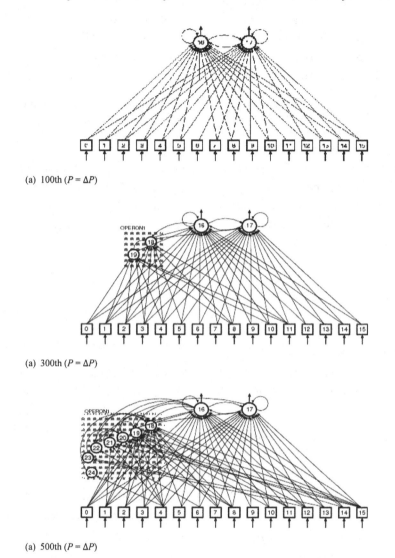

(a) 100th $(P = \Delta P)$

(a) 300th $(P = \Delta P)$

(a) 500th $(P = \Delta P)$

the robot group developed by artificial evolution. Figure 14 shows graphs for ten robots at the 500th generation in the cases where $P = 0$ and $P = \Delta P$. These graphs are drawn in a manner similar to those in Figure 8. As found in graph (a), the robot groups showed almost the same team play as in the cases of eCTRNN. On the other hand, as explained above, the effect of autonomous specialization is clearly observed in the graph (b). Robot No.1 and No.2 pushed the smallest package

in a straight line, Robot No.3, No.4, and No.9 pushed the midsize package, and the remaining five robots pushed the largest package to the goal line.

CONCLUSION

In this chapter, two approaches in EANN, called eCTRNN and MBEANN, were discussed in the context of the cooperative package pushing problem. Ten autonomous mobile robots successfully showed sophisticated collective behaviour as a result of autonomous specialization.

As the next step, the extension of the algorithm for artificial evolution in order to improve the success rate must be considered. Second, more computer simulations should be performed in order to examine the validity and the robustness of the ER approach in coordinating cooperative behaviour in an MRS since the complexity of the cooperative package pushing problem can be easily varied by changing the number of robots or the arrangement of the packages.

REFERENCES

Acerbi, A., et al. (2007). Social Facilitation on the Development of Foraging Behaviors in a Population of Autonomous Robots. In *Proceedings of the 9th European Conference in Artificial Life* (pp. 625-634).

Angeline, P. J., Sauders, G. M., & Pollack, J. B. (1994). An evolutionary algorithms that constructs recurrent neural networks. *IEEE Transactions on Neural Networks*, *5*, 54–65. doi:10.1109/72.265960

Bäck, T. (1996). *Evolutionary Algorithms in Theory and Practice: Evolution Strategies, Evolutionary Programming, Genetic Algorithms.* Oxford University Press.

Baldassarre, G., Nolfi, S., & Parisi, D. (2003). Evolving Mobile Robots Able to Display Collective Behaviours . *Artificial Life*, *9*(3), 255–267. doi:10.1162/106454603322392460

Beer, R. D. (1996). Toward the Evolution of Dynamical Neural Networks for Minimally Cognitive. In *From Animals to Animats 4: Proceedings of the Fourth International Conference on Simulation of Adaptive Behavior* (pp. 421-429).

Blynel, J., & Floreano, D. (2003). Exploring the T-Maze: Evolving Learning-Like Robot Behaviors using CTRNNs. In *Proceedings of the 2nd European Workshop on Evolutionary Robotics (EvoRob'2003)* (LNCS).

Cliff, D., Harvey, I., & Husbands, P. (1993). Explorations in Evolutionary Robotics. *Adaptive Behavior*, *2*(1), 71–104. doi:10.1177/105971239300200104

Gomez, F. J. and Miikkulainen, R. (1999). Solving Non-Markovian Control Tasks with Neuroevolution, In *Proceedings of the International Joint Conference on Artificial Intelligence* (pp. 1356-1361).

Harvey, I., Di Paolo, E., Wood, A., & Quinn, R., M., & Tuci, E. (2005). Evolutionary Robotics: A New Scientific Tool for Studying Cognition. *Artificial Life*, *11*(3/4), 79–98. doi:10.1162/1064546053278991

Harvey, I., Husbands, P., Cliff, D., Thompson, A., & Jakobi, N. (1997). Evolutionary robotics: The sussex approach. *Robotics and Autonomous Systems*, *20*, 205–224. doi:10.1016/S0921-8890(96)00067-X

Liu, Y., & Yao, X. (1996). A Population-Based Learning Algorithms Which Learns Both Architectures and Weights of Neural Networks. *Chinese Journal of Advanced Software Research*, *3*(1), 54–65.

Mondada, F., & Floreano, D. (1995). Evolution of neural control structures: Some experiments on mobile robots. *Robotics and Autonomous Systems*, *16*(2-4), 183–195. doi:10.1016/0921-8890(96)81008-6

Ohkura, K., Yasuda, T., Kawamatsu, Y., Matsumura, Y., & Ueda, K. (2007). MBEANN: Mutation-Based Evolving Artificial Neural Networks. In *Proceedings of the 9th European Conference in Artificial Life* (pp. 936-945).

Quinn, M., & Noble, J. (2001). Modelling Animal Behaviour in Contests: Tactics, Information and Communication. In *Advances in Artificial Life: Sixth European Conference on Artificial Life (ECAL 01),* (LNAI).

Stanley, K., & Miikkulainen, R. (2002). Evolving neural networks through augmenting topologies . *Evolutionary Computation, 10*(2), 99–127. doi:10.1162/106365602320169811

Triani, V., et al. (2007). From Solitary to Collective Behaviours: Decision Making and Cooperation, In *Proceedings of the 9th European Conference in Artificial Life* (pp. 575-584).

Yao, X. (1999). Evolving artificial networks. *Proceedings of the IEEE, 87*(9), 1423–1447. doi:10.1109/5.784219

Chapter 10
A Multi–Robot System Using Mobile Agents with Ant Colony Clustering

Yasushi Kambayashi
Nippon Institute of Technology, Japan

Yasuhiro Tsujimura
Nippon Institute of Technology, Japan

Hidemi Yamachi
Nippon Institute of Technology, Japan

Munehiro Takimoto
Tokyo University of Science, Japan

ABSTRACT

This chapter presents a framework using novel methods for controlling mobile multiple robots directed by mobile agents on a communication networks. Instead of physical movement of multiple robots, mobile software agents migrate from one robot to another so that the robots more efficiently complete their task. In some applications, it is desirable that multiple robots draw themselves together automatically. In order to avoid excessive energy consumption, we employ mobile software agents to locate robots scattered in a field, and cause them to autonomously determine their moving behaviors by using a clustering algorithm based on the Ant Colony Optimization (ACO) method. ACO is the swarm-intelligence-based method that exploits artificial stigmergy for the solution of combinatorial optimization problems. Preliminary experiments have provided a favorable result. Even though there is much room to improve the collaboration of multiple agents and ACO, the current results suggest a promising direction for the design of control mechanisms for multi-robot systems. In this chapter, we focus on the implementation of the controlling mechanism of the multi-robot system using mobile agents.

DOI: 10.4018/978-1-60566-898-7.ch010

INTRODUCTION

When we pass through the terminals of an airport, we often see carts scattered in the walkways and laborers manually collecting them one by one. It is a dull and laborious task. It would be much easier if carts were roughly gathered in any way before collection. Multi-robot systems have made rapid progress in various fields, and the core technologies of multi-robot systems are now easily available (Kambayashi & Takimoto, 2005). Employing one of those technologies, it is possible to give each cart minimum intelligence, making each cart an autonomous robot. We realize that for such a system cost is a significant issue and we address one of those costs, the power source. A powerful battery is big, heavy and expensive; therefore for such intelligent cart systems small batteries are desirable since energy saving is an important issue in such a system (Takimoto, Mizuno, Kurio & Kambayashi, 2007; Nagata, Takimoto & Kambayashi, 2009).

To demonstrate our method, we consider the problem of airport luggage carts that are picked up by travelers at designated points and left in arbitrary places (Kambayashi, Sato, Harada, & Takimoto, 2009). It is desirable that intelligent carts (intelligent robots) draw themselves together automatically to make their collection more efficient. A simple implementation would be to give each cart a designated assembly point to which it automatically returns when free. It is easy to implement, but some carts would have to travel a long way back to their assigned assembly point, even if they found themselves located close to another assembly point. The additional distance traveled consumes unnecessary energy since the carts are functionally identical.

To ameliorate the situation, we employ mobile software agents to locate robots scattered in a field, e.g. an airport, and make them autonomously determine their moving behavior using a clustering algorithm based on ant colony optimization (ACO). ACO is a swarm intelligence-based method and a multi-agent system that exploits artificial stigmergy for the solution of combinatorial optimization problems. Preliminary experiments yield a favorable result. Ant colony clustering (ACC) is an ACO specialized for clustering objects. The idea is inspired by the collective behaviors of ants, and Deneubourg formulated an algorithm that simulates the ant corps gathering and brood sorting behaviors (Deneuburg, Goss, Franks, Sendova-Franks, Detrain & Chretien, 1991).

We have studied a few applications of the base idea for controlling mobile multiple robots connected by communication networks (Ugajin, Sato, Sato, Tsujimura, Yamamoto, Takimoto & Kambayashi, 2007; Sato, Ugajin, Tsujimura, Yamamoto & Kambayashi, 2007; Kambayashi, Tsujimura, Yamachi, Takimoto & Yamamoto, 2009). The framework provides novel methods to control coordinated systems using mobile agents. Instead of physical movement of multiple robots, mobile software agents can migrate from one robot to another so that they can minimize energy consumption in aggregation. In this chapter, we describe the details of implementation of the multi-robot system using multiple mobile agents and static agents that implement ACO. The combination of the mobile agents augmented by ACO and mobile multiple robots opens a new horizon of efficient use of mobile robot resources. We report here our experimental observations of our simulation of our ACC implementation.

Quasi-optimal robot collection is achieved in three phases. The first phase collects the positions of robots. One mobile agent issued from the host computer visits scattered robots one by one and collects the position of each. The precise coordinates and orientation of each robot are determined by sensing RFID (Radio Frequency Identification) tags under the floor carpet. Upon the return of the position collecting agent, the second phase begins wherein another agent, the simulation agent, performs the ACC algorithm and produces the quasi-optimal gathering positions for the robots. The simulation agent is a static agent that resides in the host computer. In the third phase, a number of mobile agents are issued from the host computer. Each mobile agent migrates to a designated robot, and directs the robot to the assigned quasi-optimal position that was calculated in the second phase.

The assembly positions (clustering points) are determined by the simulation agent. It is influenced, but not determined, by the initial positions of scattered robots. Instead of implementing ACC with actual robots, one static simulation agent performs the ACC computation, and then the set of produced positions is distributed by mobile agents. Therefore our method eliminates unnecessary physical movement and thus provides energy savings.

The structure of the balance of this chapter is as follows. In the second section, we review the history of research in this area. The third section describes the agent system that performs the arrangement of the multiple robots. The fourth section describes the ACC algorithm we have employed to calculate the quasi optimal assembly positions. The fifth section demonstrates the feasibility of our system by implementing an actual multi-robot orients itself using RFID tags in its environment. The sixth section discusses our quantitative experiments and our observations from the preliminary experiments. Finally, we conclude in the seventh section and discuss future research directions.

BACKGROUND

Kambayashi and Takimoto have proposed a framework for controlling intelligent multiple robots using higher-order mobile agents (Kambayashi & Takimoto, 2005; Takimoto, Mizuno, Kurio & Kambayashi, 2007; Nagata, Takimoto & Kambayashi, 2009). The framework helps users to construct intelligent robot control software using migration of mobile agents. Since the migrating agents are of higher order, the control software can be hierarchically assembled while they are

running. Dynamic extension of control software by the migration of mobile agents enables the controlling agent to begin with relatively simple base control software, and to add functionalities one by one as it learns the working environment. Thus we do not have to make the intelligent robot smart from the beginning or make the robot learn by itself. The controlling agent can send intelligence later through new agents. Even though the dynamic extension of the robot control software using the higher order mobile agents is extremely useful, such a higher order property is not necessary in our setting. We have employed a simple, non-higher-order mobile agent system for our framework. We previously implemented a team of cooperative search robots to show the effectiveness of such a framework, and demonstrated that that framework contributes to energy savings for a task achieved by multiple robots (Takimoto, Mizuno, Kurio & Kambayashi, 2007; Nagata, Takimoto & Kambayashi, 2009). We have employed a simple non-high-order mobile agent system for our framework. Our simple agent system should achieve similar performance.

Deneuburg formulated the biologically inspired behavioral algorithm that simulates the ant corps gathering and brood sorting behaviors (Deneuburg, Goss, Franks, Sendova-Franks, Detrain, & Chretien, 1991). His algorithm captured many features of the ant sorting behaviors. His design consists of ants picking up and putting down objects in a random manner. He further conjectured that robot team design could be inspired from the ant corps gathering and brood sorting behaviors (Deneuburg, Goss, Franks, Sendova-Franks, Detrain, & Chretien, 1991). Wang and Zhang proposed an ant inspired approach along this line of research that sorts objects with multiple robots (Wang & Zhang, 2004).

Lumer improved Deneuburg's model and proposed a new simulation model that was called Ant Colony Clustering (Lumer, & Faieta, 1994). His method could cluster similar object into a few groups. He presented a formula that measures the similarity between two data objects and designed an algorithm for data clustering. Chen et al have further improved Lumer's model and proposed Ants Sleeping Model (Chen, Xu & Chen, 2004). The artificial ants in Deneuburg's model and Lumer's model have considerable amount of random idle moves before they pick up or put down objects, and considerable amount of repetitions occur during the random idle moves. In Chen's ASM model, an ant has two states: active state and sleeping state. When the artificial ant locates a comfortable and secure position, it has a higher probability in sleeping state. Based on ASM, Chen has proposed an Adaptive Artificial Ants Clustering Algorithm that achieves better clustering quality with less computational cost.

Algorithms inspired by behaviors of social insects such as ants that communicate with each other by stigmergy are becoming popular (Dorigo & Gambardella, 1996) and widely used in solving complex problems (Toyoda & Yano, 2004; Becker & Szczerbicka, 2005). Upon observing real ants' behaviors, Dorigo et al found that ants exchanged information by laying down a trail of a chemical substance (pheromone) that is followed by other ants. They adopted this ant strategy, known as ant colony optimization (ACO), to solve various optimization problems such as the traveling salesman problem (TSP) (Dorigo & Gambardella, 1996). Our ACC algorithm employs pheromone, instead of using Euclidian distance to evaluate its performance.

THE MOBILE AGENTS

Robot systems have made rapid progress in not only their behaviors but also in the way they are controlled (Murphy, 2000). Multi-agent systems introduced modularity, reconfigurability and extensibility to control systems, which had been traditionally monolithic. It has made easier the development of control systems in distributed environments such as intelligent multi-robot systems.

On the other hand, excessive interactions among agents in the multi-agent system may cause problems in the multi-robot environment. Consider a multi-robot system where each robot is controlled by an agent, and interactions among robots are achieved through a communication network such as a wireless LAN. Since the circumstances around the robot change as the robots move, the condition of each connection among the various robots also changes. In this environment, when some of the connections in the network are disabled, the system may not be able to maintain consistency among the states of the robots. Such a problem has a tendency to increase as the number of interactions increases.

In order to lessen the problems of excessive communication, mobile agent methodologies have been developed for distributed environments. In the mobile agent system, each agent can actively migrate from one site to another site. Since a mobile agent can bring the necessary functionalities with it and perform its tasks autonomously, it can reduce the necessity for interaction with other sites. In the minimal case, a mobile agent requires that the connection be established only when it performs migration (Binder, Hulaas & Villazon, 2001). Figure 1 shows a conceptual diagram of a mobile agent migration. This property is useful for controlling robots that have to work in a remote site with unreliable communication or intermittent communication. The concept of a mobile agent also creates the possibility that new functions and knowledge can be introduced to the entire multi-agent system from a host or controller outside the system via a single accessible member of the intelligent multi-robot system (Kambayashi & Takimoto, 2005).

Our system model consists of robots and a few kinds of static and mobile software agents. All the controls for the mobile robots as well as ACC computation performed in the host computer are achieved through the static and mobile agents. They are: 1) user interface agent (UIA), 2) operation agents (OA), 3) position collecting agent

Figure 1. A mobile agent is migrating from site A to site B

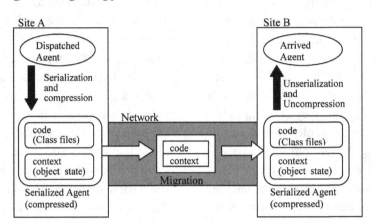

Figure 2. Cooperative agents to control a mobile robot

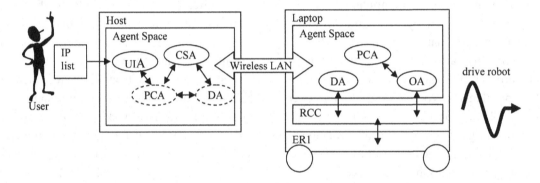

(PCA), 4) clustering simulation agent (CSA), and 5) driving agents (DA). All the software agents except UIA and CSA are mobile agents. A mobile agent (PCA) traverses robots scattered in the field to collect their coordinates. After receiving the assembly positions computed by a static agent (CSA), many mobile agents (DAs) migrate to the robots and drive them to the assembly positions. Figure 2 shows the interactions of the cooperative agents to control a mobile robot.

The functionality of each agent is described as follows:

1. User Interface Agent (UIA): The user interface agent (UIA) is a static agent that resides on the host computer and interacts with the user. It is expected to coordinate the

entire agent system. When the user creates this agent with a list of IP addresses of the mobile robots, UIA creates PCA and passes the list to it.

2. Operation Agent (OA): Each robot has at least one operation agent (OA). It has the task that the robot on which it resides is supposed to perform. Each mobile robot has its own OA. Currently all operation agents (OA) have a function for collision avoidance and a function to sense RFID tags, which are embedded in the floor carpet, to detect its precise coordinates in the field.

3. Position Collecting Agent (PCA): A distinct agent called the position collecting agent (PCA) traverses mobile robots scattered in the field and to collect their coordinates.

PCA is created and dispatched by UIA. Upon returning to the host computer, it hands the collected coordinates to the clustering simulation agent (CSA) for ACC.

4. Clustering Simulation Agent (CSA): The host computer houses the static clustering simulation agent (CSA). This agent actually performs the ACC algorithm by using the coordinates collected by PCA as the initial positions, and produces the quasi-optimal assembly positions of the mobile robots. Upon terminating the computation, CSA creates a number of driving agents (DA).

5. Driving Agent (DA): The quasi-optimal arrangement coordinates produced by the CSA are delivered by driving agents (DA). One driving agent is created for each mobile robot, and it contains the set of procedures for the mobile robot. The DA drives its mobile robot to the designated assembly position.

OA detects the current coordinates of the robot on which it resides. Each robot has its own IP address and UIA hands in the list of the IP addresses to PCA. First, PCA migrates to an arbitrary robot and starts hopping between them one by one. It communicates locally with OA, and writes the coordinates of the robot into its own local data area. When PCA gets all the coordinates of the robots, it returns to the host computer. UIA waits certain period for PCA's return. If PCA does not hear from PCA for certain period, it declares "time-out" and cancels the PCA. Then UIA re-generates a new PCA with new identification number. On the other hand, if PCA can not find a robot with one of the IP addresses on the list, it retries certain times and then declares the missing robot to be "lost." PCA reports that fact to UIA. Upon returning to the host computer, PCA creates CSA and hands in the coordinate data to CSA which computes the ACC algorithm. We employ RFID (Radio Frequency Identification) tagging to get precise coordinates. We set RFID tags in a regular grid shape under the floor carpet tiles. The tags we chose have a

small range so that the position-collecting agent can obtain fairly precise coordinates from the tag. The robots also have a basic collision avoidance mechanism using infrared sensors.

CSA is the other static agent whose sole role is ACC computation. When CSA receives the coordinate data of all the robots, it translates them into coordinates for simulation, and performs the clustering. When CSA finishes the computation and produces a set of assembly positions, it then creates the set of procedures for autonomous robot movements.

CSA creates DA that convey the set of procedures to the mobile robots. Each DA receives its destination IP address from PCA, and the set of procedures for the destination robot, and then migrates to the destination robot. Each DA has a set of driving procedures that drives its assigned robot to the destination, while it avoids collision. OA has the basic collision detection and avoidance procedures, and DA has task-specific collision avoidance guidance, such as the coordinates of pillars and how to avoid them.

We have implemented the prototype of the multi-agent system for mobile robot control using Agent Space (Satoh, 1999). Agent Space is a library for constructing mobile agents developed by Satoh. By using its library, the user can implement a mobile agent environment with Java language.

In Agent Space, mobile agents are defined as collections of call-back methods, and we implement the contents of the methods with the interfaces defined in the system. In order to create a mobile agent, the application calls the create method. An agent migrates to another site by using the move and leave methods. When an agent arrives, the arrive method is invoked. Migration by an agent is achieved by its duplication of itself at the destination site. Thus the move and leave methods are used as a pair of methods for actual migration. Figure 3 shows the move method for example. The other methods are implemented similarly. The users are expected to implement a destructor to erase the original agent in the leave

Figure 3. The move method

```
if(comeBackHost == true){
        try{
                URL url = new URL("matp://hostIP:hostPort);
                        context.move(url);
        }
        catch(InvalidURLException ex){
        ex.printStackTrace();
        }
        catch (IOException ex){
        ...
        }
```

Figure 4. The position collecting agent (PCA)

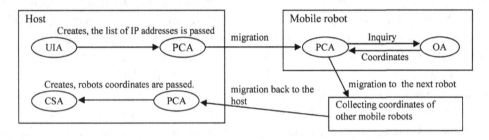

Figure 5. The destination agent (DA)

method. Agent Space also provides services in its Application Program interface (API) such as the move method to migrate agents and the invoke method to communicate to another agent. Figures 4 and 5 show how PCA and DA work, respectively.

The following is the PCA implementation:

1. UIA invokes create method to create the mobile agent PCA, and hands in the list of the IP addresses of mobile robots to PCA.
2. PCA invokes move method so that it can migrate to the mobile robot specified in the top of the list of IP addresses.
3. Invoke leave method.

4. The agent actually migrates to the specified mobile robot.
5. Invoke arrive method in the destination robot, and the PCA communicates locally to the OA in order to receive the coordinate of the robot.
6. Checks the next entry of the IP address list; if PCA visits all the mobile robots, it returns to the host computer, otherwise migrates to the next mobile robot with the IP address of the next entry in the list.

The followings are the DA implementation:

1. CSA creates the mobile agents DA as many as the number of the mobile robots in the field.
2. Each DA receives the IP address to where the DA is supposed to migrate, and the set of the procedures to drive the robot.
3. The agents actually migrate to the specified mobile robots.
4. Each DA invokes arrive method in the destination robot, constructs the sequence of commands from the given procedures, and then communicates with the robot control software called RCC (Robot Control Center) in the notebook computer on the robot in order to actually drive the mobile robot.

ANT COLONY CLUSTERING

In this section, we describe our ACC algorithm to determine the quasi-optimal assembly positions for multiple robots. Coordination of an ant colony is achieved through indirect communication through pheromones. In previously explored ACO systems, artificial ants leave pheromone signals so that other artificial ants can trace the same path (Deneuburg, Goss, Franks, Sendova-Franks, Detrain, & Chretien, 1991). In our ACC system, however, we have attributed pheromone to objects so that more objects are clustered in a place where strong pheromone is sensed. The simulation agent, CSA, performs the ACC algorithm as a simulation. The field of the simulation has the coordinates of objects and their pheromone values, so that the artificial ants can obtain all the necessary information (coordinates and pheromone) for the simulation field.

Randomly walking artificial ants have a high probability of picking up an object with weak pheromone, and putting the object where they sense strong pheromone. They are designed not to walk long distances so that the artificial ants tend to pick up scattered objects and produce many

small clusters of objects. When a few clusters are generated, those clusters tend to grow.

Since the purpose of traditional ACC is clustering or grouping objects into several different classes based on selected properties, it is desirable that the generated chunks of clusters grow into one big cluster such that each group has distinct characteristics. In our system, however, we want to produce several roughly clustered groups of the same type, and to minimize the movement of each robot. (We assume we have one kind of cart robots, and we do not want robots move long distances.) Therefore our artificial ants have the following behavioral rules.

1. An artificial ant's basic behavior is random walk. When it finds an isolated object, it picks it up.
2. When the artificial ant finds a cluster with certain number of objects, it tends to avoid picking up an object from the cluster. This number can be updated later.
3. When the artificial ant with an object finds a cluster, it put down the object so that the object is adjacent to one of the objects in the cluster.
4. If the artificial ant cannot find any cluster with certain strength of pheromone, it just continues a random walk.

By the restrictions defined in the above rules, the artificial ants tend not to convey objects for a long distance, and produce many small heaps of objects at the first stage. In order to implement the first feature, the system locks objects with certain number of adjoining objects, and no artificial ant can pick up such a locked object. The number for locking will be updated later so that artificial ants can bring previously locked objects in order to create larger clusters. When the initially scattered objects are clustered into small number of heaps, the number of objects that causes objects to be locked is updated, and the activities of the artificial ants re-start to produce smaller number

Figure 6. The behavior of the artificial ant

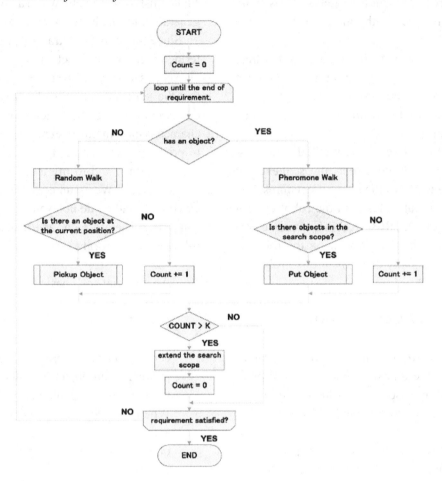

of clusters. We describe the behaviors of the artificial ants below.

In the implementation of our ACC algorithm, when the artificial ants are generated, they have randomly supplied initial positions and walking directions. An artificial ant performs a random walk; when it finds an unlocked object, it picks up the object and continues random walk. During its random walk, when it senses strong pheromone, it puts down the conveyed object. The artificial ants repeat this simple procedure until the termination condition is satisfied. Figure 6 shows the behavior of an artificial ant. We explain several specific actions that each artificial ant performs below.

The base behavior for all the artificial ants is the random walk. The artificial ant that does not

have any object is supposed to move straight in a randomly determined initial direction for ten steps at which time the ant randomly changes the direction. The ant also performs side steps from time to time to create further randomness.

When the artificial ant finds an object during its random walk, it determines whether to pick it up or not based on whether the object is locked or not, and the strength of pheromone the object has, according to the value of the formula (1) below. An artificial ant will not pick any locked object. Whether an object is locked or not is also determined by the formula (1). Here, p is the density of pheromone and k is a constant value. p itself is a function of the number of objects and the distance from the cluster of objects. The formula

Figure 7. Strengths of pheromones and their scope

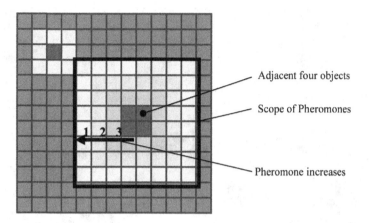

simply says that the artificial ant does not pick up an object with strong pheromone.

$$f\left(p\right) = 1 - \frac{(p+l)*k}{100} \qquad (1)$$

Currently, we choose p as the number of adjacent objects. Thus, when an object is completely surrounded by other objects, p is the maximum value, set in our experiments to be nine. We choose k to be equal to thirteen in order to prevent any object surrounded by other eight objects from being picked up. Then, the computed value of $f(p) = 0$ (never pick it up). l is a constant value at which an object is locked. Usually l is zero (not locked). When the number of clusters becomes less than two third of the number of total objects and p is greater than three, we set l to six. When the number of clusters becomes less than one third of the number of total objects and p is greater than seven, l becomes three. When the number of clusters becomes less than the number of the user setting and p is greater than nine, l becomes one. Any objects that meet these conditions are deemed to be locked. This "lock" process prevents artificial ants from removing objects from growing clusters, and contributes to stabilizing the clusters' relatively monotonic growth.

When an artificial ant picks up an object, it changes its state into "pheromone walk." In this state, an artificial ant tends probabilistically move toward a place it where it senses the strongest pheromone. The probability that the artificial ant takes a certain direction is $n/10$, where n is the strength of the sensed pheromone in that direction. Figure 7 shows the strengths of pheromones and their scope. This mechanism causes the artificial ants move toward the nearest cluster, and consequently minimizes the moving distance.

An artificial ant carrying an object determines whether to put down the object or to continue to carry it. This decision is made based on the formula (2). Thus, the more it senses strong pheromone, the more it tends to put the carried object. Here, p and k are the same as in the formula (1). The formula simply says when the artificial ant bumps into a locked object; it must put the carried object next to the locked object. Then, the value of $f(p) = 1$ (must put it down).

$$f\left(p\right) = \frac{p*k}{100} \qquad (2)$$

Conventional ACC algorithms terminate when all the objects are clustered in the field, or predefined number of steps are executed (Chen, Xu & Chen, 2004). In such conditions, however,

Figure 8. A team of mobile robots work under control of mobile agents

the clustering may be over before obtaining satisfactory clusters. Therefore we set the terminate condition of our ACC algorithm that the number of resulted clusters is less than ten, and all the clusters have three or more objects. This condition may cause longer computation time than usual the ACC, but preliminary experiments show that this produces reasonably good clustering.

THE ROBOTS

In this section, we demonstrate that the model of static and mobile agents with ant colony clustering (ACC) is suitable for intelligent multi-robot systems. We have employed the ER1 Personal Robot Platform Kit by Evolution Robotics Inc. as the platform for our prototype (Evolution Robotics, 2008). Each robot has two servomotors with tires. The power is supplied by a rechargeable battery. It has a servomotor controller board that accepts RS-232C serial data from a host computer. Each robot holds one notebook computer as its host computer. Our control mobile agents migrate to these host computers by wireless LAN. One robot

control application, called RCC (robot control center), resides on each host computer. Our mobile agents communicate with RCC to receive sensor data. Figure 8 shows a team of mobile multiple robots working under control of mobile agents.

In the previous implementation, an agent on the robot calculates the current coordinates from the initial position, and determines where computed coordinates are different from actual positions (Ugajin, Sato, Tsujimura, Yamamoto, Takimoto & Kambayashi, 2007). The current implementation employs RFID (Radio Frequency Identification) to get precise coordinates. We set RFID tags in a regular grid shape under the floor carpet tiles. The tags we chose have a small range so that the position-collecting agent can obtain fairly precise coordinates from the tag. Figure 9 shows the RFID under a floor carpet tile. The robot itself has a basic collision avoidance mechanism using infrared sensors.

For driving robots along a quasi-optimal route, one needs not only the precise coordinates of each robot but also the direction each robot faces. In order to determine the direction that it is facing, each robot moves straight ahead in the direction

Figure 9. RFID under a carpet tile

Figure 10. RFID tags in square formation

Figure 11. Determining orientation from two RFID tag positions

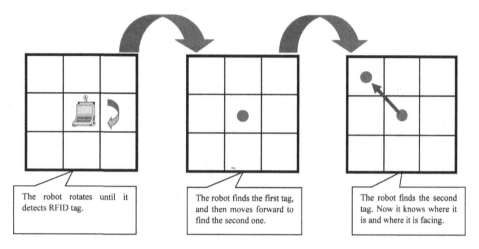

The robot rotates until it detects RFID tag.

The robot finds the first tag, and then moves forward to find the second one.

The robot finds the second tag. Now it knows where it is and where it is facing.

it is currently facing and obtains two positions (coordinates) from RFID tags under the carpet tiles. Determining current orientation is important because there is a high cost for making a robot rotates through a large angle. It is desirable for each robot be assigned rather simple forward movements rather than complex movement with several direction-changes when there are obstacles to avoid. Therefore whenever OA is awake, it performs the two actions, i.e. obtaining the current position and calculating the current direction.

Since each carpet tile has nine RFID tags, as shown in Figure 10, the robot is supposed to obtain the current position as soon as OA gets the sensor data from the RFID module. If OA can not obtain the position data, it means the RFID module can not sense a RFID tag, OA makes the robot rotate until the RFID module senses a RFID tag (usu-

ally a small degree). Once OA obtains the current position, it drives the robot a short distance until the RFID module detects the second RFID tag. Upon obtaining two positions, a simple computation determines the direction in which the robot is moving, as shown in Figure 11.

In a real situation, we may find the robot close to the wall and facing it. Then we cannot use the simple method just described above. The robot may move forward and collide with the wall. The Robot has two infrared sensors, but their ranges are very short and when they sense the wall, it is often too late to compute and execute a response. In order to accommodate such situations, we make RFID tags near the wall have a special signal to the robot that tells it that it is at the end of the field (near the wall) so that the robot that senses the signal can rotate to the opposite direction, as

Figure 12. RFID tags near the wall emit special signal

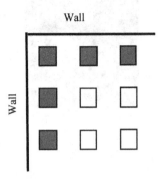

shown in Figure 12. This is only required in our experimental implementation because the current collision detection mechanism does not otherwise work well enough to avoid collision with the wall. Therefore we employ the special signal in the RFID tags. When the robot finds a wall while moving to obtain its location, it arbitrarily changes direction, and starts obtaining two positions again (Kambayashi, Sato, Harada, & Takimoto, 2009).

NUMERICAL EXPERIMENTS

We have conducted several numerical experiments in order to demonstrate the effectiveness of our ACC algorithm using a simulator. The simulator has an airport-like field with a sixty times sixty grid cut into two right equilateral triangles with fifteen unit sides at the top left and right, and one trapezoid with top side thirty, bottom side fifty-eight and height fifteen at the bottom as shown Figure 13 (a) through (c). We have compared the results of our ACC algorithm with an algorithm that performs grouping objects at predefined assembly positions. The desirable arrangements are those of low moving costs and fewer numbers of clusters. Therefore, we defined comparing fitness value that is represented in the formula (3). Here, *Clust* represents the number of clusters, and *Ave*

represents the average distance of all the objects moved.

$$Eval = Ave + Clust \qquad (3)$$

The results of the experiments are shown in Tables 1 through 3. The results of our ACC algorithm are displayed in the column "Ant Colony Clustering," and the results of gathering objects predefined assembly positions are displayed in the column "Specified Position Clustering." We have implemented both algorithms in the simulator. In the "Specified Position Clustering," we set four assembly positions. For the experiments, we set three cases that are the number of objects 200, 300 and 400. In every case, we set the number of artificial ants to be 100. We have performed five trials for each setting, and obtained the evaluation values calculated by the formula (3), as shown in Tables 1 through 3. In all three cases, the computations are over less than six seconds under the computing environment of the experiments is JVM on Pentium IV 2.8 GHz and Windows XP. Since the physical movements of robots take much more time than computation, we can ignore the computation complexity of the ACC algorithm.

In every case, our ACC algorithm produced a better result than that for predefined assembly positions. We can observe that as the number of objects increases, the average moving distance of each object decreases. We can, however, observe that increasing the number of ants does not contribute the quality of clustering as shown in Table 4 and 5.

In the first implementation, we did not employ a "lock" process. Under those conditions we observed no convergence. The number of clusters was roughly 300. When we add the "lock" process, clusters start to emerge, but they tend to stabilize at about thirty clusters. Then the artificial ants see only immediately adjacent places. When we add the feature to artificial ants that they can dynamically change their scope in the field, the small

Figure 13. Clustered objects constructed by assembling at positions computed by ACC and at predefined positions

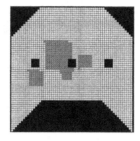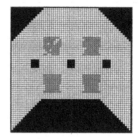

(a) Initial state   (b) Assembled by ACC   (c) Predefined positions

Table 1. Airport field (Objects: 200, Ant Agent: 100)

Airport Field	Ant Colony Clustering				Specified Position Clustering			
	Cost	Ave	Clust	Eval	Cost	Ave	Clust	Eval
1	2279	11.39	4	15.39	5826	29.13	4	33.13
2	2222	11.11	3	14.11	6019	30.09	4	34.09
3	2602	13.01	2	15.01	6194	30.97	4	34.97
4	2433	12.16	3	15.16	6335	31.67	4	35.67
5	2589	12.94	4	16.94	6077	30.38	4	34.38

Table 2. Airport field (Objects: 300, Ant Agent: 100)

Airport Field	Ant Colony Clustering				Specified Position Clustering			
	Cost	Ave	Clust	Eval	Cost	Ave	Clust	Eval
1	2907	9.69	4	13.69	8856	29.52	4	33.52
2	3513	11.71	4	15.71	9142	30.47	4	34.47
3	3291	10.97	4	14.97	8839	29.46	4	33.46
4	3494	11.64	3	14.64	8867	29.55	4	33.55
5	2299	7.66	6	13.66	9034	30.11	4	34.11

Table 3. Airport Field (Objects: 400, Ant Agent: 100)

Airport Field	Ant Colony Clustering				Specified Position Clustering			
	Cost	Ave	Clust	Eval	Cost	Ave	Clust	Eval
1	4822	12.05	1	13.05	11999	29.99	4	33.99
2	3173	7.93	6	13.93	12069	30.17	4	34.17
3	3648	9.12	4	13.12	12299	30.74	4	34.74
4	3803	9.51	3	12.51	12288	30.72	4	34.72
5	4330	10.82	5	15.82	12125	30.31	4	34.31

187

Table 4. Airport Field (Objects: 300, Ant Agent: 50)

Airport Field	Ant Colony Clustering				Specified Position Clustering			
	Cost	Ave	Clust	Eval	Cost	Ave	Clust	Eval
1	3270	10.9	5	15.9	9156	30.52	4	34.52
2	2754	9.18	5	14.18	9058	30.19	4	34.19
3	3110	10.36	4	14.36	9006	30.02	4	34.02
4	3338	11.12	3	14.12	9131	30.43	4	34.43
5	2772	9.24	5	14.24	8880	29.6	4	33.6

Table 5. Airport Field (Objects: 300, Ant Agent: 200)

Airport Field	Ant Colony Clustering				Specified Position Clustering			
	Cost	Ave	Clust	Eval	Cost	Ave	Clust	Eval
1	3148	10.49	4	14.49	8887	29.62	4	33.62
2	3728	12.42	3	15.42	8940	29.8	4	33.8
3	2936	9.78	5	14.78	8923	29.73	4	33.73
4	3193	10.64	3	13.64	9408	31.36	4	35.36
5	4131	13.77	2	15.77	9309	31.03	4	35.03

clusters converge into a few big clusters. Then the artificial ants can see ten to twenty positions ahead. As a result of our experiments, we realize that we need to continue to improve our ACC algorithm.

CONCLUSION AND FUTURE DIRECTIONS

We have presented a framework for controlling mobile multiple robots connected by communication networks. Mobile and static agents collect the coordinates of scattered mobile multiple robots and implement the ant colony clustering (ACC) algorithm in order to find quasi-optimal positions to assemble the mobile multiple robots. Making mobile multiple robots perform the ant colony optimization is enormously inefficient. Therefore a static agent performs the ACC algorithm in its simulator and computes the quasi-optimal positions for the mobile robots. Then other mobile

agents carrying the requisite set of procedures migrate to the mobile robots, and so direct the robots using the sequence of the robot control commands constructed from the given set of procedures.

Since our control system is composed of several small static and mobile agents, it shows an excellent scalability. When the number of mobile robots increases, we can increase the number of mobile software agents to direct the mobile robots. The user can enhance the control software by introducing new features as mobile agents so that the multi-robot system can be extended dynamically while the robots are working. Also mobile agents decrease the amount of the necessary communication. They make mobile multi-robot applications possible in remote site with unreliable communication. In unreliable communication environments, the multi-robot system may not be able to maintain consistency among the states of the robots in a centrally controlled manner. Since a mobile agent can bring the necessary functionalities with it and perform its tasks autonomously, it can reduce the necessity for interaction with other sites. In the minimal case, a mobile agent requires that the connection be established only when it performs migration (Binder, Hulaas & Villazon, 2001). The concept of a mobile agent also creates the possibility that new functions and knowledge can be introduced to the entire multi-agent system from a host or controller outside the system via a single accessible member of the intelligent multi-robot system (Kambayashi & Takimoto, 2005). While our imaginary application is simple cart collection, the system should have a wide variety of applications.

We have implemented a team of mobile robots to show the feasibility of our model. In the current implementation, an agent on the robot can obtain fairly precise coordinates of the robots from RFID tags.

The ACC algorithm we have proposed is designed to minimize the total distance objects are moved. We have analyzed and demonstrated the effectiveness of our ACC algorithm through simulation, performing several numerical experiments with various settings. Although we have so far observed favorable results from the experiments in the simulator, applying the results of the simulation to a real multi-robot system is difficult. Analyzing the results of the simulation, we often find the sum of the moving distances of all the robots is not minimal as we expected. We have re-implemented the ACC algorithm to use only the sum of moving distances and have found some improvement. Even though we believe that the multi-agent framework for controlling multi-robot systems is a right direction, we have to overcome several problems before constructing a practical working system.

Compared with the time for robot movements, the computation time for the ACC algorithm is negligible. Even if the number of artificial ants increases, the computation time will increase linearly, and the number of objects should not influence the computation's complexity. Because

any one step of each ant's behavior is simple, we can assume it takes constant execution time. Even though apparently obvious, we need to confirm this with quantitative experiments.

One big problem is to determine how we should include the collision avoidance behaviors of robots into the simulation. We need to quantify real robot movements more completely. Collision avoidance itself is a significant problem because the action of clustering means create a jam of moving robots. Each driving agent must maneuver its robot precisely to the destination while avoiding colleague robots as it dynamically determines its destination coordinates. This task requires much more intelligence we had expected early in this project.

During the experiments, we experienced the following unfavorable situations:

1. Certain initial arrangements of objects causes very long periods for clustering,
2. If one large cluster is created at the early stage of the clustering and the rest of the field has scarce objects, then all the objects are assembled into one large cluster. This situation subsequently makes aggregate moving distance long, and
3. As a very rare case, the simulation does not converge.

Even though such cases are rare, these phenomena suggest further avenues for research for our ACC algorithm. As we mentioned in the previous section, we need to design the artificial ants have certain complex features that changes their ability to adapt to circumstances. We defer this investigation to our future work.

On the other hand, when certain number of clusters have emerged and stabilize, we can coerce them into several (three or four) clusters by calculating the optimal assembly points. This coercion to required assembly points should be one of the other directions for our future work. We may also investigate computing the number of clusters and their rough positions prior to performing the ACC algorithm, so that we can save much computation time. In many ways, we have room to improve our assembly point calculation method before integrating everything into one working multi-robot system.

ACKNOWLEDGMENT

We appreciate Kimiko Gosney who gave us useful comments. This work is partially supported by Japan Society for Promotion of Science (JSPS), with the basic research program (C) (No. 20510141), Grant-in-Aid for Scientific Research.

REFERENCES

Becker, M., & Szczerbicka, H. (2005). Parameters Influencing the Performance of Ant Algorithm Applied to Optimisation of Buffer Size in Manufacturing. *Industrial Engineering and Management Systems*, 4(2), 184–191.

Binder, W. J., Hulaas, G., & Villazon, A. (2001). Portable Resource Control in the J-SEAL2 Mobile Agent System. In *Proceedings of International Conference on Autonomous Agents* (pp. 222-223).

Chen, L., Xu, X., & Chen, Y. (2004). An adaptive ant colony clustering algorithm. In *Proceedings of the Third IEEE International Conference on Machine Learning and Cybernetics* (pp. 1387-1392).

Deneuburg, J., Goss, S., Franks, N., Sendova-Franks, A., Detrain, C., & Chretien, L. (1991). The Dynamics of Collective Sorting: Robot-Like Ant and Ant-Like Robot. In *Proceedings of First Conference on Simulation of Adaptive Behavior: From Animals to Animats* (pp. 356-363). Cambridge: MIT Press.

Dorigo, M., & Gambardella, L. M. (1996). Ant Colony System: a Cooperative Learning Approach to the Traveling Salesman . *IEEE Transactions on Evolutionary Computation*, 1(1), 53–66. doi:10.1109/4235.585892

Evolution Robotics Ltd. Homepage (2008). Retrieved from http://www.evolution.com/

Kambayashi, Y., Sato, O., Harada, Y., & Takimoto, M. (2009). Design of an Intelligent Cart System for Common Airports. In *Proceedings of 13th International Symposium on Consumer Electronics*. CD-ROM.

Kambayashi, Y., & Takimoto, M. (2005). Higher-Order Mobile Agents for Controlling Intelligent Robots. *International Journal of Intelligent Information Technologies*, 1(2), 28–42.

Kambayashi, Y., Tsujimura, Y., Yamachi, H., Takimoto, M., & Yamamoto, H. (2009). Design of a Multi-Robot System Using Mobile Agents with Ant Colony Clustering. In *Proceedings of Hawaii International Conference on System Sciences*. IEEE Computer Society. CD-ROM

Lumer, E. D., & Faieta, B. (1994). Diversity and Adaptation in Populations of Clustering Ants. In *From Animals to Animats 3: Proceedings of the 3rd International Conference on the Simulation of Adaptive Behavior* (pp. 501-508). Cambridge: MIT Press.

Murphy, R. R. (2000). *Introduction to AI robotics*. Cambridge: MIT Press.

Nagata, T., Takimoto, M., & Kambayashi, Y. (2009). Suppressing the Total Costs of Executing Tasks Using Mobile Agents. In *Proceedings of the 42nd Hawaii International Conference on System Sciences*, IEEE Computer Society. CD-ROM.

Sato, O., Ugajin, M., Tsujimura, Y., Yamamoto, H., & Kambayashi, Y. (2007). Analysis of the Behaviors of Multi-Robots that Implement Ant Colony Clustering Using Mobile Agents. In *Proceedings of the Eighth Asia Pacific Industrial Engineering and Management System.* CD-ROM.

Satoh, I. (1999). A Mobile Agent-Based Framework for Active Networks. In *Proceedings of IEEE Systems, Man, and Cybernetics Conference* (pp. 161-168).

Takimoto, M., Mizuno, M., Kurio, M., & Kambayashi, Y. (2007). Saving Energy Consumption of Multi-Robots Using Higher-Order Mobile Agents. In *Proceedings of the First KES International Symposium on Agent and Multi-Agent Systems: Technologies and Applications* (LNAI 4496, pp. 549-558).

Toyoda, Y., & Yano, F. (2004). Optimizing Movement of a Multi-Joint Robot Arm with Existence of Obstracles Using Multi-Purpose Genetic Algorithm. *Industrial Engineering and Management Systems, 3*(1), 78–84.

Ugajin, M., Sato, O., Tsujimura, Y., Yamamoto, H., Takimoto, M., & Kambayashi, Y. (2007). Integrating Ant Colony Clustering Method to Multi-Robots Using Mobile Agents. In *Proceedings of the Eigth Asia Pacific Industrial Engineering and Management System.* CD-ROM.

Wang, T., & Zhang, H. (2004). Collective Sorting with Multi-Robot. In *Proceedings of the First IEEE International Conference on Robotics and Biomimetics* (pp. 716-720).

KEY TERMS AND DEFINITIONS

Mobile robot: In contrast to an industrial robot, which usually consist of a multi-linked manipulator and an end effecter that is attached to a fixed surface, a mobile robot has the capability to move around in its environment. "Mobile robots" often implies autonomy. Autonomous robots can perform desired tasks in unstructured environments with minimal user intervention.

Mobile agent: A piece of program that can migrate from a computational site to another computational site while it is under execution.

Multi-robots: A set of mobile robots. They are relatively small and expected to achieve given tasks by cooperating with each other.

Intelligent robot control: A method to control mobile robots. This method allows mobile robots to behave autonomously reducing user interventions to a minimum.

Swarm intelligence: the property of a system whereby the collective behaviors of (mobile) agents interacting locally with their environment cause coherent functional global patterns to emerge. A swarm has been defined as a set of (mobile) agents which are liable to communicate directly or indirectly with each other, and which collectively carry out a distributed problem solving.

Ant colony optimization: A probabilistic technique inspired by the behaviors of social insects "ants." It was proposed as a method for solving hard combinatorial optimization problems, and is known to be useful for solving computational problems which can be reduced to finding good paths through graphs.

Clustering algorithm: An algorithm that extracts similar data items from unstructured data and group them into several clusters.

Section 5
Multi-Agent Games and Simulations

Chapter 11
The AGILE Design of Reality Game AI

Robert G. Reynolds
Wayne State University, USA

John O'Shea
University of Michigan-Ann Arbor, USA

Xiangdong Che
Wayne State University, USA

Yousof Gawasmeh
Wayne State University, USA

Guy Meadows
University of Michigan-Ann Arbor, USA

Farshad Fotouhi
Wayne State University, USA

ABSTRACT

This chapter investigates the use of agile program design techniques within an online game development laboratory setting. The proposed game concerns the prediction of early Paleo-Indian hunting sites in ancient North America along a now submerged land bridge that extended between Canada and the United States across what is now Lake Huron. While the survey of the submerged land bridge was being conducted, the online class was developing a computer game that would allow scientists to predict where sites might be located on the landscape. Crucial to this was the ability to add in gradually different levels of cognitive and decision-making capabilities for the agents. We argue that the online component of the courses was critical to supporting an agile approach here. The results of the study indeed provided a fusion of both survey and strategic information that suggest that movement of caribou was asymmetric over the landscape. Therefore, the actual positioning of human artifacts such as hunting blinds was designed to exploit caribou migration in the fall, as is observed today.

DOI: 10.4018/978-1-60566-898-7.ch011

INTRODUCTION

Agile software design methodologies are a response to traditional plan-based approaches such as the waterfall model and others. There are a number of different agile methodologies. These support short term increments in the development of a software project. These increments reflect short rather than long-term planning decisions. Iterations are done in short time frames known as "time boxes". These boxes range from 1 to 4 weeks in duration. Within each time box aspects of a full software development life cycle can take place which include planning, requirements analysis, design, coding, unit and acceptance testing. At the end of each time box, there is an available release which may still not contain all of the intended functionality.

Work is done by small groups of 5 to 9 individuals whose composition is cross functional and self-organizing without consideration of organizational structure or hierarchy. The goal is to encourage face to face communication in preference to written communication. Associated with a team is a customer representative who makes a personal commitment to being available for development questions. The production of working software is the prime indicator for progress. Due to the emphasis on face to face communication the number of written documents produced is often less than other methods, although documents are viewed to rank equally with the working product and other produced artifacts.

Recently an agile approach has been applied to the development of a game to support research in locating Paleo-Indian occupation sites and hunting camps along a land bridge that spanned Lake Huron from what is now Michigan to what is now Ontario, Canada. This project received a 2008 NSF High Risk grant in order to generate evidence for possible occupation of the ancient land bridge which is now under hundreds of feet of water. Figure 1 below provides the bathymetry, or deep profile of Lake Huron, with the lighter

regions being of higher elevation. The target of the project is a stretch of land labeled the Alpena-Amberly Ridge. This runs from the Alpena region of Michigan on the west to the Godderich region of Canada on the east.

The goal of the project was to perform a sonar-based underwater survey of the area in order to see if there was evidence for human modification of the landscape. This was conducted through the University of Michigan-Ann Arbor by Dr. John O'Shea. More detailed underwater surveys would be predicted on acquired knowledge of where specifically to look. These devices can be manned or robotic but by their nature required a small area to search. With this in mind we developed a software game as part of the Computer Science Game Programming class at Wayne State University, CSC 5430. The class had a lecture and a laboratory and was both in-class and on-line. The goal of the game was to simulate the movement of animal and human *HUNTING AND FORAGING AGENTS* over the land bridge. In particular, we focused on spring and fall where it was expected that flocks of Caribou would use the bridge as part of their annual migrations. It was felt that such movements would attract hunters who would restrict aspects of the bridge in order facilitate the hunting of Caribou.

Therefore, the amount of social intelligence allocated to the caribou and the human hunters in the game will affect their movement and the number of caribou taken. In the game, the objective was to place hunters, their encampments, their hunting stands, and purposefully placed obstacles such as fence-like rock configurations in order to maximize the count of caribou taken. Those configurations of components that produced an above average yield would suggest to the underwater team where to most profitably perform further search.

Since underwater surveys are conducted in late August and September at the same time as the onset of the gaming class it was felt, that each avenue of research can provide data about human occupation, both archaeologically and via the game play. It was of interest to see how the two sources related to each other at the end of the term. The course was set up to allow the gradual layering of intelligence into each of the major agents, caribou and humans through the support of an explicitly agile methodology. In fact, we felt that the on-line aspect of the course would play an important role in the application of the methodology since it will facilitate student communication and allow the archaeologists to have access to class discussions, demonstrations, and to provide feedback. Given the physical separation of the data collection and the software development sites, a face to face approach was not possible. In this chapter, we use on-line technology to substitute for this face to face interaction required with the Agile Development methodology. Our goal will be to see if this approach can still support the agile development of a real-world game for this application.

In section 2 the Land Bridge project is briefly described. Section 3 describes how the 12 basic tenets of agile programming, its manifesto, is supported within the on-line course framework.

Then, in section 4 the basic steps in the layering of the social intelligence using Cultural Algorithms into the system through laboratory and lecture assignments is given. Section 5 provides an example of how the project developed within this context. Section 6 concludes the chapter by assessing how the project results correspond to the survey results and describes future work.

THE LAKE STANLEY LAND BRIDGE

Overview

It is difficult, if not impossible, to consider the character of the Paleo-Indian and Early Archaic occupation of the Great Lakes region without reference to the grudging withdrawal of the continental ice sheet, and the subsequent rises and drops in the waters of the Great Lakes that accompanied the region's gradual transition to its modern appearance. Archaeologists have used the sequence of high water beaches to date these early sites, although with few exceptions the sites have

Figure 1. Bathymetry of Lake Huron. The land bridge is bright yellow and labeled the Alpena Amberly Ridge on the map.

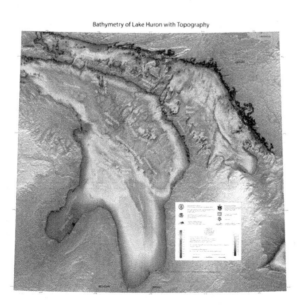

been thoroughly disturbed by later land use, and their faunal assemblages have been dissolved by the acid forest soils. For sites associated with the periods of low water, there are often no surface sites to be found at all. Some can be found deeply buried beneath later lake sediments, and many more are presumed lost forever, somewhere out under the lakes.

Newly released high resolution bathometry of the Great Lakes, coupled with advances in 3-D surface modeling, make it possible to once again view the ancient landforms from these low water periods. This in turn raises the possibility of discovering the early human settlements and activity sites that existed in this environment. In this project, we seek to explore this potential with respect to the earliest of the major low water stands in the Lake Huron basin, Lake Stanley, and the unique causeway structure that once linked Michigan with Ontario. This causeway or "land bridge" would have been available for colonization by plant, animal, and human communities. It would have also been a vehicle to support the movement of animals, specifically caribou, across Lake Huron during spring and fall migrations.

The Survey and Data Collection Methodology

The post-glacial history of the Great Lakes is characterized by a series of high and low water stands produced by the interaction of early Holocene climate, the flows of glacial melt waters and the isostatic rebound of recently deglaciated land surfaces (Lewis, et al., 1994; Moore, Rea, Mayer, Lewis, & Dobson, 1994; Larsen, 1999) as shown in Figure 2 The most extreme of the low water stands in the Lake Huron basin is referred to as the Lake Stanley stage (Lake Chippewa in the Lake Michigan basin), which spanned roughly 10,000 to 7500BP and was associated with lake levels as low as 80-90m amsl (compared to the modern standard of 176 masl (meters above sea level)) (Lewis et al., 2007).

When projected at these levels the Lake Huron basin contains two lakes separated by a ridge or causeway extending northwest to southeast across the basin from the area of Presque Isle, Michigan to Point Clark in Ontario. The causeway, termed the Alpena – Amberley Ridge, averages 16 km in width (Figure 1) and is capped with glacial till and Middle Devonian limestone and dolomite (Hough, 1958; Thomas, Kemp, & Lewis, 1973). It is represented via a topographic map where the

Figure 2. The history of the ancient Great lakes. As the glaciers retreated a precursor to Lake Huron was formed, Lake Stanley around 9000 B.B.

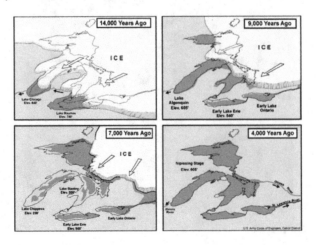

three targeted survey regions are highlighted by boxes. Region 3 in the middle of the three was the target for the first surveys here.

The earliest human occupation in the upper Great Lakes is associated with a regional fluted point Paleo-Indian tradition which conventionally ends at the start of the Lake Stanley low water stage (Ellis, Kenyon, & Spence, 1990; Monaghan & Lovis, 2005; Shott, 1999). The terminal Paleo-Indian and Early and Middle Archaic populations that inhabited the region during Lake Stanley times would have experienced an environment that was colder and drier than present with a spruce dominated forest (Croley & Lewis, 2006; Warner, Hebda, & Hahn, 1984). Sites associated with these time periods are rare. While some are found preserved beneath meters of later lake sediment (Lovis, 1989), it is generally assumed that most were lost as Lake Huron rose to its modern levels. Here we report on the first evidence for human activity on the Alpena-Amberley Land Bridge; a structure that during the Lake Stanley low water phase would have provided a land connection across the middle of modern Lake Huron linking northern Michigan with central Ontario.

Archaeologists have long recognized the potential for discovering sites of Pleistocene and early Holocene age in coastal areas that experienced repeated exposure and submergence, although these efforts have typically focused on marine environments that were subject to global changes in sea level. During the past year, investigators from the Museum of Anthropology and the Marine Hydrodynamics Laboratory at the University of Michigan have begun the task of testing whether human occupation sites are present on the Alpena-Amberley Ridge beneath Lake Huron. A particularly tantalizing possibility is the potential that stone constructions, such as caribou drive lanes, hunting blinds and habitation sites, of a kind only preserved in subarctic regions, might be visible on the lake bottom.

To discover sites within this setting, a multilayered search strategy was developed. The data used in this chapter was collected by a surface-towed side scan sonar and remote operated vehicles (ROVs). An example is shown in figure 3 below. The side scan survey was conducted using a digital side scan sonar unit (Imagenex) at a frequency of 330 kHz and a depth of 30m, mapping overlapping swaths of roughly 200m. Targets of interest, identified from acoustic survey, were examined using a remote operated vehicle (ROV). The current work utilized two mini-ROVs, a SeaBotix LBV 150, and an Outland 1000, which can be manually deployed from a small craft. Two pilot search areas have been covered, representing a total area of 72 sq km, at depths ranging from 12 to 150m.

Based upon the results of the current survey and the corresponding reality game we hope to acquire sufficient information to motivate a more detailed search using autonomous underwater vehicles (AUVs) and direct observation by archaeologists using SCUBA. The next section will provide an overview of the software development methodology that we used to develop the research game here.

THE AGILE METHODOLOGY

Why Agile?

Boehm and Turner (2004) suggest that the "home ground" for the use of Agile Program Design methodologies can be described using in terms of the following factors. First, the system to be developed has low criticality relative to existing systems. Secondly, it involves the use of senior developers. The third is that the requirements change often. Fourthly the numbers of developers is small, 5 to 10. Lastly, the target project domain is one that thrives on "chaos". That is, it is a project of high complexity.

All of the factors are true in this case. The system has a low criticality since there is no existing system that depends directly on its production. Also, the survey is currently independent of the

Figure 3. Examples of remote operated vehicles

results. Thus, results of the software are not needed at this point to determine survey agendas. Since this study is in a sense one of a kind, there is very little a priori knowledge about what to expect from the survey, as well as what the agents based game should generate.

As for the fourth point, the develop group is small, consisting of senior undergraduates and graduate students. In addition, O'Shea functioned as the expert user. Developers communicate indirectly with him via the video-streaming device. Lectures and project discussions are recorded and the web links are given to all parties for reference. Reynolds was the course instructor, and with a research position in the Museum of Anthropology he was able to facilitate a dialogue with the expert and the student developers.

Figure 4. A topographic map of the Lake Stanley Causeway extending from the mitten portion of Michigan on the left to Ontario on the right. The three target survey regions along the causeway are highlighted.

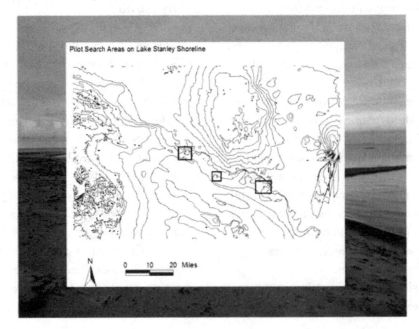

Since this is a research project that by definition is a "high risk" endeavor there is much uncertainty about what can possibly be extracted from these surveys, and inferred from the game program activities. So things tend to change from day to as the survey commenced. This was in fact an exciting thing for the students to participate in. Between the lecture and lab there were 14 assignments, about one a week which meant a constant work load for the students. However, all 10 students, both online and in class stayed on to complete the project. This is quite an unusual occurrence since during the course of the term there are many events that occur that are beyond the control of the student, yet they all delivered the final project.

Supporting the Agile Approach with Online Technology

Wood and Kleb demonstrated that agile methodologies, specifically Xtreme Programming, developed primarily to support a business customer-developer relationship can be extended to research based projects (Wood & Kleb, 2002). Here we extend the application of agile methodologies to developing research based programs using an on-line course environment within an academic setting. We feel that the addition of an online component to a game development course enhances the ability of the course to support an agile development methodology.

The best way to demonstrate how this enhances the application of agile methods is to describe how each of the basic features of the "Agile manifesto (Ambler, 2008) are supported in the classroom here. The features and their implementation are as follows:

Rapid Continuous Delivery of Useful Software

In our case this meant that there would be a lecture or lab assignment every week over the 14 weeks of the course. The duration of each assignment was on average one week. Each new assignment added complexity to the previous assignment in a gradual fashion. How the layering of complexity was performed here is described in section 4. The course instructor evaluated the submissions and passed the evaluated work and questions along to the expert through twice weekly videos and one weekly face to face meeting.

Frequently Delivered Software

The goal was to have a working program each week that contained a new feature in addition to features from previous ones. The online component allowed the instructor to critique submissions in terms of the requirements. This resulted in students "tracking" these issues and incorporating the desired features into subsequent programs. This resulted in undesirable features being eliminated quickly throughout the group. While each programmer did their own program they were able to share utility modules and graphics, which encouraged desirable features to spread. One student, for example came up with a sound package that he distributed.

Working Software is the Principle Measure of Progress

The language for the class was Python, a text based scripting language. The emphasis was on making the code as much as possible self documenting. While some documentation was required such as object-oriented diagrams etc. emphasis during the evaluations done in class was on performance, code quality, and readability. This made delivering a new program a week feasible for a student, given the reduced emphasis on additional documentation. The online component allowed the instructor to execute the submissions and provide feedback to students during the lecture as part of the critique.

Changes in Project Requirements are Welcomed at Any Time, Even Late

The video streaming component of the course supported the change of requirements. New requirements can be motivated and described as part of the lecture, and students can go back and review the lecture video if they have problems.

Close Daily Cooperation between Consumer and Developers

The online feature made it possible for the user to keep abreast of the project in the course and observe the critiques by the instructor of the students' project code. The user and instructor would meet weekly to exchange information. In this case, frequent online interaction substituted for face to face contact. Online students had to "check out" after each lecture by taking a short debriefing quiz at the end of the lecture. This worked to keep students up with the changing shape of the project.

Collocation of Developers to Support Communication

The organization of the course was three tiered; the students, the instructor, and the domain expert. The expert at the top level and the students were not collocated, the instructor was able to function as the intermediary between the two groups. The student group met with the instructors 3 times a week, and the instructor met with the expert who was not co-located one a week. While there was no physical collocation, the online framework made the interaction go smoothly since the class meetings were recorded and available for others to comment on.

Project is Built Around a Motivated Individual

Professor O'Shea is an expert in Great Lakes underwater archaeology and an enthusiastic advocate for the education of students about Michigan prehistory. This rubbed off on the students through the papers and documents that were provided as well.

Simplicity

The project was designed to start simple and to gradually add complexity to the design. The layering of complexity related primarily to adding the Artificial Intelligence components.

Self Organizing Teams

Teams were not directly supported in the classroom framework, but the weekly review of student submissions encouraged students to look to others for ideas and insights. Given that the discussions were taped, online students can get a better feel for contributions made by the in class component.

Regular Adaptation to Changing Circumstances

Since the underwater survey was ongoing as the class was taking place, the weekly assignments were adjusted to reflect student performance in the previous assignments and input from the field by the team of archaeologists. As a result, basic cognitive capabilities were added first to the agents, and then more detailed behaviors as appeared to be warranted by the survey results.

In summary, this project required a flexible methodology but since its' participants were not physically co-located the online component of the course was necessary to realize the potential of an agile approach here. In the next section we briefly describe the game and its components as they emerged over the term.

GAME DEVELOPMENT

Game Design

It was important that the students understand the genre form which the current was related. Students were introduced to popular hunting games such as Atari's "Deer Hunter", and Klaus Jurgen Wrede's "Carcassonne" a board game based upon ancient hunting and foraging agents in Europe. In our game, the idea was to position a campsite, and hunting blinds over a portion of the land bridge. Caribou, the North American reindeer, would move from northwest to southeast over the bridge in the fall, and back across the bridge in the spring. The landscape of the bridge was assumed to be that of Arctic tundra populated by spruce and grasses and containing possible small stands of water.

The player positioned the campsite and hunting stands on the land bridge. Then the hunters would autonomously move out of the campsite daily and position themselves in the nearest available blind. The hunting agents used state machines to emulate decision making using knowledge learned using Cultural Algorithms. Cultural Algorithms are an extension of Genetic Algorithms that used to generate social intelligence to guide the interaction of multiple agents. When a caribou came within a certain distance of the blind, the hunter became aware of it, and tried to kill it. At the end of the day, hunters would return to the camp with a portion of their kill. Additionally, other predators co-existed on the landscape with humans, in this case wolves. Wolves hunted in packs and generally kept away from humans unless they met head on via a chance encounter, or the wolf becomes so hungry that they will attack humans.

Each of the object categories had both an individual and a social intelligence. For example, caribou moved in herds, wolves moved in packs, and humans lived as a group and hunted in a distributed manner. The placement of hunting blinds will determine whether they hunt individually or in a collective fashion. The result of a players' placement of hunting blinds and campsites on the land bridge will be a function of the intelligence that each of the interacting species has. While these Paleo-Indians were hunter-gatherers our focus is on the hunting component. Gathering will be added later.

Game Object Organization

The game supports the coupling or interaction of both human and natural systems. The basic object hierarchy supported by the game is given in figure 5. The key is that the terrain object is constantly being updated through the survey process. Changes in its composition then need to be reflected through changes to the information used in the construction of the state machines for the three different intelligent agent object classes, human, caribou, and wolf. Each state machine encoded basic individual behaviors based on its environmental context. Thus, the key to the use of the agile technology here is to synchronize changes between the terrain and game objects. In this section we focus on the game objects, and in the following section we focus on the terrain objects.

Each group of agents therefore possesses both individual and social intelligence. Each component will now be described. The individual cognitive capabilities supported by the system are as follows.

Sensing their immediate physical environment: Every object class has a certain set of cognitive capabilities in terms of being able to identify objects within its' local area. The capabilities vary from class to class, and can be adjusted over game levels if necessary.

Sensing their immediate social environment: Each object can sense its relative position in its social group, whether it is a herd, a pack, or a family. We added behaviors for the caribou such alignment, flee, and separation in order to reflect the ability of the caribous to move as a herd in a

Figure 5. An example object hierarchy for the game agent classes

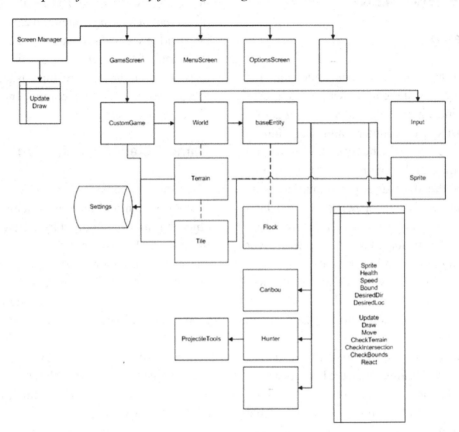

line, to avoid obstacles, wolves, hunters, etc, and to separate into more than one herd if they were attacked by the hunters or the wolves.

Basic goals: Caribou, wolves, and humans all have a goal of a survival which requires a certain amount of food, water, and other resources daily. Humans for example will also need firewood for cooking and warmth, as well as stone for tool making.

Path Planning: A basic set of waypoints are located within the region. Individual agents from each category, human, caribou, or wolf can plan their path in order to avoid obstacles, moving in herds, and to achieve goals such as attaining food and water.

State Machines: Each category of agents inherits a state machine from its class. The state machine keeps track of its current state in terms of its food and resource needs, its' active goals, and its perceived environment.

Learning capabilities: Cultural Algorithms have been successfully used to acquire social intelligence in a variety of application and will be used to support organizational learning here (Reynolds & Ali, 2008; Reynolds, Ali, & Jayyousi, 2008).

The Terrain Object

At the onset of the project, there was some information about the general topography of the region as shown by the computer generated GIS representation in Figure 6. However, prior to the survey it was not clear what level of detail might actually be found on the bottom. That is, will there be evidence of non-natural structures made by humans as well as other evidence suggesting the past distribution of resources. As the survey proceeded it became clear that there was sufficient evidence for campsite and hunting blinds to allow them to be included in the model, as well as behaviors that would utilize them. Therefore, as more information was garnered from the survey, the requirements for the game changed since there was now more information that the game could be used to predict.

At the beginning of the game design, hunting blinds and campsites were not considered but once it was clear that there was sufficient evidence for them they were added into the game and the objects behaviors adjusted to include them. With an agile approach, synchronizing the project requirements with the survey results would have been very difficult to achieve. And, without the online component it would have been difficult to effectively implement the agile technology in a situation that lacked aspects of collocation that are critical to traditional agile approaches.

When the development of the game first began campsites and hunting blinds were not considered to be used. The preliminary results of the survey suggested that remains of such structures were present, so we added in those features along with the AI required to exploit them. We then can observe how adding this new information in can affect the behavior of the various agents. Figure 7 gives a screen shot of main menu of the game containing three options. The three options are as follows: "Play Game" to load and start playing the game; "Options" to change the difficulty of

the game; and "Exit" to close the game application. The options screen gives the user a chance to configure the game and to change the difficulty level of the game. The menu has four main options that can be useful in the modification of the number of wolves, caribou, and hunters.

RESULTS

Figure 8 gives a screen shot of the game environment containing spruce forest, water features, rock formations along with hunting blinds and campsites. One can see caribou herds, hunters, and wolves distributed over the landscape as well. In this figure wolves are attacking the caribou. Hunters are running to find ambush sites ahead of the herd. Notice the emergent property of the herd to split into smaller herds as result of an attack. This behavior is observed in real caribou and emerged here as a result of the interaction of herd members. The semi-circle of rocks corresponds to a man-made hunting blind.

Figure 9 presents another emergent behavior produced by the model. In the figure the herd of caribou moves in a line, with one individual following another other. This behavior is observed in real caribou and merges here as a result of the individual movements of the herd members.

Figure 6. A GIS (geographical information system) representation of the topography of the land bridge

Figure 7. A screen shot of the main menu of the game. The user can start playing the game, change the difficulty level, and exit the game.

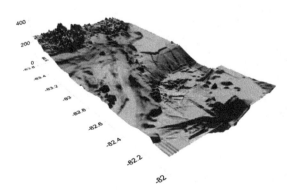

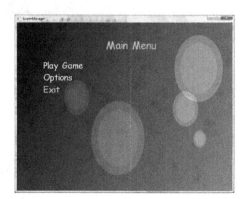

Figure 8. A screen shot of the game as caribou move from the northwest to the south east along a portion of the land bridge

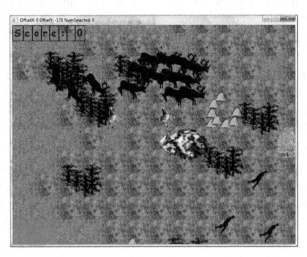

Figure 10 is a screen shot of the caribou herd moving in a line as they head south while avoiding the water and rocks. They are also able to avoid hunters and wolves. This avoidance of objects gives the hunters an opportunity to produce drive lanes. These lanes force the herd into a linear pattern within a narrowing area which makes hunting easier.

This emergent asymmetry provides us with new information that we can use to answer the following questions:

1. Is there a positioning of blinds and sites that exhibits equal kill probabilities for both north and south migrations?
2. How does this compare to the optimal positioning blinds and sites for north migration or south migration alone?

Figure 9. In this screen shot the caribou are moving in a line across the land bridge. Notice that caribou avoid the hunters in their path.

Figure 10. The caribou herd moves along the water's edge while avoiding the hunter

3. If the north and south optimal locations are different, what are the differences and why?
4. For an observed archaeological configuration of sites and blinds on the land bridge can we infer when they were used during the year?

For example, we placed the hunting blinds discovered by the archaeological survey in their proper positions on the land bridge. We then had the herd simulate southern and northern migrations respectively. The question of interest is whether the positioning of the blinds produces a better kill count for northern or southern migration. The presence of a significant difference may allow us to infer the season in which the blinds were most useful. As it turned out, blind placement was significantly more effective in north to south migration rather than from south to north. This suggests that the primary hunting season was in the fall and that positioning of the currently observed hunting blinds was to support fall hunting.

These are new insights that the fusion of survey and gaming approaches through an agile technology online has produced. The agile approach online has taken advantage of the synergy of the two different technologies to produce insights that neither one could have produced on its own.

Figure 11. An autonomous underwater vehicle to be used for future surveys

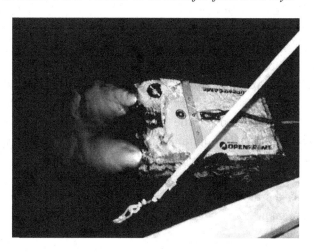

FUTURE WORK

In this study we used an agile software design methodology to produce a real-world game that can be used to help predict the location of Paleo-Indian sites on a submerged prehistoric land bridge. It was critical to the success of the project that results of the ongoing underwater survey be synchronized with the current requirements of the developing game. Specifically, as new information about environments contents was obtained, adjustments would be needed to the Artificial Intelligence of the game games in order to exploit them. However, some key requirements for an agile approach such as co-location of the expert and the developers were missing. With the use of online technology we were able to work around these obstacles.

As a result we have produced a computer game that fuses the most recently collected data with the requisite AI to exploit it. In order to support the incremental addition of computational intelligence into our game an Agile Program Design methodology was employed. The results of playing the game suggest that there is an inherent asymmetry to the migration pathways of the caribou. That is, the positioning of blinds and the related campsites is a function of migration direction. This means that we are relating the positioning of non-natural objects with the different seasons of the year. That is an additional level of understanding that we had not bargained for at the onset of the project.

Future work will involve the use of manned and autonomous underwater vehicles such as that shown in figure 11. Since sites are between 60 and 200 meters below the surface, this means that for the deepest sites the US navy recommends at most 20 minutes exposure at those levels with 44 minutes of decompression in three stops on the way up. Therefore, it will be important to pinpoint areas that will be most likely to yield important information about prehistoric occupants in order to make diving expeditions the more effective. The results produced by the game so far suggest that it will be an important factor in making good decisions as to where to deploy these devices.

In addition, we will consider adding the ability of agents to learn to coordinate their own hunting activities using the Cultural Algorithm, a socially motivated hybrid learning algorithm developed by Reynolds (2008) We can then compare the best learned results with those of the human player. Likewise, we can allow caribou herd and wolf packs to adjust their dynamics in order to achieve their own goals of survival.

ACKNOWLEDGMENT

This project was supported in part by NSF High-Risk Grant #BCS-0829324.

REFERENCES

Ambler, S. (2008). Scaling Scrum – Meeting Real World Development Needs. *Dr. Dobbs Journal.* Retrieved April 23, 2008 from http://www.drdobbsonline.net/architect/207100381.

Boehm, B., & Turner, R. (2004). *Balancing Agility and discipline: A Guide for the Perplexed.* Addison-Wesley Press.

Croley, T., & Lewis, C. (2006)... *Journal of Great Lakes Research, 32*, 852–869. doi:10.3394/0380-1330(2006)32[852:WADCTM]2.0.CO;2

Ellis, C., Kenyon, I., & Spence, M. (1990). Occasional Publication of the London Chapter . *OAS, 5*, 65–124.

Hough, J. (1958). Geology of the Great Lakes. [Univ. of Illinois Press.]. *Urbana (Caracas, Venezuela)*, IL.

Larsen, C. (1999). Cranbrook Institute of Science. *Bulletin, 64*, 1–30.

Lewis, C. (1994)... *Quaternary Science Reviews, 13*, 891–922. doi:10.1016/0277-3791(94)90008-6

Lewis, C. (2007).. . *Journal of Paleolimnology*, *37*, 435–452. doi:10.1007/s10933-006-9049-y

Lovis, W. (1989). *Michigan Cultural Resource Investigations Series 1*, East Lansing.

Monaghan, G., & Lovis, W. (2005). *Modeling Archaeological Site Burial in Southern Michigan.* East Lansing, MI: Michigan State Univ. Press.

Moore, T., Rea, D., Mayer, L., Lewis, C., & Dobson, D. (1994).. . *Canadian Journal of Earth Sciences*, *31*, 1606–1617. doi:10.1139/e94-142

Reynolds, R. G., & Ali, M. (2008). Computing with the Social Fabric: The Evolution of Social Intelligence within a Cultural Framework. *IEEE Computational Intelligence Magazine*, *3*(1), 18–30. doi:10.1109/MCI.2007.913388

Reynolds, R. G., Ali, M., & Jayyousi, T. (2008). Mining the Social Fabric of Archaic Urban Centers with Cultural Algorithms. *IEEE Computer*, *41*(1), 64–72.

Shott, M. (1999). Cranbrook Institute of Science . *Bulletin*, *64*, 71–82.

Thomas, R., Kemp, A., & Lewis, C. (1973).. . *Canadian Journal of Earth Sciences*, *10*, 226–271.

Warner, G., Hebda, R., & Hahn, B. (1984). *Palaeogeography, Palaeoclimatology, Palaeoecology*, *45*, 301–345. doi:10.1016/0031-0182(84)90010-5

Wood, W., & Kleb, W. (2002). Extreme Programming in a research environment . In Wells, D., & Williams, L. (Eds.), *XP/Agile Universe 2002* (pp. 89–99). doi:10.1007/3-540-45672-4_9

Chapter 12
Management of Distributed Energy Resources Using Intelligent Multi-Agent System

Thillainathan Logenthiran
National University of Singapore, Singapore

Dipti Srinivasan
National University of Singapore, Singapore

ABSTRACT

The technology of intelligent Multi-Agent System (MAS) has radically altered the way in which complex, distributed, open systems are conceptualized. This chapter presents the application of multi-agent technology to design and deployment of a distributed, cross platform, secure multi-agent framework to model a restructured energy market, where multi players dynamically interact with each other to achieve mutually satisfying outcomes. Apart from the security implementations, some of the best practices in Artificial Intelligence (AI) techniques were employed in the agent oriented programming to deliver customized, powerful, intelligent, distributed application software which simulates the new restructured energy market. The AI algorithm implemented as a rule-based system yielded accurate market outcomes.

INTRODUCTION

The electricity grid is the backbone of the power network and is at the focal point of technological innovations. Utilities need to introduce distributed intelligence into their existing infrastructure to make them more reliable, efficient, and capable of exploiting and integrating alternative sources of energy. The intelligent grid includes the infrastructure and technologies required to allow distributed generation of energy with increasing

the operational efficiency through distributed control and monitoring of resources.

Intelligent grid should be self-healing and reconfigurable to guard against man-made and natural disasters. One way to assure such characteristics in an electric power grid is to design small and autonomous subsets of the larger grid. These subsets are called intelligent microgrids which are used as a test bed for conglomerate innovations in communication technologies, smart metering, co-generation, and distributed intelligence and control. The test bed serves to showcase the capabilities of the developed systems, thereby ac-

DOI: 10.4018/978-1-60566-898-7.ch012

celerating the commercialization of technologies and solutions for smart grids all over the world.

Multi-agent system is one of the most exciting and fastest growing domain in agent oriented technology which deals with modeling of autonomous decision making entities. Multi-agent based modeling of a microgrid is the best choice to form an intelligent microgrid (Rahman, Pipattanasomporn, & Teklu, 2007; Hatziargyriou, Dimeas, Tsikalakis, Lopes, Kariniotakis, & Oyarzabal, 2005; Dimeas & Hatziargyriou, 2007), where each necessary element in a microgrid is represented by an intelligent agent that uses a combination of AI-based and mathematical models to decide on optimal actions.

Recent developments (Rahman, Pipattanasomporn, & Teklu, 2007; Hatziargyriou, Dimeas, Tsikalakis, Lopes, Kariniotakis, & Oyarzabal, 2005; Sueyoshi & Tadiparthi, 2007) in multi-agent system have shown very encouraging results in handling multi-player interactive systems. In particular, multi-agent system approach has been adopted to simulate, validate and test the open deregulated energy market in some recent works (Sueyoshi & Tadiparthi, 2007; Bagnall & Smith, 2005; Praça, Ramos, Vale, & Cordeiro, 2003; Logenthiran, Srinivasan, & Wong, 2008). Each participant in the market is modeled as an autonomous agent with independent bidding strategies and responses to bidding outcomes. They are able to operate autonomously and interact pro-actively within their environment. Such characteristics of agents are best employed in situations where the role identities are to be simulated as in a deregulated energy market simulation.

The dawn of the 21st century has seen numerous countries de-regulating or lobbying for deregulation of their vertically integrated power industry. Electric power industry has seen an evolution from a regulated to a competitive industry. The whole industry of generation, transmission and distribution has been unbundled into individual competing entities. Although the journey has been far from seamless as observed in the California'

electricity crisis (Budhraja, 2001), many critics have agreed that deregulation is indeed a noble endeavour. The problem associated with deregulation can be solved with structural adjustments to the markets and learning from past mistakes.

This chapter shows the development and implementation of multi-agent application to deregulated energy market. The developed application software is a testament of the multi-agent framework implementation and effectiveness of dynamic modeling of multi-agent environment where the internal tasks of each agent are executed concurrently with external inputs from the agent world. Successful deployment of the application software coupled with high degree of robustness indicates the relevance and operational level of multi-agent system based application software development. User can use the software for any size of power system by defining the number of agents in the system and inserting the associated information.

The structure of the remaining chapter is as follows: Section 2 provides the introduction of microgrid and Distributed Energy Resource (DER), and Section 3 gives an introduction of restructured electricity market. Section 4 describes the implementation of multi-agent system based application software for PoolCo energy market simulation. Section 5 demonstrates the flow of simulation of the implemented application software. Section 6 discusses results of PoolCo outcome of a sample microgrid. Finally, it is concluded in the seventh section.

BACKGROUND

Microgrid and Distributed Energy Resource

Over the years, the computer industry has been evolving continuously and the power industry has remained relatively stable. In the past few years, the power industry also has seen many revolutionary changes. The deregulated energy environment has favoured a gradual transition from centralized power generation to Distributed Generation (DG) where sources are connected at the distribution network. Several technologies, such as diesel engines, micro turbines, fuel cells, wind turbines and photovoltaic systems can be part of a distributed generation system. The capacity of the DG sources varies from few kWs to few MWs. Distributed systems can also bring electricity to remote communities which are not connected with the main grid. Such multiple communities can create a microgrid of power generation and distribution.

Microgrids can be defined as low voltage intelligent distribution networks comprising various distributed generators, storage devices and controllable loads which can be operated as interconnected system with the main distribution grid, or as islanded system if they are disconnected from the main distribution grid. The common communication structure and distributed control of DG sources together with controllable loads and storage devices such as flywheels, energy capacitors and batteries, are central to the concept of microgrids (Lasseter, Akhil, Marnay, Stephens, Dagle, Guttromson, Meliopoulos, Yinger, & Eto, 2002). From the grid's point of view, a microgrid can be regarded as a controlled entity within the power system that can be operated as a single aggregated load and a small source of power or ancillary services supporting the network. From the customers' point of view, microgrids are similar to traditional low voltage distribution networks which provide local thermal and electricity needs. In addition, microgrids enhance the local reliability, reduce emissions, improve the power quality by supporting voltage, and potentially lower the cost of energy supply.

Deregulated Energy Market

Around the world, the electricity industry which has long been dominated by vertically integrated utilities is experiencing major changes in the structure of its markets and regulations (Lasseter, Akhil, Marnay, Stephens, Dagle, Guttromson, Meliopoulos, Yinger, & Eto, 2002; Shahidehpour & Alomoush, 2001). The power industry has become competitive because the traditional centralized operation is replaced with an open market environment. This transformation is often called as the deregulation of electricity market. Market structure varies from country to country depending on the policies adopted in the country. For example, the Independent System Operator (ISO) and the Power Exchange (PX) are separate entities in some countries' markets like California's market, although the PX functions within the same organization as the ISO, while they are under the same structure with control of the ISO in some other markets.

To implement a competition, vertically integrated utilities are required to unbundle their retail services into generation, transmission and distribution. Generation utilities will no longer have a monopoly. Even small business companies will be free to sign contracts for buying power from cheaper sources. Many new concepts (Shahidehpour & Alomoush, 2001; Shahidehpour, Yamin, & LI, 2002) have appeared to facilitate the way of dealing with restructuring. A few critical roles of these entities and concepts which are instrumental for understanding the multi-agent system based modeling of restructured energy markets are discussed here.

Independent System Operator (ISO)

ISO is an independent entity of individuals in market energy market such as generation, transmission, distribution companies and end users. The ISO administers transmission tariffs, maintains the system security, coordinates maintenance scheduling, and has a role in coordinating long-term planning. The main purpose of an ISO is to ensure fair and non discriminatory access of the grid, transmission lines and ancillary services. ISO manages the power flow over the transmission system and facilitates reliability requirements of the power system. The ultimate role of ISO is to ensure that the total generation meets the demand by taking congestion and ancillary services into the account. This function is carried out by controlling dispatch of flexible plants and giving orders to adjust the power supply levels or curtail loads to ensure loads matching with the available power generation in the system.

Power Exchange (PX)

PX monitors and regulates the economic operation of the interconnected grid. It operates as an independent, non-government, non-profit entity which provides schedules for loads and generators. In the PoolCo model, PX establishes an auction like spot market where energy bids of the generators and consumers are matched anonymously based on the bidding quantities and prices. PX finds the market clearing quantities and the market clearing price from market equilibrium point. PX operates a day-ahead market as well as an hour-ahead market separately depending on the marketing period.

Transmission Companies (TRANSCOs)

The transmission system is the most essential element in electricity markets. The secure and efficient operation of the transmission system is necessary for efficient operation of these markets. A TRANSCO has the role of building, owning, maintaining, and operating the transmission system in a certain geographical region to provide services for maintaining the overall reliability of the electrical system. The use of TRANSCOs assets comes under the control of the ISO and they are regulated to provide non-discriminatory connections and comparable services for cost recovery. The ISO oversees the operation and scheduling of TRANSCOs' facilities.

Generation Companies (GENCOs)

Generation companies are formed once the generation of electric power is segregated from existing utilities. They take care of the operation and maintenance of existing generating plants. Electricity from them is either sold to short term markets or provided directly to the entities that have contracts with them for purchase of electricity. Besides real power, they may sell reactive power and operating reserves. GENCOs include Independent Power Producers (IPP).

Customers

They are the end users of electricity with different load requirements. They may be broadly categorized into industrial, commercial and residential. In a restructured market, customers are no longer obligated to purchase any services from their local utility company. Customers have rights to make direct access to generators or other power providers and choose best packages of services that meet their needs.

Physical Power System Constraints and Transmission Pricing

The agreements between trading parties are made based on the outcome of the energy market, which does not represent the actual power flow in the system. Constraints in the power system, for example transmission losses and contract transmission paths, affect the operation of the transmission system. Due to transmission line losses, the power injected at any node in the system to satisfy a certain demand at another node depends on the loss factors between the nodes. Transmission losses will affect the actual power injection pattern and quantity to the network.

Another issue that affects transmission pricing is Contract Path which has been used between transacted parties as dedicated paths where power flows are assumed to flow through pre-defined paths. However, physically electrons could flow in a network over parallel paths owned by several utilities that may not be through the contract path. As a result, transmission owners need to be compensated for the actual use of their facilities.

The above are just two of the many implications that power system constraints affect pricing in a restructured market. Though it is beyond the scope of this discussion and also beyond the scope of this application development, managing such constraints and their impacts in pricing is essential.

Market Models

The main objectives of an electricity market are to ensure the secure efficient operation and to decrease the cost of electricity through competition. Several market structure (Praça, Ramos, Vale, & Cordeiro, 2003; Shahidehpour & Alomoush, 2001; Shrestha, Song, & Goel, 2000) models exist all over the world. These market models would differ in terms of marketplace rules and governance structure. Generally they can be classified into three types such as PoolCo model, Bilateral contract model and Hybrid model.

The PoolCo market model is a marketplace where power generating companies submit their production bids, and consumer companies submit their consumption bids. The market operator uses a market clearing tool to find the market clearing price and accepted production and consumption bids for every hour. The bilateral contracts are negotiable agreements between sellers and buyers about power supply and reception. Bilateral contract model is very flexible because negotiating parties can specify their own contract terms and conditions. Finally, the third market model is hybrid model which is a combination of PoolCo and Bilateral contracts models. It has the features of PoolCo as well as Bilateral contracts models. In this model, customers can either negotiate with a supplier directly for a power supply agreement or accept power from the pool at the pool market clearing price. For this software development,

ISO and PX are modeled as separate entities like in the California's energy market (Shahidehpour & Alomoush, 2001; Shahidehpour, Yamin, & LI, 2002) to illustrate their individual roles in the energy market and the typical day-ahead PoolCo model is chosen because of its simplicity.

The PoolCo model (Shrestha, Song, & Goel, 2000) consists of competitive independent power producers, vertically integrated distribution companies load aggregators and retail marketers. The PoolCo does not own any part of the generation or transmission utilities. The main task of PoolCo is to centrally dispatch and schedule generating units in the service area within its jurisdiction. The operating mechanism of the PoolCo model is described in Figure 1. In a PoolCo market operation, buyers (loads) submit their bids to the pool in order to buy power from the pool and sellers (generators) submit their bids to the pool in order to sell power to the pool. All the generators have right to sell power to the pool but they can not specify customers.

During PoolCo operation, each player will submit their bids to the pool which is provided by PX. The PX sums up these bids and matches interested demand and supply of both sellers and buyers. The PX then performs economic dispatch to produce a single spot price for electricity for the whole system. This price is called the Market Clearing Price (MCP) which is the highest price in the selected bids of the particular PoolCo simulation hour. Winning generators are paid the MCP for their successful bids while successful loads are obliged to purchase electricity at MCP. Generators compete for selling power. If the bids submitted by generator agents are too high, they have low possibility to sell power. Similarly, loads compete for buying power. If bids submitted by load agents are too low, they have low possibility to get power. In such a model, generators bids with low cost and load bids with high cost would essentially be rewarded.

Figure 1. PoolCo market model

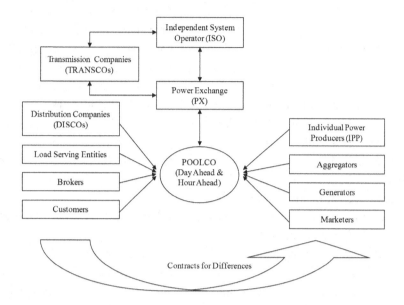

Market Operation

The ISO and The PX handle the market operation which can be a day-ahead market or an hour-ahead market (Shahidehpour & Alomoush, 2001; Shahidehpour, Yamin, & LI, 2002). In the day-ahead market, for each hour of the 24-hour window, sellers bid a schedule of supply at various prices, buyers bid a schedule of demand at various prices, and market clearing price and market clearing quantities are determined for each hour. Then the PX schedules supply and demand with help of the ISO. The ISO finalizes the schedules without congestion. An hour-ahead market is similar to a day-ahead market, except the total scheduling period is an hour instead of a day. Typical market operation in a restructured power system is shown in Figure 2.

Figure 2. Flow of PX-ISO in market operation

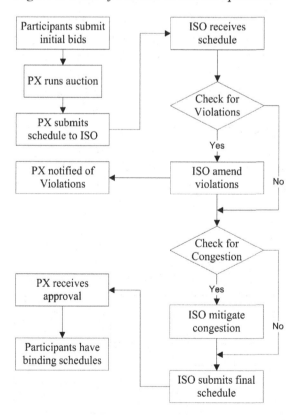

IMPLEMENTATION OF MULTI AGENT SYSTEM

Multi-Agent Platform

A multi-agent platform provides a platform for implementing agent world and managing agents' execution and message passing. JADE (Java Agent DEvelopment) is a multi-agent platform which is chosen for this development software. JADE (http://jade.tilab.com/) aims for developing multi-agent systems and applications conforming to Foundation for Intelligent Physical Agents (FIPA) standards (http://www.fipa.org/) for intelligent agents. JADE is a middleware which means that JADE provides another layer of separation between the software and the operating system. In this implementation, the underlying operating system is the Java virtual machine. JADE is fully coded in Java which provides an attractive features of this implementation because Java has several attractive features over the other programming languages such as extensibility, security and cross-platform.

JADE platform provides Agent Management Service (AMS), Directory Facilitator (DF) and Message Transport System (MTS) which are the necessary elements in a multi-agent system as specified by FIPA standards. Typical FIPA compliant agent platform is illustrated in Figure 3. Agent management service is responsible for managing the agent platform which maintains a directory of Agent Identifiers (AIDs) and provides white page and life cycle services. Directory facilitator provides the default yellow page services in the platform which allows the agents to discover the agents in the network based on the services they wish to offer. Finally, message transport system provides a channel for agent communication and is responsible for delivering messages between the agents.

Structure of Software Packages

A generic service component software packages have been designed and developed for this software implementation which is shown in Figure 4. Agents, Behaviours, Ontology, Tools and Power System are the main packages developed in this agent framework.

The Agents package consists of a collection of different types of agents. The Behaviour package details a collection of different types of tasks assigned to various agent entities. The Ontology package specifies a set of vocabulary for agent language which agents understand and use for their communication with each other in the framework. The Tools package implements a collection of tools which are used by the agents in the framework for managing and processing the auction.

Figure 3. FIPA compliant agent platform

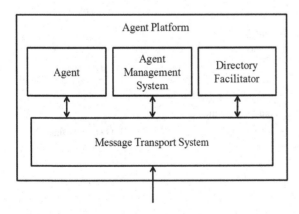

Figure 4. Structure of software packages in the framework

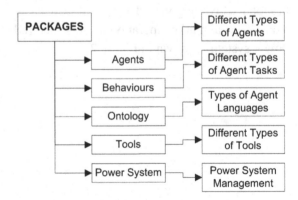

Figure 5. Different types of agents in the agent world

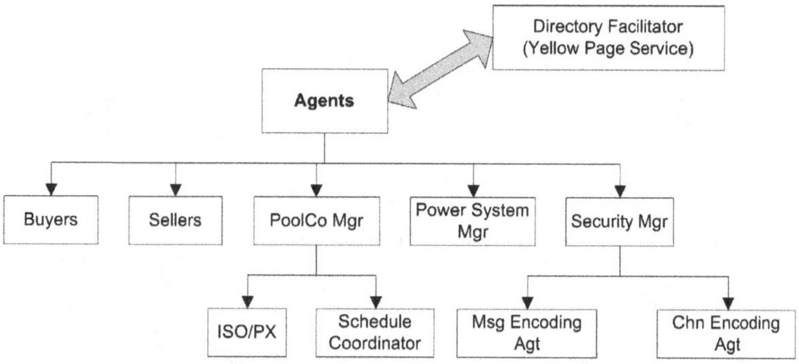

Finally, the Power System package comprises of data structures used to represent the state of the physical power system network.

Agents Package

The Agent package in this framework consists of several agents such as ISO, PX, Schedule Coordinator, PoolCo Manager, Power System Manager, Security Manager, Sellers and Buyers. This application software focuses only on the restructured energy market simulation. The different entities would interact in this framework to simulate a day-ahead market. A test system with three sellers and five buyers is considered for the case study, however it can be extended for any number of market participants. The Figure 5 shows some of the main agents implemented in this framework.

PX Agent: PX agent has been customized for purpose of modeling the restructured energy market. PX acts as a middle man between the various market participants and ISO. Market participants will submit their bids to the pool. PX performs scheduling through Schedule Coordinator (SC) agent. The schedule coordinator will collate all the bids and determine a schedule using the market clearing engine.

For any particular day schedule, PX will also scans for any violation bids which are sent back by the ISO. If any vectorized violated bids are

received, PX disseminates this information back to the relevant market participants.

ISO Agent: ISO in this framework performs the roles of a regulatory body. ISO seeks to ensure the authenticity of the bids and the stability of the network. Bids are checked for violations and acted upon accordingly. ISO would conduct these checks with the help of a rule based engine customized for ISO. In addition, network simulations are carried out on the day schedules to ensure stability of the network with power world simulator. ISO also maintains a database of day schedules. As mentioned earlier, ISO has the broad role of seeing to the system operation and the stability of the network in this project.

Security Manager Agent: The security manager is an overall central information security hub that provides all encryption, decryption, encryption keys generation, issue of digital certificates and other security related services to the agent world. All agents have to register with security manager to make use of the security services in the network. As all message transmission is done through the Secured Socket Layer (SSL) protocols, agents which do not register with the security manager will have no access to the SSL service thus they will not be able to communicate with any other agents in the network.

Authorized agents will have valid ID in the agent world and these IDs are used to register with security manager. In this application soft-

215

Figure 6. Security architecture of the agent world

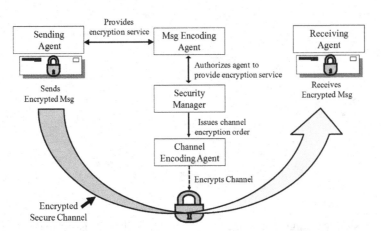

ware, security architecture for intelligent agents is employed in the network as shown in Figure 6.

Security manager has exclusive access to message encoding agent and channel encoding agent which are mainly responsible for providing encryption of messages and channels services to all agents. All agents who wish to send messages across the network will have to engage their service in encrypting the message and channel which they are going to send through. The agents need to contact the security manager agent, upon the authentication; message encoding agent and channel encoding agent provide security services to the agents. Message encoding agent will provide encryption service for the sending agent after receiving encryption order from security manager and channel encoding agent will encrypt the message channel between the sending and receiving agent after receiving encryption order from security manager. When the message is successful encrypted, sending agent will send the encrypted message through the encrypted channel.

Such architecture provides double redundancy in communications security for extra security. For any hacker to successfully decrypt the messages sent between two agents, it is needed to break the encryption of the channel and then the message itself. This is difficult to achieve for the following reasons.

- The encryption process is done by two separate entities (Message encoding agent and channel encoding agent). Unlike systems with only one encrypting entity where all encryption are done centrally, it takes twice the effort to search or guess a key same as the generation key is done by two separate entities.

- The channel encryption provides dual level of security. Every time a message is sent between two agents, a new secured channel is established. The encryption key used to establish the secured channel is always different. Since the channel encryption is always different, the key value for decryption is also always different. This makes it even harder for unauthorized interception of messages to succeed.

Behaviour Package

In multi agent systems, agents with different nature are endowed with different attributes, beliefs and objectives. These behaviours are added to internal tasks of agents. Behaviour is basically an event handler. Behaviours in the agents are executed in an order as arranged by the JADE internal behaviour scheduler. This scheduler is responsible for organizing and arranging all the

Figure 7. Schematic of behaviours package

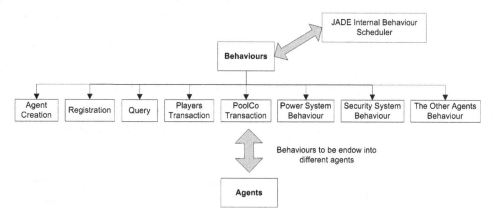

behaviours of each and every agent. The Figure 7 shows some of the main behaviours implemented in this framework.

Ontology Package

The concept of ontology can be imagined as vocabulary in an agent world for the communication. By defining any information as a specific ontology, it will look like our speaking language. Once the agent determines what language this information is coded in, it will try to retrieve an ontology "decoder" from the ontology package. With the help of these "decoders", the receiving agent though not speaking the native language will be able to understand the other agent. The Figure

8 shows some of the main ontologies implemented in this framework.

Back to the software programming of the implementation; when an agent sends information to another agent, it will be serialized in a fashion which is normally not understood by another agent unless it knows what ontology is used. Once the agent knows which class of ontology does this information belongs to, it will be able to "decode" and "reassemble" the information in a meaningful way such that it can read useful information from it. Some of the ontologies implemented in this framework are given in details below.

Bid: This is a class for all bids submitted by sellers and buyers to the PoolCo. These bids contain the bid price, quantity and much other

Figure 8. Schematic of ontology package

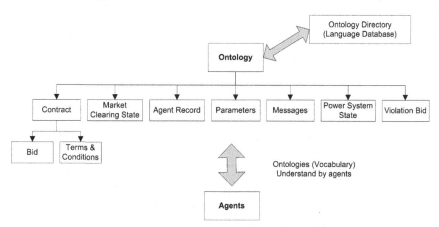

Figure 9. Structure of contract ontology

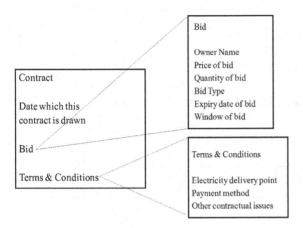

peripheral information like the owner of bid, date of submitted, date of expiry, type of bid (seller bid or buyer bid) and originating bid address. Owner of the bid refers to the name of the agent in the agent world and it also includes information about the physical address of the agent where it is residing on. Information on the date which a bid is submitted is useful when tracking the bids' order for processing. Information on the expiry date of a bid is very important because it is used by ISO for finding the violation bids.

Terms and Conditions: This ontology is an abstract object that defines the specific terms and conditions. This object will be embedded together with the bid object to form the contract object. This object specifies contractual issues pertaining to the agreement on the electricity sale between

buyers and sellers like location of electricity purchased, location of electricity injected, payment method and the other related issues pertaining to the conditions of sale.

Contract: This is the integrated ontology of bid and, terms and conditions ontologies. The bid defines the price and quantity of the electricity sale. The terms and condition object defines all contractual issues pertaining to the electricity sale agreement. As the name suggest, these information will be embedded in the contract object defining all scopes of the electricity transaction details. Figure 9 shows the structure of this ontology.

Agent Record: This ontology implements a data structure record of every agent with regards to its participation in the market network. Every agent holds a copy of its own record. The PoolCo also maintains a collection of all the agents on the network in the form of agent record object. The PoolCo keeps these agent record objects in a secure hashtable for its own reference and management of players. Each agent record contains the information about the agent such as subscription status with PoolCo, proposals status with the PoolCo, owner of this agent, current bid submitted to the PoolCo, successful bid by the PoolCo, record of its own message history with all other agent entities and state of the owner agent in current negotiations. Figure 10 shows the structure of this ontology.

Market Clearing State: This ontology is a data structure detailing outcome of an auction. It in-

Figure 10. Structure of agent record ontology

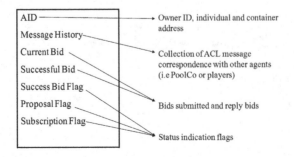

Figure 11. Structure of market clearing state

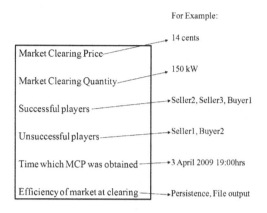

cludes information on market clearing price, total successful volume of electricity transacted, successful buyers, successful sellers and details about the market is experiencing excess demand or supply at the going market clearing price. Schematic of this object is as shown in the Figure 11.

Violation Bid: This is a class that deals with violated hour bids. An hour bid that violates the rules and conditions laid out by ISO which is carried out by the ISO rule engine will be created as a violation bid. The violation bid object will contain details of the faulty, amended bids and the type of violation. In a particular day schedule, all violation bids will be collated into a vector form and the vectorized violated bids will be sent to PX as an encoded message. Table 1 shows examples for these three rules of violation of bidding implemented in this software.

Time violation is shown for Buyer 1 because it sent a delayed bid which can be seen in table. Seller 2 sent a bid with 420kW even though it agreed to sell up to a maximum of 400kW and

Seller 1 sent a bid with 95kW even though it agreed to sell at least minimum of 100kW. Therefore Seller 2 violated its maximum limit and Seller 1 violated its minimum limit.

Tools Package

This package provides a set of generic and specific tools used in this framework. They are provided as public type services to all agents. Some of these tools like ISO rule engine provide provisions for expansions. The necessary extensions can be applied in future. The Figure 12 shows the tools implemented in this framework and some of them are given in details below.

ISO Rule Base System: These rules belong to the operation of ISO and implemented to check the hourly bids of the day schedule based on the set of user defined rules. In this framework, only three main rules are developed. First rule is to ensure that the date and time of bid submission for every hour is not violated. Market participants in this simulated environment are required to submit their 24-hour bids for a particular day on or before at 12:00 PM of the day before the simulation day. Any time submission violation would result in a monetary penalty for the concern participant. After sieving out the time violation, the violated bid would then be vectorized into a vector of violation bids. The second and third rules are being carried out on the quantity of sellers' bids. Sellers in the network are allowed to produce pre-determined maximum and minimum quantities. Seller quantity bids are checked to ensure that the maximum and minimum allowable limits are not violated.

Table 1. Different types of violated bids

Market Participants	Bid Received Time	Bid Received for	Quantity	Price
Buyer 1	01-04-2009 12:00 PM	01-04-2009 11:00 PM	379.0	27.7
Seller 2	31-04-2009 11:00 AM	01-04-2009 03:00 PM	420.0	28.7
Seller 1	01-04-2009 12:00 PM	02-04-2009 11:00 PM	95.0	32.0

Figure 12. Tools in the framework

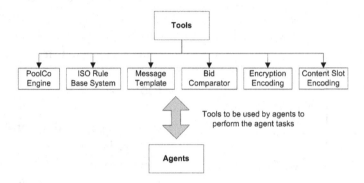

Content Slot Coding: This is a main encoding and encryption engine used for scrambling messages sent from one agent to other. The message is first serialized and then encoded using the Base64.jar which is the encoding file used for JADE serialization and transmitting sequences of bytes within an ACL Message. Further, it is encrypted using the OpenSSL package to apply strong encryption on the message which will be used in the RSA algorithm.

Coordination between Agents

The coordination between agents is an important issue in the MAS. In an energy market model, the agents coordinate (Koritarov, 2004; Krishna, & Ramesh, 1998; Krishna & Ramesh, 1998a) among themselves in order to satisfy the energy demand of the system accomplish with the distributed control of the system. The coordination strategy defines the common communication framework for all interactions between agents. Simple contract-net coordination is chosen for the processing of wholesale market because of its simplest coordination strategies. All discussions between agents are started simply by a requesting agent asking the other agents for a proposed contract to supply some commodity, and then awarding contracts from the returned proposals in a fashion that minimizes cost or fulfils some other goal. The disadvantage of simple contract-net coordination is only simple negotiation without allowing for

counter proposals. Effectively, the initiating agent has to pick from the presented contracts and cannot negotiate the price. The advantage of contract-net is that it distributes computing, allowing the specific agent which started a contract net process to be responsible for evaluating bids and deciding based on its own rules which contracts to accept. It also separates internal agent information from one another, since agents only communicate through the defined contract-net protocol and all calculations are done internally to each agent. Since the agents can change at every contract-net cycle, there is no dependence on a specific agent. A system with more complex negotiation might lead to lower costs for the system. However, simple contract- net is sufficient to demonstrate a distributed coordination framework.

A directory service allows agents to register themselves and publish their capabilities. By using a directory service, agents do not have to be aware of the other agents. For example, a load agent will look up sources in the directory every time it wishes to secure a new supply contract. This allows for agents to be added or removed from the system at any time since agents are included in contract-net negotiations once they register themselves with the directory service. The coordination layer that the approach defines is the strategic layer above the real time layer. Because of the time required for a contract-net interaction to complete, and since contracts are assigned in discrete time intervals, this coordination layer

Figure 13. Communication between the agents

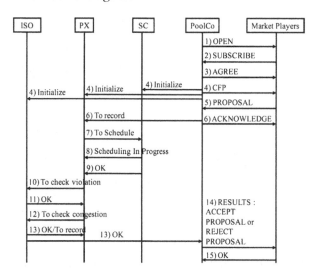

cannot address real time issues. The coordination layer allows for the distributed agents to plan how resources should be applied for satisfying demand. The actual operation of the system components self regulates through negative feedback since the system cannot produce more energy than is consumed. Figure 13 shows the overall communication between the agents in this simulation.

SIMULATION OF DEVELOPED SOFTWARE

The Multi-agent framework and generic service components implemented in the software are integrated to deploy a simulation application of modeling of restructured energy market. This simulation framework consist of four main states namely, agent world creation and initialization,

non-contractual binding and SSL communications, contractual binding communications and finalization and sealing of contracts. The general flow of the software simulation can be seen in Figure 14.

The Figure 15 shows multi-agent system launching. It is started via agent launch pad by which all the administrative agents such as PoolCo manager agent and its subordinate agents, security manager agent and its subordinate agents, and power system manager agent are launched. Buyer agents and seller agents are created and launched as static agents in a local machine.

After seller agents and buyer agents are created, they will execute their own thread to initialize their generation capacities, load requirements and bidding price by obtaining these values from a centrally held database in the simulation environment. When all the parameters of the agents are properly initialized, each agent will autono-

Figure 14. General flow of software simulation

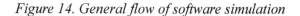

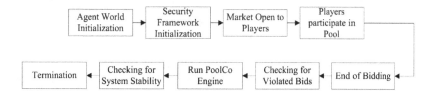

Figure 15. Initialization of the agent world

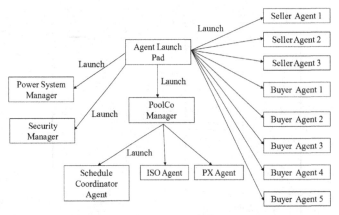

mously register itself with the DF as their first task. The process is illustrated in Figure 16.

As soon as the agents registers themselves with the DF, the agents will query the DF for a complete listing of agents and their services in the network using a certain search constraints. These search constraints are usually queries to the DF for agents with a certain types of services or agents with certain types of names. Sellers will send a query to the DF on all other buyer agents and PoolCo manager agent. Buyers will also end a query to the DF on all other seller agents and PoolCo manager agent. The DF will respond with listing of all agents that match their search constraints, and all the physical addresses of these agents.

With this information, agents will be able to autonomously contact them at their own thread of execution. After retrieving the necessary directory listing of various agents in the network, each agent will contact the security manager for allocation of ID keys to be used for encryption purpose and SSL algorithm engine as shown in Figure 17.

As soon as the agents registered for security services, all further communication on the network will be encrypted. When PoolCo is ready to communicate with all player agents, it will broadcast an OPEN message as shown in Figure 13. All the player agents who wish to take part in this round of bidding, will respond by sending a SUBSCRIBE message to subscribe PoolCo manager agent. PoolCo will close the subscription window after everyone in the network has subscribed or when the subscription date is expiry, whichever is earlier.

Figure 16. Registration and query of the agents

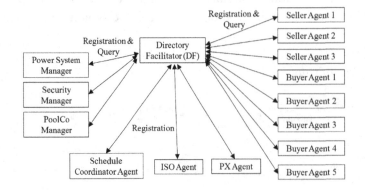

Figure 17. Role of the security manager in the agent world

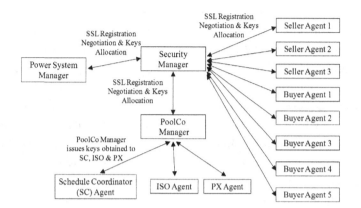

Once everyone in the network has subscribed PoolCo manager agent, the PoolCo manager agent issues an AGREE message to agree all agents who have signed up for the subscription service. This reply is to confirm their subscription. When the AGREE message arrives, player agents will stop their own internal execution of whatever task they are involving with (e.g. receiving updates from the DF for listing of agents in the network, manipulating input data from DF) to handle this newly arrived message and they will record this correspondence in their internal message history database. All message exchanges are recorded by each and every agent in their own internal message history database. After that, they will resume their operation at whatever they were doing before. At the same time, they will continue to listen for new messages.

After PoolCo manager agent sends out AGREE message to every agent who sent subscription message, PoolCo manager agent will also update its own internal message history database and proceed to prepare for a call for proposal broadcast. Once it prepared Call For Proposal (CFP), it will retrieve the list of subscribed agents in the network and send a CFP message to all the subscribers. After PoolCo manager agent sent CFP message, it will also send a message to ISO agent, PX agent and SC agent to initialize them and prepare for eminent auction and scheduling task. PX agent

and SC agent are the entities used to model the internal division of PoolCo management. PoolCo manager agent is the front-door communication entity for representing the bidding system.

CFP message will arrive at the player agents who have previously subscribed to the PoolCo service. Upon receiving this message, agents will stop their execution as before and handle this newly arrived message. The player agents prepare themselves for bidding if they are interested to participate in this round of bidding. In this stage, player agents will submit the formal bids to the PoolCo manager agent. PoolCo manager agent will process these bids and send the results to them. These submissions of bids and replies from PoolCo manager agent are legally binding contracts. Buyers who submitted bids to buy are legally obligated to buy the quantity of power at bided price. The same things are applied for sellers too. Agents, who are interested for submitting bids, will have access to their internal bidding records. They will prepare the necessary parameters like price and quantity of electricity to buy or offer in the market and the prepared parameters will be encoded as a bid object. When encoding is completed, they will send a PROPOSAL message to PoolCo manager agent with the bid object enclosed. PoolCo manager agent receiving up on the PROPOSAL message will re-directed these messages to PX agent for recording. PoolCo manager agent will only close the proposal window after everyone in the network has submitted their proposals or proposal window expiry date is due, whichever is earlier. The proposal expiry date is by default one day after PoolCo manager agent sent out its CFP message. After the proposal window is closed, PX agent will process the bids collected by a series of sorting and tagging. These whole set of data will be hashed into a hashtable. Then this hashtable will be sent to SC agent. At the same time, SC agent will send a message to the PoolCo manager agent to notify scheduling in progress.

SC agent has an algorithm which computes a data structure that represents the aggregated demand and the aggregated supply with respect to the price component using a rule based system. These sets of data will be processed to produce a single spot price at market equilibrium where the demand meets the supply. This price is called as the Market Clearing Price (MCP). It will also calculate the quantity of electricity transacted at this price. PX agent will also determine the successful buyer agents and seller agents in this round of bidding based on the MCP and quantity of electricity transacted. Then the whole set of data will be sent to ISO agent to check for violation of bidding as well as to check for congestion of scheduling. If any bidding is violated or the scheduling is congested, ISO will do the necessary actions. On the other hand, the whole set of data comprising of MCP, quantity of electricity transacted, list of successful buyer agents and seller agents, list of unsuccessful buyer agents and seller agents will be sent to PoolCo manager agent.

After receiving this data, PoolCo manager agent extracts out the relevant information and sends to power system manager agent so that power system manager can update the power system state. PoolCo will also extract the list of successful bidders from the set of data and sends a ACCEPT PROPOSAL message to successful bidders embedded with details of the successful bids. PoolCo will also extract the list of unsuccessful bidders from the data and sends a REJECT

PROPOSAL message to unsuccessful bidders. All bidders will be notified of their bidding outcomes at the end of every bidding round.

Agents, who receive an ACCEPT PROPOSAL message, will record their successful bid object and update their internal records. Then they send an OK message to the PoolCo manager agent to acknowledge the contract. Agents, who receive a REJECT PROPOSAL message, will record their unsuccessful attempt and make changes to their internal records.

This whole process is a round of bidding in the PoolCo model for one slot. In case of day-ahead market, it is for one hour of the 24 hour slots. Agents usually submit a complete scheduling of 24 bids, representing their bids for the day-ahead market.

RELIABILITY CHECKING OF MICROGRID

Once the market is simulated, before the scheduling is proposed, the stability and reliability of the power network is checked using power world simulator in order to ensure that the scheduling does not undermine the stability and reliability of the system. Power world simulator is a commercial power system simulating package based on comprehensive, robust power flow solution engine which is capable of efficiently solving a system up to 100000-bus problems. It also allows the user to visualize the system through the use of animated diagrams providing good graphical information about the technical and economic aspects of the network. A snapshot of power world simulator is shown in Figure 18. It has several optional add-ons. OPF and SimAuto add-ons are integrated in this software development.

The optimal power flow (OPF) provides the ability to optimally dispatch the generation in an area or group of areas while simultaneously enforcing the transmission line and interface limits. The advantages of this OPF over other commercially available optimal power flow packages are

Figure 18. A snapshot of the distributed power system

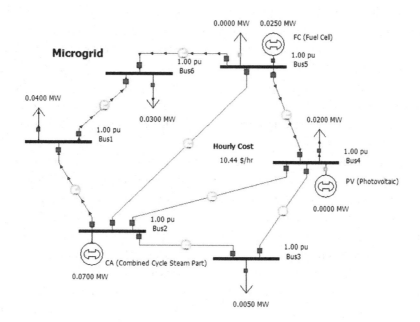

its ability to display the OPF results on system one-line diagrams and contour the results for ease of interpretation, and the ease with which the users can export the OPF results to a spreadsheet, a text file, or a power world AUX file for added functionality. SimAuto is an automated server that enables user to access the functionalities from a program written externally by Microsoft Com-

Figure 19. Demonstration of successful implementation of MAS

ponent Object Model (COM) technology. Even though Java does not have COM compatibility, Java integrates the Java Native Interface (JNI) which is a standard programming interface between Java programming and COM objects. JNI allows Java virtual machine to share a process space with platform native code.

If the schedule generated for the microgrid results in congestion, ISO would employ the power world simulator to mitigate congestion. The purpose of the OPF is to minimize the cost function by changing system controls and taking into account both equality and inequality constraints which are used to model the power balance constraints and various operating limits. It functionally combines the power flow with economic dispatch. In power world simulator, the optimal solution is being determined using linear programming. Once congestion has been mitigated, the new network schedule and relevant network information will be extracted from the power world simulator.

RESULTS AND DISCUSSION

Development and implementation of multi-agent application to restructured energy markets is pre-

sented in this chapter. The developed multi-agent application software simulates the restructured energy markets with accurate results. Further, this is a testament of the multi-agent framework implementation and effectiveness of dynamic modeling of multi-agent environment where internal tasks of each agent are executed concurrently with external inputs from the agent world. Successful deployment of the application software coupled with high robustness indicates the relevance and operational level of multi-agent system based application software development. User can use the software for any size of power system by defining the number of agents in the system and inserting the associated information.

The Figure 19 shows a demonstration of developed software simulation. The Remote Monitoring Agent (RMA) console can be run in the JADE runtime environment where developed agents in the frame work can be monitored and controlled. The other graphical tools such as dummy agent, sniffer agent and introspector agent, which are used to monitor, debug and control the MAS programming, can be activated from RMA. In the figure, the sniffer agent is activated and the successful implementation of agents' communication is observed.

Figure 20. Excess demand at the MCP

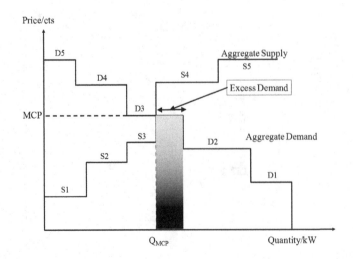

Figure 21. Excess supply at the MCP

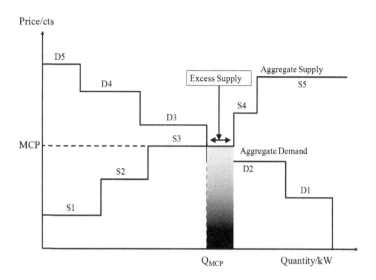

Several different scenarios of double sided bidding PoolCo market are simulated and the scenarios are defined as follows: scenario 1 is defined for a case where excess demand is available at the MCP which is illustrated in Figure 20; scenario 2 is defined for a case where excess supply is available at the MCP which is illustrated in Figure 21; and scenario 3 is defined for a case where supply and demand are matched at the MCP as illustrated in Figure 22.

Table 2 shows the above scenarios numerically. In scenario 1, at the market equilibrium, the bidding quantity of Load 1 is 10kW whereas the successful market output is only 5kW. Therefore additional 5kW power is necessary for Load 1. Here, the excess demand of 5kW is available at

Figure 22. Perfect supply and demand matching at the MCP

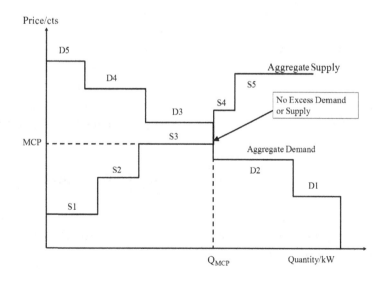

Table 2. Results of different scenarios

Agents	Scenario 1				Scenario 2				Scenario 3			
	Input		Output		Input		Output		Input		Output	
	P	Q	P	Q	P	Q	P	Q	P	Q	P	Q
Pgen1	11	70	11	70	11	**70**	11	**65**	11	70	12	70
Pgen2	12	20	0	0	12	20	0	0	12	20	0	0
Pgen3	10	25	11	25	10	35	11	35	10	20	12	20
Load1	11	**10**	11	**5**	11	10	11	10	11	10	0	0
Load2	12	20	11	20	12	20	11	20	12	**20**	12	**20**
Load3	10	10	0	0	10	10	0	0	10	10	0	0
Load4	14	30	11	30	14	30	11	30	14	30	12	30
Load5	13	40	11	40	13	40	11	40	13	40	12	40

P- Price in cents and Q- Quantity in kW.

the market equilibrium. In scenario 2, at the market equilibrium, the bidding quantity of Pgen 1 is 70kW whereas the successful market output is only 65kW. Therefore 5kW power is available at Pgen 1. Here, the excess supply 5kW is available at the market equilibrium. In scenario 3, at the market equilibrium, the bidding quantity of Load 2 is 20kW and the successful market output is also 20kW. Here, the supply and the demand are exactly matched at the market equilibrium.

The agent platform (JADE) used in the software development is a FIPA compliant platform. In the implementation of agent oriented designing, it strictly follows the FIPA standards compliance to ensure interoperability with future systems of FIPA standards as well. JADE is fully Java coded platform. It shows the complete cross platform portability on all the systems when tested on UNIX, Linux, Windows 95, 98, 2000, XP and Vista machines.

CONCLUSION

This chapter presents a multi-agent software development to simulate the ISO/PX operations for a restructured energy markets. This is done

with agent oriented programming methodology to deliver customized and powerful application software. The simulation of this software demonstrates the successful development and implementation of the multi-agent framework, the feasibility and effectiveness of a multi-agent platform to model the restructured energy markets, and the roles of ISO and PX in particular for carrying out the market operations.

This application is a fully cross platform, FIPA compliant software written in Java language. The application is made by various Java packages, giving future programmers to work with both readymade pieces of functionality and abstract interfaces of custom and application tasks. Further, the attractive features of Java, in particular its cross platform deployment, security policies and provisions for distributed computing through Remote Method Invocation (RMI) and sockets, have benefited this software development.

ACKNOWLEDGMENT

The funding for this project was received from SERC IEDS programme grant R-263-000-507-306.

REFERENCES

Bagnall, A. J., & Smith, G. D. (2005). A Multi agent Model of UK Market in Electricity Generation. *IEEE Transactions on Evolutionary Computation*, 522–536. doi:10.1109/TEVC.2005.850264

Budhraja, V. S. (2001). California's electricity crisis. *IEEE Power Engineering Society Summer Meeting*.

Dimeas, A. L., & Hatziargyriou, N. D. (2007). Agent based control of Virtual Power Plants. *International Conference on Intelligent Systems Applications to Power Systems*.

Hatziargyriou, N. D., Dimeas, A., Tsikalakis, A. G., Lopes, J. A. P., Kariniotakis, G., & Oyarzabal, J. (2005). Management of Microgrids in Market Environment. *International Conference on Future Power Systems*.

Jayantilal, A., Cheung, K. W., Shamsollahi, P., & Bresler, F. S. (2001). Market Based Regulation for the PJM Electricity Market. *IEEE International Conference on Innovative Computing for Power Electric Energy Meets the Markets* (pp. 155-160).

Koritarov, V. S. (2004). *Real-World Market Representation with Agents* (pp. 39–46). IEEE Power and Energy Magazine.

Krishna, V., & Ramesh, V. C. (1998). Intelligent agents for negotiations in market games. Part I. Model. *IEEE Transactions on Power Systems*, 1103–1108. doi:10.1109/59.709106

Krishna, V., & Ramesh, V. C. (1998a). Intelligent agents for negotiations in market games. Part II. Application. *IEEE Transactions on Power Systems*, 1109–1114. doi:10.1109/59.709107

Lasseter, R., Akhil, A., Marnay, C., Stephens, J., Dagle, J., Guttromson, R., et al. (2002, April). White paper on Integration of consortium Energy Resources. The CERTS MicroGrid Concept. CERTS, CA, Rep.LBNL-50829.

Logenthiran, T., Srinivasan, D., & Wong, D. (2008). Multi-agent coordination for DER in MicroGrid. *IEEE International Conference on Sustainable Energy Technologies* (pp. 77-82).

Praça, I., Ramos, C., Vale, Z., & Cordeiro, M. (2003). MASCEM: A Multi agent System That Simulates Competitive Electricity Markets. *IEEE International conference on Intelligent Systems* (pp. 54-60).

Rahman, S., Pipattanasomporn, M., & Teklu, Y. (2007). Intelligent Distributed Autonomous Power System (IDAPS). *IEEE Power Engineering Society General Meeting*.

Shahidehpour, M., & Alomoush, M. (2001). *Restructured Electrical Power Systems: Operation, Trading, and Volatility*. Marcel Dekker Inc.

Shahidehpour, M., Yamin, H., & LI Z. (2002). *Market Operations in Electric Power Systems: Forecasting, Scheduling, and Risk Management*. Wiley-IEEE Press.

Shrestha, G. B., Song, K., & Goel, L. K. (2000). An Efficient Power Pool Simulator for the Study of Competitive Power Market. *Power Engineering Society Winter Meeting*.

Sueyoshi, T., & Tadiparthi, G. R. (2007). Agent-based approach to handle business complexity in U.S. wholesale power trading. *IEEE Transactions on Power Systems*, 532–543. doi:10.1109/TPWRS.2007.894856

KEY TERMS AND DEFINITIONS

Multi-Agent System: A distributed network of intelligent hardware and software agents that work together to achieve a global goal.

Restructuring of Power System: Reform the vertically integrated utility monopoly power system to transform distributed control power system which provide competition and open access to all user in the interconnection.

PoolCo Market Model: One of the market models in restructured power system. PoolCo is a centralized marketplace that clears the market for buyers and sellers according to the bids of sellers and buyers.

Rule-Based System: One of the ways to store and manipulate knowledge to interpret information in a useful way.

Distributed Energy Resource: Distributed generation technology with distributed storage and controllable loads and their combination is referred to as Distributed energy resource.

Coordination Between Agents: In multi-agent systems, an agent usually plays a role with cooperative or completive behaviours with other agents. Therefore communication between agents is necessary in a multi-agent system.

Reliability of Power System: Concerns sufficient generation and transmission resources are available to meet projected demand and status of system after outages or equipment failures. Reliable power system operation must satisfy voltage constraints and power flows within thermal limits.

Section 6
Multi-Agent Learning

Chapter 13
Effects of Shaping a Reward on Multiagent Reinforcement Learning

Sachiyo Arai
Chiba University, Japan

ABSTRACT

The multiagent reinforcement learning approach is now widely applied to cause agents to behave rationally in a multiagent system. However, due to the complex interactions in a multiagent domain, it is difficult to decide the each agent's fair share of the reward for contributing to the goal achievement. This chapter reviews a reward shaping problem that defines when and what amount of reward should be given to agents. We employ keepaway soccer as a typical multiagent continuing task that requires skilled collaboration between the agents. Shaping the reward structure for this domain is difficult for the following reasons: i) a continuing task such as keepaway soccer has no explicit goal, and so it is hard to determine when a reward should be given to the agents, ii) in such a multiagent cooperative task, it is difficult to fairly share the reward for each agent's contribution. Through experiments, we found that reward shaping has a major effect on an agent's behavior.

INTRODUCTION

In reinforcement learning problems, agents take sequential actions with the goal of maximizing a time-delayed reward. In this chapter, the design of reward shaping for a continuing task in a multiagent domain is investigated. We use an interesting example, *keepaway* soccer (Kuhlmann, 2003; Stone, 2002; Stone, 2006), in which a team

tries to maintain ball possession by avoiding the opponent's interceptions. The keepaway soccer problem, originally suggested by Stone (2005), provides a basis for discussing various issues of multiagent systems and reinforcement learning problems(Stone, 2006). The difficulties of this problem are twofold, i.e., the state space is continuous and the sense-act cycle is triggered by an event, such as a keeper (learner) getting the ball. Since the learner selects a macro-action which

DOI: 10.4018/978-1-60566-898-7.ch013

requires a different time period, it is appropriate to model this problem as a semi-Markov decision process.

To our knowledge, *designing the reward function* has been left out of reinforcement learning research, even though the reward function introduced by Stone (2005) is commonly used. However, designing the reward function is an important problem (Ng, 2000). As an example, the following are difficulties of a designing reward measure for keepaway. First, it is a continuing task that has no explicit goal to achieve. Second, it is a *multiagent cooperative task*, in which there exists a reward assignment problem to elicit desirable teamwork. Because of these two features of keepaway, it is hard to define the reward signal of each keeper to increase the time of ball possession by a team. It should be noted that the reward for increasing each keeper does not always lead to increased possession time by a team.

In the case of a continuing task, we can examine a single-agent continuing task such as the *pole balancing task*, in which one episode consists of a period from the starting state to the failure state. If the task becomes a failure, a penalty is given, and this process can be used to evaluate teamwork and individual skills. In contrast, in the case of a *multiagent task*, which includes both a teammate and at least one opponent, it is hard to tell who contributes to the task. In a multiagent task such as keepaway, it is not always suitable to assign positive rewards to agents according to the amount of time cycles of each agent. Appropriately assigning an individual reward for each agent will have a greater effect on cooperation than sharing a common reward within the team. But, if the individual reward is not appropriate, the resulting performance will be worse than that after sharing a common reward. Therefore, assigning an individual reward to each agent can be a double-edged sword. Consequently, our focus is on assigning a reward measure that does not have a harmful effect on multiagent learning.

The rest of this chapter is organized as follows. In the next section, we describe the keepaway soccer domain, and discuss its features from the viewpoint of reinforcement learning. In Section 3, we introduce the reinforcement learning algorithm we applied and our reward design for keepaway. Section 4 shows our experimental results, including the acquired behavior of the agents. In Section 5, we discuss the applicability of our reward design on reinforcement learning tasks. We state our conclusion and future work in Section 6.

PROBLEM DOMAIN

Keepaway Soccer

Keepaway (Stone, 2002) is known as a subtask of RoboCup soccer, and it provides a great basis for discussion on important issues of multiagent systems. Keepaway consists of *keepers* who try to keep possession of the ball, and *takers* who attempt to take possession of the ball within a limited region. The *episode* terminates whenever takers take possession or the ball runs out of the region, and then players are reset for a new episode. When takers keep the ball for more than four cycles of simulation time, they are judged to have gained ball possession successfully.

Figure 1 shows the case of three keepers and two takers (3 vs. 2) playing in a region of size 20×20[m]. Here, keeper K_1 currently has the ball, K_2 is the closest to K_1, and K_3 is the next closest, and so on, up to K_n when n keepers exist in the region. In a similar way, T_1 is the closest taker to K, T_2 is the next closest one, and so on, up to T_m, when m takers exist in the region.

Macro-Actions

In the RoboCup soccer simulation, each player executes a primitive action, such as a turn (*angle*), dash (*power*) or kick (*power, angle*) every 100[ms].

However, it is difficult to employ these primitive actions when we take a reinforcement learning approach to this domain, because the parameters of actions and state variables have continuous values that make the state space very huge and complicated. To avoid the state representation problem, macro-actions proposed by Stone[1] are very helpful, and we employ the following macro-actions.

HoldBall(): Remain stationary while keeping possession of the ball in a position that is as far away from the opponents as possible.

PassBall(): Kick the ball directly towards keeper k.

GetOpen(): Move to a position that is free from opponents and open for a pass from the ball's current position.

GoToBall(): Intercept a moving ball or move directly towards a stationary ball.

BlockPass(k): Move to a position between the keeper with the ball and keeper k.

Since each macro-action consists of some primitive actions, it requires more than one step (100[ms/step]). Therefore, the keepaway task can be modeled as a *semi-Markov* decision process (SMDP). In addition, in the case of the RoboCup soccer simulation, it assumed that noise affects the visual information during the keepaway task. Considering the above features of the task model, the distance to an object from a player is defined in the following equations.

$$d' = Quantize\left(\exp\left(Quantize\left(\log\left(d\right), q\right), 0.1\right)\right)$$
(1)

$$Quantize\left(V, Q\right) = rint\left(V \,/\, Q\right)Q$$
(2)

Here, d' and d are the quantized value and the exact value of the distance, respectively. The function $rint(x)$ truncates an x number of decimal places. Parameter q is set as 0.1 and 0.01 when an object is moving and when it is fixed, respectively. The noise parameter ranges from 1.0 to 10.0. For example, when the distance between players is less than 10.0[m], the noise parameter is set as 1.0, and it is set as 10.0 when the distance is 100.0[m], the most noisy case.

Takers' Policy

In keepaway soccer, the two types of takers' policies shown in Figure 2 have been generally used to see the effects of the reward design. In the case of policy-(i), the takers select GoToBall() to interrupt the keeper's HoldBall() whenever either taker is a certain distance from the ball. In the case of policy-(ii), the taker who is nearest to the ball selects GoToBall(), and the other taker selects BlockPass(k). Because each taker plays a distinct role in policy-(ii) to intercept a pass, policy-(ii) is more strategic than policy-(i).

Figure 1. 3 vs. 2 keepaway task in 20 [m] × 20 [m]. (a) Object names; (b) Initial positions

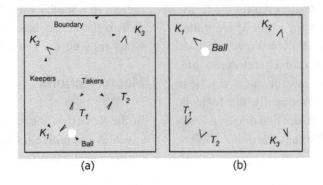

(a)   (b)

Issues of Reward Shaping

Figure 3 shows the task classification of testbeds for multiagent and reinforcement learning from the viewpoints of designing a reward problem. As we mentioned in the previous section, the difficulties of designing a reward are the lack of an explicit goal and the number of agents involved in a task.

First, in a *continuing task* there is no explicit goal to achieve, so the designer cannot tell when the reward should be given to the agent/s. Second, in a multiagent cooperative task, there exists a reward assignment problem of deciding the amount of reward be allotted to each agent for achieving desirable teamwork. The keepaway task contrasts with a *pursuit game*, which is a traditional multiagent research testbed that has an explicit common goal. Because the *pursuit game* is an episodic task, the reward just has to be given when hunters achieve their goal. In addition, it is easier to assign a hunter's (learner's) reward than the reward of a keeper (learner), because all four hunters definitely contribute to capture the prey. Therefore, keepaway is classified as a harder task in terms of designing a reward because we have no clues to define an explicit goal beforehand and to assign a reward to each agent.

From the aspect of a continuing task, we can refer to the case of single-agent continuing tasks, e.g., *pole balancing*, in which one episode consists of a period from the starting state to the failure state. In such a continuing task, an episode will always end with failure, and the penalty can help the design of a good reward measure to improve both teamwork and individual skills. In contrast, from the aspect of a *multiagent task* that includes both teammates and opponents, as in keepaway, it is hard to tell who contributed to keeping possession of the ball within the team. In other words, we should consider the case where some keepers contribute and others may not, or an opponent (taker) contributes by taking a ball. What has to be noted is that the episode of a multiagent's continuing task end with someone's failure. This problem has been discussed as a *credit assignment* in time-extended single-agent task and multiagent task domains (Agogino, 2004).

Though the *credit assignment* issue is closely related to our research here, we design a reward function to evaluate the "last" state-action pair of each agent in the SMDP (semi Markov Decision Process) domain, where each agent's action takes a different length of time, instead of assigning each state-action pair of each agent's whole state-action sequence. Here, we consider the *reward design* problem that consists of setting the *amount of reward value* and *time of reward assignment* so that we can optimize design issues of the reward measure in the multiagent learning process.

Figure 2. Takers' policy: 3 vs. 2 keepaway task in a 20 [m] × 20 [m] region. (i) Always GoToBall; (ii) GOTOBall and BlockPass

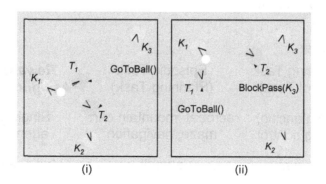

COMPONENTS OF REINFORCEMENT LEARNING

Learning Algorithm

Sarsa (λ) with replacing eligibility trace (Singh, 1996) has been generally employed as the multiagent learning algorithm because of its robustness within non-MDPs. Here, we take Sarsa (λ) with replacing eligibility trace as the learning algorithm of a keeper by following Stone's approach (Stone, 2005). The list below shows the learning algorithm that we use.

Initialize $\theta(i,a)$, $e(i,a)$

1. each episode:
2.   each SMDP step:
3.   keeper gets s_t from environment
4.   make up a feature vector $\mathbf{F}_t(i)$ from s_t
5. $\quad Q_a = \sum_{i=0}^{N1} \theta(i,a) F_t(i)$ for all a
6.   select action a_t using Q_a
7.   senses s_{t+1}, r_{t+1}
8.   make up a feature vector $\mathbf{F}_{t+1}(i)$ from s_{t+1}
9. $\quad Q_a = \sum_{i=0}^{N1} \theta(i,a) \mathbf{F}_{t+1}(i)$ for all a
10.   select action a_{t+1} using Q_a
11.   for all i
12.
$$\delta = r_{t+1} + \gamma Q(s_{t+1}, a_{t+1})\} - Q(s_t, a_t)$$
13. $\theta(i,a) = \theta(i,a)\ e(i,a)\delta$, for all a
14. $e(i,a) = \lambda e(i,a)$, for all a
15. $e(i,a) = \mathbf{F}_{t+1}(i)$, if $a = a_{t+1}$
16. $e(i,a) = 0$, if $a \neq a_{t+1}$
17    $s_t \leftarrow s_{t+1}$
18. if episode ends, go to line 2

State Representation: Tile-coding (Arai, 2006): For the state representation, we introduce a revised version of tile-coding (Sutton, 1998) as function approximations for continuous state variables, as shown in Figure 4. A linear tile-coding function approximation that have been introduced to avoid a state explosion within many reinforcement learning applications. It allows to divide the continuous state space by using the arbitral sizes of tiling, and be able to generalize the similar situations by using the multiple overlapping tilings. The number of tiles and each size of it are previously defined by the designer. This definition affects learning performance materially.

Figure 3. Classification of testbeds

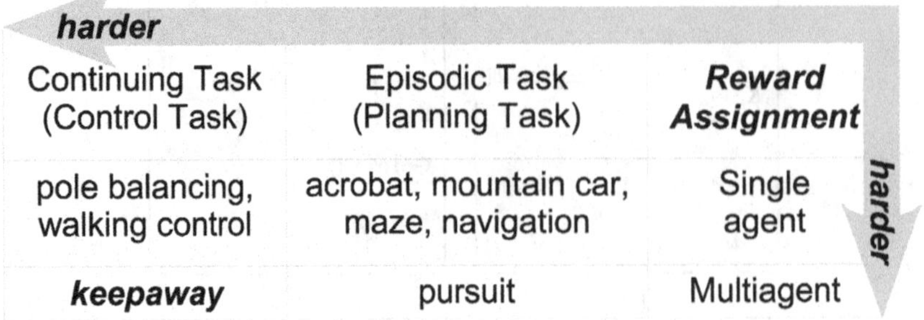

harder		
Continuing Task (Control Task)	**Episodic Task (Planning Task)**	***Reward Assignment***
pole balancing, walking control	acrobat, mountain car, maze, navigation	Single agent
keepaway	pursuit	Multiagent

harder

Figure 4. Function approximations for the state representation

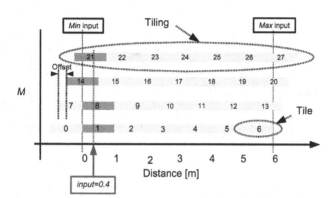

In Figure 4, from lines 4 to **8**, feature vector $\mathbf{F}_t$ and $\mathbf{F}_{t+1}$ are made by tile-coding. In our experiments, we use primarily single-dimensional tilings: 32 tilings are overlaid with a 1/32 offset, and each tiling is divided into n_{tile} segments. Each segment is called a *tile* and each state variable lies in a certain tile, which is called an *active tile*.

In keepaway, the distances between players and angles are represented as state variables. In the case of 3 vs. 2, the total number of state variables is 13, which consists of 11 for distance variables and 2 for angle variables. In our experiments, n_{tile} = 10 for each of the 11 distance variables, and n_{tile} =18 for each of the 2 angle variables.

Accordingly, the number of total tiles N_{tile} is 4672. Each value of the state variable i is represented as a feature vector, $\mathbf{F}(i)$, $\Sigma_{i=tile}^{N} \mathbf{F}(i) = 1$ and. Each value of i is shown as follows:

$$i = \begin{cases} \dfrac{n_{tile}}{N_{tile}} & (i\text{ th tile is active}) \\ 0 & (\text{otherwise}) \end{cases} \qquad (3)$$

Previous Work in Reward Design

For the reward r appearing in line 7 of Figure 2, r_s defined by Equation (4) (Stone, 2005) is commonly used for keepaway. Here, *CurrentTime* is the simulation time when the keeper holds the ball or the episode ends, and *LastActionTime* is the simulation time when the keeper selects the last action. We hereafter refer to function Equation (4) as r_s. In this approach, a reward is defined as the amount of time between the last action and the current time (or end time). That is, as the amount of time increases after taking the ball, the amount of reward given to the keeper also increases.

$$r_s = CurrentTime - LastActionTime \qquad (4)$$

This approach seems reasonable and proper. However, there are some problematic cases by using r_s, as shown in Figure 5, for example. In Figure 5(b), K_1, who is the current ball holder, gets a larger reward than the one in Figure 5(a). Consequently, the keeper passes the ball to the intercepting takers on purpose, because the ball will bounce back directly to K_1, and then K_1 is paid some reward for selecting this action. This action seems to yield a larger reward than other actions.

K_1 in Figure 5(d) gets a larger reward than the one in Figure 5(c) when the reward is defined by r_s. Consequently, the keeper is likely to pass to the teammate (keeper) who is in the farthest position to get a larger reward than pass to the nearer one. Although it seems reasonable, we cannot tell which one is the better strategy because it depends on the amount of noise and the keepers' skill.

Figure 5. Problematic situations. (a) K_1 takes HoldBall(); (b) K_1 takes PassBall(); (c) K_1->K_2->K_1; (d) K_1->K_3->K_1

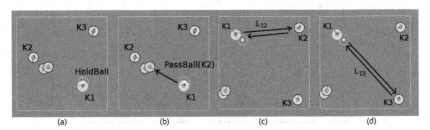

These examples show the difficulty of designing a reward function for a continuing task, as previously mentioned.

Reward Design for Collective Responsibility

In the well-known continuing task, *pole balancing*, the amount of reward is defined by Equation (5). The agent receives either 0 or -1 for a successful and failure condition, respectively. We follow the scheme that the reward makes an agent do what it takes to achieve, not how it achieves (Sutton, 1998). From this standpoint, the reward value has to be constant during a successful situation such as *pole balancing*, because we do not know which action achieves the task. While r_s by Equation (4) (Stone, 2005) provides a differentially

programmed reward for each step, as shown in Figure 6, it is usually difficult to say whether these values become an appropriate indicator to keep a successful situation. Therefore, we introduce a novel reward function based on a constant reward sequence (Figure 6) to reduce the harmful effects of the reward design on emerging behavior.

$$r = \begin{cases} -1 & \text{(under a failure condition)} \\ 0 & \text{(otherwise)} \end{cases} \qquad (5)$$

The major difference between the domain of *pole balancing* and keepaway is the number of agents involved. Unlike the single-agent case, in which one agent is responsible for the failure or success of the task, responsibility is diffused in the multiagent case. However, in the keepaway

Figure 6. Reward sequence in a continuing task

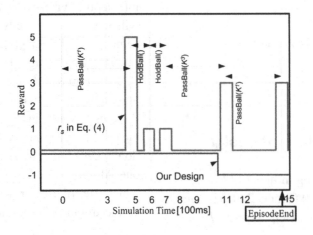

task, specifying which agent causes failure seems much easier than specifying which one contributes success. Therefore, we design reward functions of agent j, as shown in Equation (6), where t^j is given by $TaskEndTime\text{-}LastActionTime(K^j)$ when the task fails. Otherwise, the agent receives 0 constantly during a successful situation. The reason for sharing the same reward ($= 0$) among the agents in the successful situation is that we cannot tell which agent contributes to the task solely from the length of keeping time. Also, we know that if one agent keeps a task longer locally, good teamwork does not always result.

In our design, each agent receives the reward $f(t^j)$ according to its amount of time, i.e., from taking the ball ($LastActiontime$) to the end of the task ($TaskEndTime$) at the end of each episode. We make function $f(t^j)$ fulfill $f(t^j) \leq 0$ to reflect the degree of success for the agents' joint action sequences. Figure 7 shows some examples of function $f(t^j)$ by which the keeper who terminates the episode receives the largest penalty (i.e., the smallest reward). Here, the x and y-axes indicate the length of t^j defined by Equation (6) and the value of $f(t^j)$, respectively. In our

experiments, we use $f(t^j) = -\beta^{t^j}$ and $f(t^j) = -1/(t^j)$ as reward functions. For simplicity, we refer to $f(t^j) = -1/(t^j)$ as r_f in the following sections.

$$r(t^j) = \begin{cases} f(t^j) & \text{(under failure condition)} \\ 0 & \text{(otherwise)} \end{cases} \quad (6)$$

$$t^j = TaskEndTime \text{ - } LastActionTime(K^j)$$

EXPERIMENT

In this section, we show the empirical results in the keepaway domain. The learning algorithm is shown in Figure 4, and the parameters are set as $\alpha = 0.125$, $\gamma = 0.95$, $\lambda = 0$, and $\varepsilon = 0.01$ for ε-greedy. For the noise parameter q, mentioned in Section 2.2, we set q as 10^{-5}, which is the same setting used by Stone (Stone, 2005) to represent the behavior of a noise-free environment.

Figure 8 shows the learning curves of the five different reward functions introduced in Figure 7. Here, the x- and y-axes indicate the length of training time and length of keeping time, respectively. We plot the moving average of 100 episodes.

Figure 7. Reward function under different β

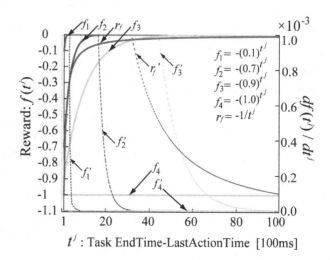

t^j : Task EndTime-LastActionTime [100ms]

Figure 8. Learning curve under the various reward functions (moving average of 100 episodes) against takers' policy (i)

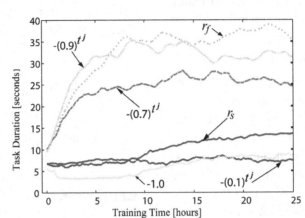

Performances

In the case of $f(t^j) = 1$, which is the same reward function given in the *pole balancing* problem, the performance declines in the early learning stage, as shown in Figure 8. The major reason for the decline is multiagent learning problems, such as *reward assignment* and simultaneous learning.

By giving -1 to all keepers under failure conditions, keepers selecting both appropriate and inappropriate actions equally receive the penalty of -1. This causes a harmful effect on learning, especially in the initial stage. However, after considerable learning, all keepers learned better actions. We find that the cases of $\beta = 0.1$ and $\beta = 1.0$ have similar curves in Figure 8, though each value of $f(t^j)$ (Figure 7) is totally different. The likely explanation for this similarity lies in the domain of $f(t^j)$. When *TaskEndtime* $t^j > 1$, the case of $\beta = 0.1$ reaches 0, and the case of $\beta = 1.0$ reaches 1, as shown in Figure 7. This indicates that all keepers receive the same reward value, and the reward assignment has harmful effects in both cases.

Figure 9 shows the learning curves for two cases with different reward functions. One is r_s and the other is $r_f = -1/T$. Because r_f shows the best performance of the keepers, we focus on this reward function, which is based on the time of failure. Figure 7 shows the episode duration of

keepers who experienced 25 hours of keepaway training. In Figure 7, keepers who learned by r_s possess the ball for up to approximately 100 seconds, while keepers who learned by r_f possess the ball for more than 300 seconds.

Emerging Behavior

Effect of Reward Design

We compare the acquired behaviors of both cases, r_s and r_f after 25 hours of training. One of the notable differences between the two cases is the timing of the pass. Figure 11(a) and (b) show the behaviors of K_1 reinforced by r_s and r_f respectively. The keeper reinforced with r_f shown in (b) does not pass the ball until the takers become quite close to him. Whereas, the keeper with r_s seems to pass regardless of the takers' location. To examine the difference of pass timing between (a) and (b), we compare the distance from the keeper having the ball to the nearest taker of each reward function. The distance in the case when using r_f is approximately 2 [m] shorter than that in the case using r_s, as shown in Table 1. The behavior (i.e., keeper holds the ball until takers becomes quite close to the keeper) means that keepers often select HoldBall(). Then, we focused on the change of the pass frequency within 1 seconds halfway through the learning process, as shown in Figure 12. Table

Figure 9. Learning curves with the best-performance function and the existing function (moving average of 1000 episodes) against takers' policy (i)

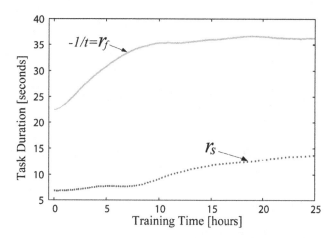

2 shows the pass frequency after learning. Here, we find that the frequency of passing a ball was smaller for r_f than for r_s.

Effect of Takers' Policy

Here, we discuss the emerged behavior from the viewpoint of the takers' policies introduced in Figure 2(i) and (ii). It seems that the emerged behavior shown in Figure 11(a) is especially effective against the takers with policy-(i). Because both takers always select GoToBall(), keeper K_2

and K_3 are always free from these takers. Thus, to examine the availability of our reward function, r_p in the different situation where the takers have a more complicated policy, we apply our reward function to the situation where takers act with policy-(ii). Figure 13 shows the comparison between the two learning curves with our keepers using r_f and r_s.

When the takers act with policy-(ii), the performance of keepers with r_s is worse than for takers with policy-(i). As mentioned in Section 3.2, reward function r_f has adverse effects on

Table 1. Pass timing: distance to the nearest taker from the keeper with the ball after 25 hours

	Distance [m]	
	Taker's policy-(i)	Taker's policy-(ii)
r_s	5.60	3.50
r_f	7.44	7.38

Table 2. Pass frequency: (number of passes during 1 second)

	Frequency [times/second]	
Reward function	Takers' policy-(i)	Takers' policy-(ii)
r_f	1.4 ± 0.1	2.1 ± 0.2
r_s	2.3 ± 0.3	2.5 ± 0.2

Figure 10. Results after 25 hours learning, 0-180 episode duration

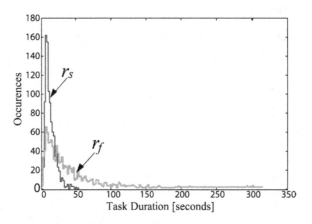

Figure 11. Pass timing. (a) r_s: Equation(4); (b) $r_f = -1/t_j$

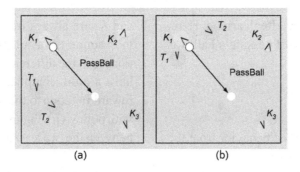

Figure 12. Pass frequency halfway through learning

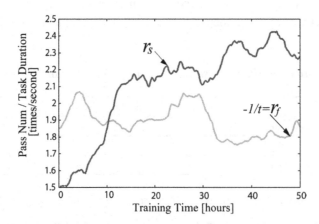

learning. Also, when the takers act with policy-(ii), r_f makes keepers possess the ball longer than function r_s does. Figure 14 shows the results of the episode duration after keepers experienced 25 hours of training against takers acting with policy-(ii). We found that keepers reinforced by r_f could possess the ball at least twice as long as keepers reinforced by r_s. However, episode dura-

Figure 13. Learning curves with the best reward function and the existing function (moving average of 1000 episodes):against takers' policy (ii)

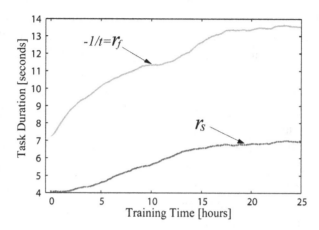

Figure 14. Results after 25 hours learning, 0-350 episode duration

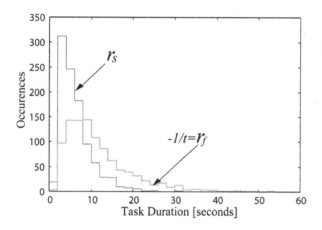

tion becomes shorter than that against takers with policy-(i), shown in Figure 2(i), because of the sophisticated takers' policy. There is not large difference of pass frequencies between the cases of r_f and r_s because the takers play distinct roles by using policy-(ii); in this case, one of the takers can immediately reach the keeper who receives the ball, and so the keeper with the ball must pass the ball as soon as possible. Therefore, the pass frequency increases. As for the emerged behavior, we find the same tendency (i.e., keepers do not pass the ball until the takers become closer to K_1) for both takers' policy-(ii) and for takers' policy-(i), as shown. in Table 1.

DISCUSSION

This section presents some of the problems in designing the function for the keepaway task through experiments. Since keepaway is a continuing and multiagent task, it is hard to decide when and what amount of reward should be given to the learners. From the aspect of a continuing task, we refer to the case of the single-agent continuing task, pole balancing, where failure terminates the task. However, because keepaway is a multiagent task, simultaneous learning problems occur in the early learning stage and we must consider getting high-performance cooperative behavior. The amount of reward that is given to the agent is defined by Equation (5). The agent receives -1 or 0 for the failure or success condition, respectively.

The difference between pole balancing and keepaway is the number of agents involved. Unlike the single-agent case, in which one agent is responsible for the failure or success of the task, responsibility is diffused in the multiagent case. However, in the keepaway task, specifying the agent causing a failure seems much easier than specifying the agent contributing to the success. Therefore, we design reward functions for the keepaway task so that a keeper who terminates the episode receives a larger penalty (i.e., smaller reward). Table 3 shows the comparison among three cases of keepaway (hand-coded, and two reward functions). Though the empirical results show that keepers using our function can possess the ball for approximately three times longer than those hand-coded and learned by the other reward function (Stone, 2005), the reason for this high performance has not been qualitatively analyzed yet.

First, we discuss the reward functions that we introduced. We introduce *T=TaskEndTime-Last-ActionTime* and give -1/T to the agent when the task fails. Otherwise, the agent receives 0 constantly during a successful situation, as mentioned in Section 3.4. The reason for sharing the same reward (= 0) among agents in the successful situation is that we cannot identify which agent contributes to the task solely by the length of keeping time. Since it is not always good for a team when one agent keeps a task longer locally, we do not introduce the predefined value of each agent's reward individually. In our design, each agent receives the reward $f(t^j)$ according to its amount of time from taking the ball (*LastAction-time*) to the end of the task (*TaskEndTime*) at the end of each episode. For the introduced reward functions, $r_f = -1/t^j$ provides relatively better performance than that of the other functions. Though function $f(t^j) = 0.7^{t^j}$ has a similar curve to r_f when $t^j < 20$, as shown in Figure 7, it doesn't perform as well as the case of r_f. The main reason for this result is due to the value range of T. The range of T is always larger than 20, and so the similarity in $T < 20$ does not have much effect on the performance.

Second, the keeper with r_f passes the ball more frequently than the keepers with r_s in the earlier stage of learning, as shown Figure 12. Because the keeper with r_s can receive some reward when selecting HoldBall in the early stage, the keeper tends not to pass the ball so many times to the other keeper. Meanwhile, our keepers reinforced by r_f do not receive any penalty when they pass the ball; that is, they receive a penalty only when they are intercepted or miss the pass. So, our keepers are not afraid to pass to other keepers. In the middle and late learning stages, the keepers with r_s pass the ball frequently because they experience a larger reward using PassBall(k) than

Table 3. Comparison of average possession times (in simulator seconds) for hand-coded and learned policies against two types of takers in region 20 [m] Ã × 20 [m].

	Keep Time [seconds] ($\pm 1\sigma$)	
Reward function	Takers' policy-(i)	Takers' policy-(ii)
r_f	35.5 ± 1.9	14.0 ± 2.3
r_s	14.2 ± 0.6	7.0 ± 0.6
Hand-coded	8.3 ± 4.7	-

using HoldBall. However, the pass frequency of our keepers decreases because they experience having their ball intercepted or missing the pass after considerable training.

Third, as we described in Section 2.2, the visual information contains some noise. The passed ball often fails to reach the intended destination because of the noise, and so the noise has a large effect on the emerging behavior. Since the action of passing carries some probability of missing or being intercepted, the frequency of the pass of our keepers learned with r_f becomes small. This is considered reasonable and proper behavior in a noisy environment and against takers' policy-(i), shown in Figure 2(i).

Fourth, we look at the effects of the takers' policy. We found in Figure 13 that our keepers with r_f against takers' policy-(ii) (Figure 2(ii)) possess the ball less than in the case against takers' policy-(i) (Figure 2(i)). It seems that, as the frequency of the pass increases, the duration of the episode decreases. We found in Table 2 that the frequency of the pass becomes smaller when our keepers learn about takers' policy-(ii) in comparison with takers' policy-(i).

Last, we discuss the macro-actions we currently use. As the pass frequency increases, keepers do not have enough time to move to a position that is free from the opponent and cannot clear a path to let the ball pass from its current position. Consequently, the probability of missing a pass seems to increase. This problem might be resolved by introducing more sophisticated macro-actions such as **GetOpen()**, and so forth.

CONCLUSION

In this chapter, we discuss the issue of the revised version of tile-coding as state representation and reward design for multiagent continuing tasks, and introduce an effective reward function for the keepaway domain. Though our experimental results show better performance than that of pre-

vious studies of keepaway (Stone 2006), we are not yet able to provide a theoretical analysis of the results. At present, we have been examining the problem peculiar to a continuing task that terminates at failure, such as a pole balancing, and a reward assignment within a *multiagent task* in which simultaneous learning takes place. For the continuing task case, we show that a certain penalty causes an agent to learn successfully. Whereas, for the multiagent case, we avoid the harmful effect of the agents' simultaneous learning by parameter tuning. It is necessary to analyze the breakdown of the reward, such as which agent gets a greater penalty, and which gets a lesser penalty, when designing a multiagent system.

REFERENCES

Agogino, A. K., & Tumer, K. (2004). Unifying Temporal and Structural Credit Assignment Problems. In *Proceedings of the Third International Joint Conference on Autonomous Agents and Multi-Agent Systems* (pp. 980-987).

Arai, S. & Tanaka, N. (2006). Experimental Analysis of Reward Design for Continuing Task in Multiagent Domains. *Journal of Japanese Society for Artificial Intelligence, in Japanese, 13*(5), 537-546.

Kuhlmann, G., & Stone, P. (2003). Progress in learning 3 vs. 2 keepaway. In *Proceedings of the RoboCup-2003 Symposium*.

Ng, A. Y. Ng & Russell, S. (2000). Algorithms for Inverse Reinforcement Learning. In *Proceedings of 17th International Conference on Machine Learning* (pp. 663-670). Morgan Kaufmann, San Francisco, CA.

Singh, S. P., & Sutton, R. S. (1996). Reinforcement Learning with Replacing Eligibility Traces. *Machine Learning, 22*(1-3), 123–158. doi:10.1007/BF00114726

Stone, P., Kuhlmann, G., Taylor, M. E., & Liu, Y. (2006). Keepaway Soccer: From Machine Learning Testbed to Benchmark . In Noda, I., Jacoff, A., Bredenfeld, A., & Takahashi, Y. (Eds.), *Robo-Cup-2005: Robot Soccer World Cup IX*. Berlin: Springer Verlag. doi:10.1007/11780519_9

Stone, P., & Sutton, R. S. (2002). Keepaway Soccer: a machine learning testbed . In Birk, A., Coradeschi, S., & Tadokoro, S. (Eds.), *RoboCup-2001: Robot Soccer World Cup V* (pp. 214–223). doi:10.1007/3-540-45603-1_22

Stone, P., Sutton, R. S., & Kuhlmann, G. (2005). Reinforcement Learning for RoboCup Soccer Keepaway. *Adaptive Behavior, 13*(3), 165–188. doi:10.1177/105971230501300301

Sutton, R., & Barto, A. G. (1998). *Reinforcement Learning: An Introduction*. Cambridge, MA: MIT Press.

ADDITIONAL READING

Agogino, A. K., & Tumer, K. (2004). Efficient evaluation functions for multi-rover systems. In *Proceedings of the Genetic and Evolutionary Computation Conference (GECCO-2004)* (pp. 1-12).

Agogino, A. K., & Tumer, K. (2005). Multi-agent reward analysis for learning in noisy domains. In *Proceedings of the fourth international joint conference on Autonomous agents and multiagent systems* (pp. 81-88).

Agogino, A. K., & Tumer, K. (2008). Efficient Evaluation Functions for Evolving Coordination. *Evolutionary Computation, 16*(2), 257–288. doi:10.1162/evco.2008.16.2.257

Erez, T., & Smart, W. D. (2008). What does shaping mean for computational reinforcement learning? In *Proceedings of 7th IEEE International Conference on Developing and Learning* (pp. 215-219).

Grzes, M., & Kudenko, D. (2008). Multigrid Reinforcement Learning with Reward Shaping. (. *LNCS, 5163*, 357–366.

Konidaris, G., & Barto, A. (2006). Autonomous shaping: Knowledge transfer in reinforcement learning. *Proceedings of the 23rd international conference on Machine learning* (pp. 489-496).

Marthi, B. (2007). Automatic shaping and decomposition of reward functions. In *Proceedings of the 24th International Conference on Machine Learning* (pp. 601–608).

Mataric, M. J. (1994). Reward functions for accelerated learning. In *Proceedings of the 11th International Conference on Machine Learning* (pp. 181-189).

Ng, A., Harada, D., & Russell, S. (1999). Policy invariance under reward transformations: Theory and application to reward shaping. In *Proceedings of the 16th International Conference on Machine Learning* (pp. 278-287).

Taniguchi, T. & Sawaragi, T. (2006). Construction of Behavioral Concepts through Social Interactions based on Reward Design: Schema-Based Incremental Reinforcement Learning. *Journal of Japan Society for Fuzzy Theory and Intelligent Informatics, 18*(4), (in Japanese), 629-640.

Tumer, K., & Agogino, A. K. (2006). *Efficient Reward Functions for Adaptive Multi-Rover Systems Learning and Adaptation in Multi Agent Systems* (pp. 177–191). LNAI.

Wolpert, D. H., & Tumer, K. (2001). Optimal payoff functions for members of collectives. *Advances in Complex Systems, 4*(2/3), 265–279. doi:10.1142/S0219525901000188

Wolpert, D. H., Tumer, K., & Bandari, E. (2004). Improving search algorithms by using intelligent coordinates. *Physical Review E: Statistical, Nonlinear, and Soft Matter Physics, 69*, 017701. doi:10.1103/PhysRevE.69.017701

KEY TERMS AND DEFINITIONS

Reward Shaping: A technique to make reinforcement learning agent converge to the successful policy for rational behavior.

Continuing Task: Has no explicit goal to achieve, but task requires agent to keep the desirable state(s) as long as possible.

RoboCup Soccer: See http://www.robocup.org/

Keepaway: Consists of *keepers* who try to keep possession of the ball, and *takers* who attempt to take possession of the ball within a limited region. The *episode* terminates whenever takers take possession or the ball runs out of the region, and then players are reset for a new episode.

ENDNOTE

[1]   Learning To Play Keepaway: http://www.cs.utexas.edu/users/AustinVilla/sim/keepaway/

Chapter 14
Swarm Intelligence Based Reputation Model for Open Multi Agent Systems

Saba Mahmood
School of Electrical Engineering and Computer Science (NUST-SEECS), Pakistan

Azzam ul Asar
Department of Electrical and Electronics Eng NWFP University of Engineering and Technology, Pakistan

Hiroki Suguri
Miyagi University, Japan

Hafiz Farooq Ahmad
School of Electrical Engineering and Computer Science (NUST-SEECS), Pakistan

ABSTRACT

In open multiagent systems, individual components act in an autonomous and uncertain manner, thus making it difficult for the participating agents to interact with one another in a reliable environment. Trust models have been devised that can create level of certainty for the interacting agents. However, trust requires reputation information that basically incorporates an agent's former behaviour. There are two aspects of a reputation model i.e. reputation creation and its distribution. Dissemination of this reputation information in highly dynamic environment is an issue and needs attention for a better approach. We have proposed a swarm intelligence based mechanism whose self-organizing behaviour not only provides an efficient way of reputation distribution but also involves various sources of information to compute the reputation value of the participating agents. We have evaluated our system with the help of a simulation showing utility gain of agents utilizing swarm based reputation system. We have utilized an ant net simulator to compute results for the reputation model. The ant simulator is written in c# and utilizes dot net charting capabilities to graphically represent the results.

DOI: 10.4018/978-1-60566-898-7.ch014

INTRODUCTION

Interactions in Human Societies

Agent based systems share a number of characteristics with human societies in terms of interactions, communications and various other factors. Human beings create a perception about another human based upon several factors. For example if someone bought a product from a seller and that product proves to be good enough on various parameters, the buyer would rate that seller as good as compared to any other available seller of the same product. So, in future shopping the buyer will keep this rating in mind before making a decision from whom to buy and not to buy. But if a buyer is to experience for the first time the interaction with certain seller and has no prior knowledge, then knowledge of peers can be utilized in this scenario to rate a particular seller. For example, if Mr. X bought a product from a seller and was satisfied, another buyer who has no knowledge about the product of the seller can use this information. Thus human beings are using notion of trust and reputation of other humans with whom they want to interact.

Trust in Computer Systems

Multiagent systems (MAS) are composed of individual agents working towards a certain goal. These agents need to interact with one another in order to achieve the goal. However in open systems it's very difficult to predict the behaviour of the agents. *Open systems* are characterized by high degree of dynamism. Thus interactions among the agents require some degree of certainty. The notion of *Trust* is used in recent years in the field of computer systems to predict the behaviour of the agents based upon certain factors. Another term *Reputation* is also being used and sometimes both trust and reputation are used interchangeably but they do differ from one another. Reputation is defined as a collected and processed information about one entity's former behaviour as experienced by others while Trust is the measure of willingness to proceed with action (decision) which places parties at risk of harm and based on an assessment of the risks, rewards and reputation associated with all the parties in involved in the given situation.

Several computational and empirical models have been suggested in recent years trying to address various issues of open multiagent system. Earlier work involved model and mechanism developed for centralized multiagent systems. However, with the evolution of distributed computing of decentralized nature, those models proved to be incomplete in addressing certain issues. Models like REGRET and FIRE take in to account various sources of information to compute the final reputation value of the agent under observation designed specifically to address issues of open MAS.

Trust

Trust is a fundamental concern in open distributed systems. Trust forecast the outcome of interaction among the agents in the system. There are basically two approaches to trust in multiagent systems; firstly trust requires that the agents should be endowed by some knowledge in order to calculate the trust value of the interacting agent. A high degree of trust would mean most probable selection of that agent for interaction purposes. Second approach to trust revolves around the design of protocols and mechanisms of interaction i.e. the rules of encounter. These interaction mechanisms need to be devised to ensure that those involved can be sure they will gain some utility if they rightly deserve it and malicious agent cannot tamper with the correct payoff allocation of the mechanism(Schlosser, Voss, Bruckner 2004). This definition of reputation and trust above depicts that for trust management, reputation is required. Reputation creation and distribution is an issue especially in case of open multiagent systems.

Thus we can say that the two approaches to trust come under reputation creation and distribution.

Trust has evolved as the most recent area in the domain of Information systems. A wide variety of trust and reputation models have been developed in the past few years. Basically we have divided the models in to two areas, centralized and decentralized.

Centralized Reputation Mechanism

Online electronic communities manage reputation of all the users in a centralized manner, for example eBay and SPORAS.

eBay Model

It is implemented as a centralized system where users can rate the interactions of other agents in the past and also leave some textual comments about their behaviour. For example in eBay, an interaction a user can rate its partner on the scale −1, 0 or +1 that means positive neutral or negative ratings respectively. These ratings are stored centrally and the reputation value is computed as the sum of those ratings over six months(Huynh, Jennings and Shadbolt 2006).

SPORAS

SPORAS (Maes & Zacharia, 2000) extends the online reputation systems by introducing a new method for rating aggregation. Specifically instead of storing all the ratings, each time a rating is received it updates the reputation of the involved party using the following algorithm:

1.   New users start with a minimum reputation value and they build up reputation during their activity on the system.
2.   The reputation value of a user never falls below the reputation of a new user.
3.   After each transaction the reputation values of the involved users are updated according to the feedback provided by the other parties, which reflect their trustworthiness in the latest transaction.
4.   Users with very high reputation values experience much smaller rating changes after each update.
5.   Ratings must be discounted over time so that the most recent ratings have more weight in the evaluation of a user's reputation.

SPORAS provide a more sophisticated model as compared to previous model. However, it is designed for a centralized system that is unable to address the issues related with open MAS.

Decentralized Systems

In decentralized systems, each agent can carry out trust evaluation itself without a central authority. Following section gives details of some of the decentralized models.

Jurca and Faltings

Jurca and Faltings introduce a reputation system (Jurca & Faltings 2003) where agents are incentivised to report truthfully about their interactions results. They define a set of broker agents called R agents whose tasks are buying and aggregating reports from other agents and selling back reputation information to them when they need it. All reports about an agent are simply aggregated using the averaging method to produce the reputation value for that agent. Though the agents are distributed in the system, each of them collects and aggregate reputation reports centrally.

TRAVOS

TRAVOS(Huynh, Jennings and Ramchurn 2004) is a model of trust and reputation that can be used to support informed decision making to assure good interactions in a Grid environment that supports virtual organizations. If a group of agents are to form a Virtual Organization, then it's important for them to choose the most appropriate partner. This model is built upon the probability theory. TRAVOS equips an agent (the trustier) with three methods for assessing the trustworthiness of another agent (the trustee).

First, the trustier can make the assessment based on the direct interactions it had with the trustee. Second, the trustier can assess the trustworthiness of the trustee based on the opinions provided by others in the system. Third, the trustier can assess the trustworthiness of another based on a combination of the direct interactions with and the reputation of the trustee. TRAVOS considers the behaviour of an agent as a probability that it will participate in a successful interaction and a probability that it will perform an unsuccessful interaction (untrustworthy behaviour). This abstraction of agent behaviour means that in this model the outcome of an interaction is a binary value (successful or not).

The FIRE Model

The FIRE model uses wide variety of sources to compute the reputation value of an agent. These sources include IR, RR, WR and CR. IR is based on an agent's personal experience, while WR and CR are reputation information reported by neighboring agents. RR is a role-based reputation and involves some rule-based evaluation of an agent's repute. The model has considered a scenario of producers and consumers with an assumption of only one type of service availability. The producers (Providers) are categorized as good, ordinary, intermittent and bad, depending upon the quality of service they are rendering. The Intermittent and Bad providers are the most random ones. If consumer agent needs to use the service, it can contact the environment to locate nearby provider agents. The consumer agent will then select one provider from the list to use its services. The selection process depends upon the reputation model of the agent. Consumer's

satisfaction is calculated in terms of Utility Gain (UG). UG for bad and intermittent providers is less then the good and average ones. The model has incorporated the factor of dynamism in order to address the changing environment of open MAS. The dynamism in FIRE is based on population of agents that cannot exceed a particular threshold and location of agents with an assumption that some agents cannot change their location.

SWARM INTELLIGENCE

Swarm intelligence (SI) is an artificial intelligence technique based around the study of collective behaviour in decentralised, self-organised, systems. The expression "swarm intelligence" was introduced by Beni & Wang in 1989(Beni, Wang 1989), in the context of cellular robotic systems.

SI systems are typically made up of population of agents interacting locally with one another and the environment. There is no central authority to dictate the behaviour of the agents. In fact, the local interactions among agents lead to emergence of a global behaviour. Examples of systems like this can be found in nature, including ant colonies, bird flocking, animal herding, bacteria molding and fish schooling.

When insects work together collaboratively there is no apparent communciation between them. Infact, they utilize envoirnment as carrier of information. They make certain changes in the envoirnement that are sensed by the insects .

There are two popular techniques in Swarm intelligence, i.e *Ant Colony Optimization* (ACO) and *Particle Swarm Optimization* (PSO). Ant colony algortihm mimic the behvior of simple ants trying to locate their food source. Ants while following a certain path lay down a special chemical called pheromone, other ants follow the same path by sensing the pheromone concentration.

Stigmergy is altering the state of environment in a way that it will effect the behavior of others for whom environment acts as a stimulus. "Ant

Colony Optimization" is based on the observation that ants will find the shortest path around an obstacle separating their nest from a target such as a piece of candy simmering on a summer sidewalk.

As ants move around they leave *pheromone trails*, which dissipate over time and distance. The pheromone intensity at a spot, that is, the number of pheromone molecules that a wandering ant might encounter, is higher either when ants have passed over the spot more recently or when a greater number of ants have passed over the spot. Thus ants following pheromone trails will tend to congregate simply from the fact that the pheromone density increases with each additional ant that follows the trail. By exploitation of the positive feedback effect, that is, the strengthening of the trail with every additional ant, this algorithm is able to solve quite complicated combinatorial problems where the goal is to find a way to accomplish a task in the fewest number of operations. Research on live ants has shown that when food is placed at some distance from the nest, with two paths of unequal length leading to it, they will end up with the swarm following the shorter path.

If a shorter path is introduced, though, for instance, if an obstacle is removed they are unable to switch to it. If both paths are of equal length, the ants will choose one or the other. If two food sources are offered, with one being a richer source than the other, a swarm of ants will choose the richer source if a richer source is offered after the choice has been made, most species are unable to switch but some species are able to change their pattern to the better source. If two equal sources are offered, an ant will choose one or the other arbitrarily. Particle Swarm Optimization is the technique developed by Dr. Eberhart and Dr. Kennedy (Eberhart & Kennedy 2001) inspired by social behavior of bird flocking or fish schooling. In PSO, the potential solutions called particles fly through the problem space by following the current optimum particles. Each particle keeps track of its coordinates in the problem space, which are

associated with the best solution (fitness) it has achieved so far. (The fitness value is also stored.) This value is called *pbest*. Another "best" value that is tracked by the particle swarm optimizer is the best value, obtained so far by any particle in the neighbors of the particle. This location is called *lbest*. When a particle takes all the population as its topological neighbors, the best value is a global best and is called *gbest*(Covaci 1999).

Implementation of SI

Agent Behavior

It is very hard to predict the behaviour of agents in an open multiagent systems that is characterized by high degree of uncertainty. In our system we have limited our model to the successful or unsuccessful interaction among the agents. Swarm based model considers behaviour of an agent as its willingness to carry out a particular interaction. This behaviour is then propagated throughout the system. If the agent is unable to fulfil certain interaction, it automatically gets isolated and is no longer called for the interaction in future.

Basic Concept

Our model is based upon the swarm intelligence paradigm. Ant colony algorithm that is the subset of Swarm Intelligence is utilized in our model (Figure 1). Ants and insects communicate with one another indirectly by making changes in the environment a process called as *stigmergy*. Ants while following a particular path towards the food source lay down a special chemical called as pheromone. If it's the valid path, more ants follow the same path thereby increasing the pheromone concentration. So the newer generation of ants automatically given an option automatically follow the path with higher pheromone concentration.

In our model there are basically two types of ants: unicast and broadcast ants. If no pheromone concentration is available then the broadcast ant

Figure 1. Swarm intelligence reputation model

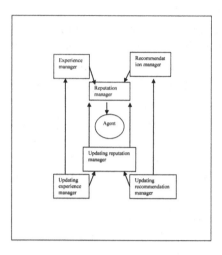

is sent to all the agents in the MAS. The ant that is able to find the valid path towards the food source then returns back to the source to complete the process unicast ants are sent to the path in order to accumulate the pheromone concentration.

Each agent in our model is equipped with certain modules. Reputation calculation requires basically two sources of information that are agent's direct experience and agent's peers or neighbours experience. Our model consists of three modules that serve to maintain these sources of information: namely, *experience manager*, *recommendation manager* and the *reputation manager*. Each one of these is discussed in detail below.

Modeling Direct Experience

This component basically involves the personal experiences of the agent with the intended agent. The experience is quantified in terms of the pheromone value. If the agent interacted with certain agent X and that interaction was successful, the experiencing agent will automatically record a value against pheromone variable. This value is used as agents own personal experience with the agent under consideration.

Modeling Peer Experience

This component basically involves the reputation information from the peers. The peers maintain special tables that contain pheromone information of its neighbors. If the agent does not contain any direct experience information. It has to utilize the information from the peers or neighbors. For doing so, it broadcasts the request to the neighbors and records a value against pheromone variable for the replying agent.

Updating

The changing levels of pheromone concentration capture the dynamic behaviour of agents in Swarm Intelligence based system. If a path is no longer followed by the agents the pheromone level starts weakening with the passage of time.

Now the pheromone concentration might be high at some other path. If an agent leaves the current best path, then the agent may opt for some other path based upon available reputation information and comes up with another path towards agents that requires evaluation. Based on this phenomenon the updating components of the model carry out the process. The final updating value is calculated at reputation update component.

Reputation Distribution

The proposed mechanism involves basically three modules: i.e., *collector*, *processor* and *transmitter*. Each agent in our model is equipped with these three modules. The collector module is responsible to collect reputation information from different sources, the processor basically computes the reputation value and the transmitter provides the reputation requestor with the required information (Figure 2).

The collector module receives reputation information of a target agent from the neighbors, as well as contains some personal reputation value computed by the containing agent based on some past experience with the agent under consideration. The reputation value from the neighbors is placed in a special table.

The processor module carries out some computations on the reputation values gathered by the collector module according to the algorithm discussed in the following section and thus finds the final reputation value of the agent under consideration. Basically the experience manager and recommendation manager and their updating counter parts together make up the collector module in any agent.

The transmitter module gets invoked whenever it receives reputation request from other requesting agents about a particular agent.

The collector module basically consists of two functions, i.e., to get information from different sources plus to update them as well so as to reflect recent changes in the system. The updating process follows the pheromone concentration philosophy of the ant colony algorithm (Ramchurn, Huynh and Jennings 2004). If an agent is no longer used for particular service etc, means that there is some other better service provider. Thus the value of pheromone variable decreases with the passage of time according to some factor. For example consider a scenario. Let's say an agent has to reach to agent six according to Figure 3. There are two paths to reach the destination but one of them provides an optimum solution to reach the intended destination. The agent would use path from node two or node three depending upon the reputation of each node. According to philosophy of Swarm if the pheromone concentration on the path two is found or is greater then the other option; agents would automatically follow that path.

Now the question arises about different levels of pheromone concentration on the two paths. The answer is very simple in a way that from the biological behavior of ants. If ants reach to the food source, on its way back it lays down the chemical on the path so that other ants could sense it. If for example the ant is unable to reach to the destination, its choice of the path has failed and it goes for some other path and carries out the same process. An important point to be noted over here is that with the passage of time if the current optimum path is not being followed the concentration of pheromone weakens thereby letting ants to find some other path.

Figure 2. Reputation distribution modules

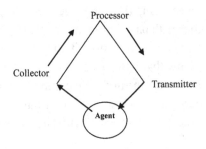

Figure 3. A swarm example

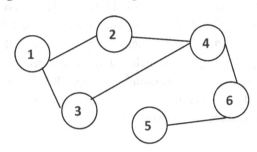

The Reputation Distribution Phenomenon

We describe the reputation distribution phenomenon with the help of connected nodes scenario, where each node is connected with other, and is also subject to dynamism. In this condition, if the node or agent is directly connected with the target agent, then the peer experience is equal to the personal experience. But if it is not directly connected with the target node or agent the information of the neighbours or peers is utilized in order to compute the reputation value of the agent. Each agent maintains a table containing the pheromone information of the immediate connected nodes to it. The node learns through the personal experience of the neighbouring nodes. Thus the two important sources of information in building of the reputation value of an agent are done in a very novel manner. This phenomenon in fact also addresses issue of distribution of reputation information. The two types of messages are generated, called as forward message and the backward message.

The forward message invokes the transmitter module of the nodes which in turn calls the collector module to find if the table contains information for the target agent. If yes, the transmitter module immediately generates the backward message. This backward message basically retraces the same path updating the pheromone information of all the intermediate nodes maintained at the collector module. The processor module carries out the calculations based upon hop count and thus assigns the *pheromone* value before and also after the updating processes (Figure 4).

From the above discussion we can further elaborate the functionality of the Experience manager and the Recommendation manager.

Experience Manager contains personal information of each node with its immediate connected agent or node.

Recommendation manager basically captures the learning behavior of the ACO; where by reputation of the target agent or node is evaluated on the basis of personal experience of all the intermediate nodes with one another, thus making the target agent as the best possible source of the service.

If for example the target under observation moves away as it's the characteristic of the highly dynamic environment, the backward message won't be generated. Instead, the pheromone information contained at the last connected node or agent would go on decreasing with the passage of time, thus this information would be propagated to the source agent or node that the current service provider is no longer available as its reputation has decreased. So the source agent or node has to now consult any other group of surrounding neighbours or group of nodes in order to evaluate and propagate the reputation information, starting with the broadcast message.

Through literature review we found that latest model FIRE incorporated personal, certificate and the witness reputation information. About certificate information it is the node or neighbor that the service providing agent delegates its self to the evaluating agent. It is just like the reference we mention in our CV. The biggest dilemma of this approach is that certificate reputation can

Figure 4. Swarm Algorithm example

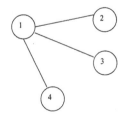

Next Node	% chance
2	33.33333%
3	33.33333%
4	33.33333%

sometimes propagates wrong full information. If we look at the ant colony algorithm we come across the conclusion that personal experience that is the most highly rated reputation information is used to build the witness or peer information. In addition, the certificate reputation comes into play in a different fashion with the generation of the backward message. If the target agent replies back and generates the backward message, it is in fact delegating a certificate to the last connected agent or node to be used by other neighboring nodes.

An important concern in trust and reputation models is how to capture the wrongful or deceitful behaviour of the agents. In our case the deceitful action can be in the form of generation of illegal backward message, thereby attracting the source agent to the wrong path towards the destination agent. Proposed by (Lam and Leung 2004) where by a unique id is assigned to each forward ant, backward ant with the same id is generated that retraces the given path and only the recorded id at each node is entertained as the backward message so the problem of lie in reputation mechanism can be easily dealt. Lie detection has been one of the most important aspects of any *reputation model*. However, our algorithm can solve this problem and certain other issues related to false reputation propagation (Figure 5).

For example the legitimate way to reach to agent 4 is through 1. But somehow agent 2 starts generating wrong backward messages making agent 0 to realize agent 4 through it. The problem can be solved if nodes only entertain the backward messages with the id's recorded in special tables

Figure 5. Lie detection example

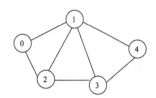

maintained at each node. Once a backward message is accepted by the agent, it deletes its id form the table to avoid duplication. Also, if it is unable to receive the backward message within a specified period of time, its id is treated as stale.

THE ALGORITHM

The algorithm captures the phenomena of reputation distribution using ant colony algorithm. The collector module basically searches for the pheromone information of the requested agent. If found, it generates the backward message; while in other case, it repeats the same process and generates the broadcast message for all the agents in search of the pheromone information.

Processing module receives the reputation values from the collector and carries out the required calculations as mentioned in the algorithm. Transmitter module responses back to any request generated for the agent to acquire the reputation value.

```
Collector Module
{
        Check if the requested
node's pheromone info available
        If available then
                Call Backward()
        Else
                Call Broadcast
()
}
Transmitter Module
{
        Call Collector Module
}

        Broadcast()
{
        Send hello message in
search of pheromone information
to all the agents in the domain
```

```
}
forward()
{
        Assign id to the for-
ward message
        Call transmitter
}
Backward()
{
        Retrace nodes in back-
ward direction from the calling
agent node
        Call Processor Module
}
Processor Module
{
        Check the id of back-
ward message
        If legal then do the
following steps
                Calculate hop-
count
                Divide hopcount
by 100        -a-
                Add 100 to
proabaility value of the node
                Add that pro-
abability value of  the tar-
get node  with the probabil-
ity value of other connected
nodes                -b-
                Calculate the
ratio a/b
                Set the prob-
ability of the node to its cur-
rent value multiplied by the
ratio
}
```

We can evaluate our proposed scheme through two parameters: i.e. hop counts and success rate. Hop count depicts the number of nodes or agents needed to be followed in order to reach to the required destination, while success rate is the percentage of the requests of which the requestor successfully obtains the evidence, or the reputation value. The FIRE (Patel 2007) model has considered a scenario of producers and consumers with an assumption of only one type of service availability. The producers (Providers) are categorized as good, ordinary, intermittent and bad, depending upon the quality of service they are rendering. The Intermittent and Bad providers are the most random ones. If consumer agent needs to use the service, it can contact the environment to locate nearby provider agents. The consumer agent will then select one provider from the list to use its services. The selection process depends upon the reputation model of the agent. Consumer's satisfaction is calculated in terms of *Utility Gain (UG)*. We apply our mechanism to the simulation environment used by the FIRE model. Consisting of consumers and producers, FIRE model incorporates various sources of information in order to compute the final value. But the issue of this reputation information distribution is still a big question mark. Also there is no mechanism defined in the model that can make it adaptable to the changes taking place in the open environment.

RESULTS

Experimental Methodology

We have utilized an ant net simulator to compute results for the reputation model. The ant simulator is written in c# and utilizes dot net charting capabilities to graphically represent the results. ANT Net uses virtual pheromone tables much like when an ant follows a path dropping pheromones to re-enforce it. The quicker the ants move down a path the more throughput of ants thus a greater concentration of pheromones. In the same way pheromone tables in ANT Net allow fast routes to score a higher chance of being selected whilst the less optimal route scores a low chance of being selected.

Every node holds a pheromone table for all other nodes of the network. Each pheromone table holds a list of table entries containing all the connected nodes of the current node.

Simulation Results

The simulation results are computed in terms of the average hop counts in various scenarios. The Reputation Model how ever shows its efficiency in different scenarios in terms of the Utility Gain (UG) value. *Utility Gain* can be defined as the level of satisfaction an agent gains after interacting with any other agent to consume certain services. UG is inversely proportional to the average number of hop counts.

Experimental Test Bed

To evaluate the proposed reputation model for multiagent systems, we required an open system whose dynamic nature is captured by the model. Therefore, we chose mobile adhoc network as a case study to find the effectiveness of our model. An adhoc network involves collection of mobile nodes that are dynamically and arbitrary located. The interconnections between the nodes change on continual basis.

Each node in Mobile adhoc networks(*Manets)* decides about its route. In fact, there are no designated routers and nodes forward packets from node to node in multi hop fashion. Nodes movement implies that current routes become invalid in future instances (Huynh, Jennings, Shadbolt 2006, Patel 2007). In such an environment quality of Service (QoS) is a big issue in terms of secure paths, bandwidth consumption, delays and etc. Our reputation model can be applied to Manets in order to show its applicability and robustness. The nodes in Manets could be thought as agents in mutliagent systems. Let's take a simple problem of route discovery in Manets with reputation mechanism. Mobile adhoc networks are open system and constantly undergo change in agent topology.

This inherent uncertainty in Manets makes it a very suitable candidate for the evaluation of the proposed reputation model. The reputation model for such environment needs to posses following properties.

- It should take variety of sources of information to have a more robust and reliable reputation value.
- Since there is no central authority in the system each agent should be equipped with the reputation model in order to interact with one another.

The swarm intelligence based model not only provides a mechanism to calculate the reputation value of an agent but in fact also provides the mechanism that is utilized to disseminate that reputation information to all the agents in the system, under high dynamism.

We have evaluated our system on the basis of hop counts only. The basic phenomenon behind our algorithm is the concentration level of pheromone. As the level of trust upon agents increases so as the concentration of the pheromone variable contained by every agent in the system. The reputation value degrades by the weakening level of pheromone variable. The number of hop counts efficiently captures this thing. As the reputation gets built around certain agents, the probability of selecting them for any future service becomes higher. Therefore, the same destination that would have been achieved by trial and error can now be consulted for any service efficiently and effectively by the inherent capability of the proposed algorithm. Previous trust and reputation models have incorporated various methods to measure the UG attained by agents in the system that depicts the level of satisfaction that an agent gains from another after the interaction. We have made number of hop counts the very basis of the UG in our proposed scheme. As the number of hops decreases in different scenarios UG gets increased thus they are inversely proportional to each other.

Figure 6. Ant algorithm set to off

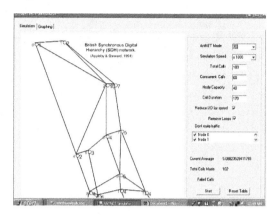

Figure 7. Ant algorithm set to on

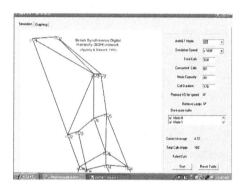

Since we are utilizing special case of Manets, for evaluation we take average hop counts as the measure of effectiveness of our model. The lesser the number of hops required in any particular scenario, the more adaptable our system is under certain conditions.

We tested the scenario with different parameters. For example, first we evaluated the system with ant net algorithm set to off (Figure 6). The average hop counts found are 5.08.

As compared to when the ant algorithm is set to on in the same scenario, the average hop counts becomes 4.72 (Figure 7). The reduction in the number of hop counts is the result of the pheromone variable value. The agents in the system learn about each other's reputation based on the previously discussed phenomena.

When no ant algorithm is utilized, meaning that there is no learning capability in the process, thus in order to utilize the same services, agents have to seek more agents for further information, thereby increasing the average hop counts.

The adaptability of the system to the dynamic environment is captured by making certain nodes unavailable. Let's say we have removed node 5, 9, and 11 from the system which no longer can be utilized to propagate the reputation information. Under such circumstances, the average hop counts comes up to be 4.89. As compared to the same situation when the ant algorithm is switched off the average hop count comes up to be 5.43

Capturing dynamism in the open environment is one of the important targets of the interactions among agents in the open multi agent systems (Figures 8 and 9). However, even the previous

Figure 8. Capturing dynamism with ant algorithm set to on

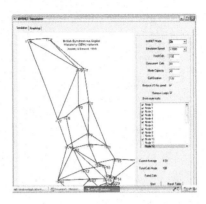

Figure 9. Capturing dynamism with ant algorithm set to off

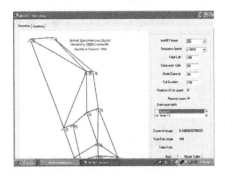

Figure 10. Graphical comparisons between ant algorithm on and off

recent model FIRE was unable to accurately capture the highly dynamic environment. When certain nodes (agents) are removed from the system the average number of hop counts increases if no swarm algorithm utilized as compared to the fact when swarm algorithm is utilized. The average hop counts is less showing the system is able to adapt itself to the changing conditions (Figure 10).

Looping is an important issue in case of Manets in order to detect and correct the duplicate infor-

Figure 11. Looping allowed

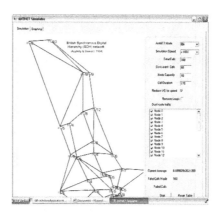

Figure 12. Graphical representation of allowed Looping

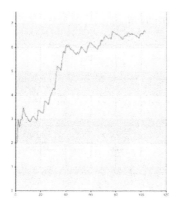

Figure 13. Looping removed

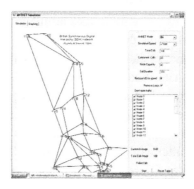

mation travelling in the network. If looping is allowed the average hop counts results to be 6.69; while if it is set to off the average hop counts computes to be 5.43 (Figures 11 through 14).

The above results show that the swarm intelligence based reputation mechanism is able to learn with the changes taking place in the system. And can quickly adapt itself to the changing

Figure 14. Graphical comparison between allowed and is allowed looping

Figure 15. Utility gain with no ant algorithm

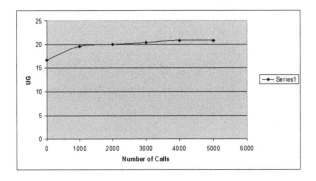

conditions as opposed to the system when no ant net algorithm is utilized.

Compared to previous recent work in the domain of reputation models, the FIRE model states that in order to have a robust reputation the model should be able to gather information from diverse sources. FIRE model has incorporated personal experience, witness reputation, rule based and the certificate reputation. The swarm intelligence reputation model inherently incorporates different types of information in order to come up with the final reputation value of the agent. On top of that, these sources of information have learning capability and can adapt themselves to the changing nature of open MAS. This fact is apparent from the simulation results. Thus we can further depict

the effectiveness of the proposed model in terms of utility gain (Figures 15 and 16).

From the charts it is visible that while ant algorithm is utilized the utility gain increases with the same number of interactions as opposed to the fact when no ant algorithm is utilized. Similarly UG in case of changing conditions also shows a marked increase showing that the system is able to learn and adapt itself. Comparing our results with the research aims set, we find that the proposed reputation model truly captures the dynamism in the agent environment. The reputation information is not held centrally and exists in distributed form and finally captures information from various sources in order to compute the final reputation value. In previous work, weights were

Figure 16. Utility gain with ant algorithm

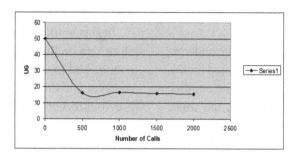

Table 1. Percent change in UG of Swarm Model

Number of interactions	21	41	61	81
UG	4	4.2	4.5	4.4
%Gain		5%	7.1%	-2.2%

Table 2. Percent change in UG of FIRE Model

Number of interactions	21	41	61	81
UG	6.3	6.8	6.5	6.4
%Gain		7%	-4.4%	-1.5%

assigned to the different sources of information, particularly information gained from the neighbors is weighed lower due to risk associated with it. In Swarm based system, the weight assigned to the peer information is very logical in a way if the agent doesn't reply back in particular period of time, the agent will loose its reputation value in the form of evaporating pheromone concentration, thereby making it less attractive for the agents to consult for any information.

Over All Performance

We evaluated the over all performance of the proposed model by computing the percentage gain in special case of dynamism. And compared the result with that of the percentage gain in the FIRE model. We computed our data from the simulation results obtained after removing certain agents from

the system finding the average hop counts; at 21, 41, 61 and 81 instances of interactions. The data is given in Table 1.

As compared to the percentage change in the Utility Gain of the interactions among agents in the system using FIRE model to compute the reputation value of the agents given in Table 2.

By analyzing the percentage change in the utility gains of the two systems given same number of interactions, we find that Swarm based system is more adaptable to the dynamism in the environment as compared to the previous FIRE model. Thus from above results we can deduce that utility gain based on hope counts yields more value as compared to the Utility Gained by agents utilizing FIRE model to compute the reputation values.

CONCLUSION AND FUTURE RESEARCH DIRECTIONS

Open multiagent systems continuously undergo changes. In such environment there is need for the participating agents to interact with one another in a trustworthy manner regarded as the trust. Trust requires creation of the reputation of the participating agents and the distribution of this reputation information that can adapt itself to the dynamism in the environment. We have proposed a swarm intelligence based mechanism for reputation creation and dissemination, specifically utilizing the ant colony algorithm. Our model is novel in a sense that no existing model of trust has as yet addressed the issue of reputation creation an information distribution with minimum overheads. Our model has the capability of utilizing environment itself as the carrier of information. The inherent optimizing capability of the Swarm Algorithm makes process of reputation creation and distribution highly adaptable to the changing environment.

Research in the field of Trust Management is still at its infancy stage, newer models are coming up however no standard has been widely accepted as such. Certain models try to compute their efficiency in terms of the empirical evaluation while others restrict their use to certain domains. So there is a requirement of one such model that can capture the basic characteristic of the dynamism of the open Multiagent Systems and also fulfils the requirements of a reputation model. For the robust performance of the reputation models, they should incorporate information about an agent from various sources so as to achieve a comprehensive value that does not depends only upon one factor but takes into account various dimensions. Swarm based model, however successfully achieves this by incorporating information from different sources with minimum overheads because the environment itself is used as the carrier of information. Environment acts as the stimuli to

the agents to change value of certain parameters that are used to compute the reputation values.

In this work we have evaluated our proposed algorithm on the basis of average number of hop counts, and making certain changes in the environment and then comparing the results with the previous recent reputation model called that FIRE judges the learning capability of the algorithm. The output of reputation models are judged on the basis of Utility Gain that shows the level of satisfaction an agent gets while interacting with one another. It is found that in Swarm based algorithms the percentage utility gain is higher as compared to previous works.

Future work can focus on additional parameters for performance evaluation such as lie detection, which will determine the authenticity of reputation information of the agents.

REFERENCES

Beni, G., & Wang, J. (1989). Swarm Intelligence in Cellular Robotic Systems. In *Proceed. NATO Advanced Workshop on Robots and Biological Systems*, Tuscany, Italy, June 26–30

Covaci, S. (1999). Autonomous Agent Technology. In *Proceedings of the 4th international symposium on Autonomous Decentralized Systems*. Washington, DC: IEEE Computer Science Society.

Huynh, T. D., Jennings, N. R., & Shadbolt, N. R. (2006). An integrated trust and reputation model for open multi-agent systems. *Journal of Autonomous agents and multi agent systems*.

Jurca, R., & Faltings, B. (2003). Towards Incentive-Compatible Reputation Management. *Trust, Reputation and Security: Theories and Practice* (LNAI 2631, pp. 138-147).

Kennedy, J., & Eberhert, R. C. (2001). *Swarm Intelligence*. Morgan Kaufmann.

Lam, K., & Leung, H. (2004). An Adaptive Strategy for Trust/ Honesty Model in Multi-agent Semi- competitive Environments. In *Proceedings of the 16th IEEE International Conference on Tools with Artificial Intelligence*(ICTAI 2004)

Patel, J. (2007). *A Trust and Reputation Model For Agent-Based Virtual Organizations.* Phd thesis in the faculty of Engineering and Applied Science School of Electronics and Computer Sciences University of South Hampton January 2007.

Ramchurn, S. D., Huynh, D., & Jennings, N. R. (2004). Trust in multiagent Systems. *The Knowledge Engineering Review, 19*(1), 1–25. doi:10.1017/S0269888904000116

Schlosser, A., Voss, M., & Bruckner, L. (2004). *Comparing and evaluating metrics for reputation systems by simulation.* Paper presented at RAS-2004, A Workshop on Reputation in Agent Societies as part of 2004 IEEE/WIC/ACM International Joint Conference on Intelligent Agent Technology (IAT'04) and Web Intelligence (WI'04), Beijing China, September 2004.

Zacharia, G., & Maes, P. (2000). Trust management through reputation mechanisms. *Applied Artificial Intelligence Journal, 14*(9), 881–908. doi:10.1080/08839510050144868

ADDITIONAL READING

Abdui-Rahman, A., & Hailes, S. (1997). A Distributed Trust Model. Paper presented at the 1997 New Security Paradigms Workshop. Langdale, Cumbria UK. ACM.

Amir Pirzada, A., Datta, A., & McDonald, C. (2004). Trusted Routing in Ad-hoc Networks using Pheromone Trails. In *Congress of Evolutionary Computation, CEC2004.* IEEE.

Botely, L. (n.d.). Ant Net Simulator. *University of Sussex, UK.*

Boukerche, A., & Li, X. (2005). An Agent-based Trust and Reputation Management Scheme for Wireless Sensor Networks. In *Global Telecommunications Conference, 2005.* GLOBECOM '05. IEEE.

Fullam, K., Klos, T., Muller, G., Sabater, J., Schlosser, A., Topol, Z., et al. (2005). A Specification of the Agent Reputation and Trust (ART) Testbed. In *Proc. AAMAS.*

Gunes, M., Sorges, U., & Bouazizi, I. (2002). ARA- the ant Based Routing Algorithm for MANETs. *International workshop on Adhoc networking (IWAAIN 2002) Vancouver British Columbia Canada, August 18-21 2002.*

Hughes, T., Denny, J., & Muckelbauer, P. A. (2003). *Dynamic Trust Applied to Ad Hoc Network Resources.* Paper presented at 6th Work shop on Trust, Privacy Deception and Fraud In Agent Societies Melbourne 2003

Kagal, L., Cost, S., Finin, T., & Peng, Y. (2001). A Framework for Distributed Trust Management. Paper presented at the Second Workshop on Norms and Institutions in MAS, Autonomous Agents.

Kennedy, J., & Eberhart, R. C. (2002). Book Review Swarm Intelligence. *Journal of Genetic Programming and Evolvable Machines, 3*(1).

Li, X., Hess, T.J., & Valacich, J.S. (2006). Using Attitude and Social Influence to Develop an Extended Trust Model for Information Systems. *Database for advances in Information Systems.*

Liu, J., & Issarny, V. (2004). Enhanced Reputation Mechanism for Mobile Ad Hoc Networks. In *Proceedings of iTrust 2004, Oxford UK.*

Mars, S. P. (1994). Formalising Trust as a Computational Concept. Department of Computing Science and Mathematics University of Stirling April 1994 PhD thesis

Marsh, S., & Meech, J. (2006). *Trust in Design*. National Research Council Canada Institute of Information Technology.

Nurmi, P. (2005). *Bayesian game theory in practice: A framework for online reputation systems*. University of Helsinki Technical report, Series of Publications C, Report C-2005-10.

Pujol, J. M., Sangüesa, R., & Delgado, J. (2002). Extracting Reputation in Multi Agent Systems by Means of Social Network Topology. In *Proceedings of the First International Joint Conference on Autonomous Agents and Multiagent Systems: Part 1* (pp. 467-474). ACM.

Sabater, J. (2003). *Trust and Reputation for Agent Societies*. PhD thesis, University Autonoma de Barcelona.

Sabater, J. (2004). Toward a Test-Bed for Trust and Reputation Models. In R. Falcone, K. Barber, J. Sabater, & M. Singh (Eds.), *Proc. of the AAMAS-2004 Workshop on Trust in Agent Societies* (pp. 101-105).

Schlosser, A., & Voss, M. (2005). Simulating Data Dissemination Techniques for Local Reputation Systems. In Proceedings of the Fourth International Joint Conference on Autonomous Agents and Multiagent Systems (pp. 1173-1174). ACM.

Sierra, C., & Debenham, J. (2005). An Information-based model for trust. *AAMAS 05*. Utrecht Netherlands.

Smith, M.J., & desJardins, M. (2005).A Framework for Decomposing Reputation in MAS in to Competence and Integrity. *AAMAS'05*.

Song, W. (2004). Neural Network-Based Reputation Model in a Distributed System. In *Proceedings of the IEEE International Conference on E-Commerce Technology*. IEEE.

Teacy, W.T.L., Huynh, T.D., Dash, R.K., Jennings, N.K., & Patel, J. (2006). *The ART of IAM: The Winning Strategy for the 2006 Competition*.

Teacy, W.T.L., Patel, J., Jennings, N.R., & Luck, M. (2005). Coping with Inaccurate Reputation Sources: Experimental Analysis of a Probabilistic Trust Model. *AAMAS'05*.

Theodorakopoulos, G., & Baras, J. S. (2006). On Trust Models and Trust Evaluation Metrics for Ad Hoc Networks. *IEEE Journal on Selected Areas in Communications*, *24*(2). doi:10.1109/JSAC.2005.861390

Xiong, L., & Liu, L. (2003). A Reputation-Based Trust Model for Peer-to-Peer eCommerce Communities. In *Proceedings of the IEEE International Conference on E-Commerce (CEC'03)*.

Yamamoto, A., Asahara, D., Itao, T., Tanaka, S., & Suda, T. (2004). Distributed Pagerank: A Distributed Reputation Model for Open Peer-to-Peer Networks. In *Proceedings of the 2004 International Symposium on Applications and the Internet Workshops (SAINTW'04)*.

Zheng, X., Wu, Z., Chen, H., & Mao, Y. (2006). Developing a Composite Trust Model for Multi agent Systems. In *Proceedings of the Fifth International Joint Conference on Autonomous Agents and Multiagent Systems* (pp. 1257-1259). ACM.

Chapter 15

Exploitation–Oriented Learning XoL:
A New Approach to Machine Learning Based on Trial–and–Error Searches

Kazuteru Miyazaki
National Institution for Academic Degrees and University Evaluation, Japan

ABSTRACT

Exploitation-oriented Learning XoL is a new framework of reinforcement learning. XoL aims to learn a rational policy whose expected reward per an action is larger than zero, and does not require a sophisticated design of the value of a reward signal. In this chapter, as examples of learning systems that belongs in XoL, we introduce the rationality theorem of profit Sharing (PS), the rationality theorem of reward sharing in multi-agent PS, and PS-r*. XoL has several features. (1) Though traditional RL systems require appropriate reward and penalty values, XoL only requires an order of importance among them. (2) XoL can learn more quickly since it traces successful experiences very strongly. (3) XoL may be unsuitable for pursuing an optimal policy. The optimal policy can be acquired by the multi-start method that needs to reset all memories to get a better policy. (4) XoL is effective on the classes beyond MDPs, since it is a Bellman-free method that does not depend on DP. We show several numerical examples to confirm these features.

INTRODUCTION

The approach, called *reinforcement learning* (RL), is much more focused on goal-directed learning from interaction than are other approaches to machine learning (Sutton and Barto, 1998). It is very attractive since it can use *Dynamic Programming* (DP) to analyze its behavior in the *Markov Decision Processes* (MDPs) (Sutton, 1988; Watkins and Dayan, 1992; Ng et al., 1999; Gosavi, 2004; Abbeel and Ng, 2005). We call these methods that are based on DP *DP-based RL methods*. In general, RL uses a *reward* as a teacher signal for its learning. The DP-based RL method aims to optimize its behavior under the values of reward signals that are designed in advance.

We want to apply RL to many real world problems more easily. Though we know some

DOI: 10.4018/978-1-60566-898-7.ch015

important applications (Merrick et al., 2007), generally speaking, it is difficult to design RL systems to fit on a real world problem. We think that the following two reasons concern with it. In the first, the interaction will require many trial-and-error searches. In the second, there is no guideline how to design the values of reward signals. Though they are not treated as important issues on theoretical researches, they are able to be a serious issue in a real world application. Especially, if we have assigned inappropriate values to reward signals, we will receive an unexpected result (Miyazaki and Kobayashi, 2000). We know the *Inverse Reinforcement Learning* (IRL) (Ng and Russell, 2000; Abbeel and Ng, 2005) as a method related to the design problem of the values of reward signals. If we input our expected policy to the IRL system, it can output a *reward function* that can realize the policy. IRL has several theoretical results, i.e. *apprenticeship learning* (Abbeel and Ng, 2005) and *policy invariance* (Ng et at., 1999).

On the other hand, we are interested in the approach where reward signals are treated independently and do not require a sophisticated design of the values of them. Furthermore, we aim to reduce the number of trial-and-error searches through strongly enhancing successful experiences. We call it *Exploitation-oriented Learning* (*XoL*). As examples of learning systems that can belong in XoL, we know *the rationality theorem of Profit Sharing* (PS) (Miyazaki et al., 1994), *the Rational Policy Making algorithm* (Miyazaki et al., 1998), *the rationality theorem of PS in multi-agent environments* (Miyazaki and Kobayashi, 2001), *the Penalty Avoiding Rational Policy Making algorithm* (Miyazaki and Kobayashi, 2000) and *PS-r** (Miyazaki and Kobayashi, 2003).

XoL has several features. (1) Though traditional RL systems require appropriate values of reward signals, XoL only requires an order of importance among them. In general, it is easier than designing their values. (2) XoL can learn more quickly since it traces successful experi-

ences very strongly. (3) XoL may be unsuitable for pursuing the optimality. It can be guaranteed by the *multi-start method* (Miyazaki et al., 1998) that resets all memories to get a better policy. (4) XoL is effective on the classes beyond MDPs such that the *Partially Observed Markov Decision Processes* (POMDPs), since it is a method that does not depend on DP called a *Bellman-free method* (Sutton and Barto, 1998).

In this chapter, we focus on the POMDPs environments where the number of types of a reward is one. As examples of learning systems that belong in XoL at the environments, we introduce the rationality theorem of PS, the rationality theorem of PS in multi-agent environments, and PS-r*. We show several numerical examples to support how to use these methods.

PROBLEM FORMULATIONS

Notations

Consider an agent in some unknown environment. The agent senses a set of discrete attribute-value pairs and performs an action in some discrete varieties. The environment provides a reward signal to the agent as a result of some sequence of an action. We denote the sensory inputs as x, y,… and actions as a, b,…. A sensory input and an action constitute a pair that is termed as a *rule*. We denote the rule "if x then a" as xa. In Profit Sharing (PS), a scalar weight, that indicates the importance of the rule, is assigned to each rule. The weight of the rule xa is denoted as ω_{xa}. The function that maps sensory inputs to actions is termed a *policy*. We call a policy *rational* if and only if the expected reward per an action is larger than zero. Furthermore, a *useful rational policy* is a rational policy that is not inferior to the *random walk* (RA) where the agent selects an action based on the same probability to every action in every sensory input. The policy that

Figure 1. A conflict structure

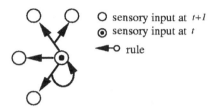

○ sensory input at *t+1*
◉ sensory input at *t*
◄○ rule

maximizes the expected reward per action is termed as the *optimal policy*.

The environment is treated as consisting of stochastic processes, where a sensory input corresponds to some state and an action to some state transition operator. We show an environment represented by a state transition diagram in Figure 1. The node with a token denotes a sensory input at time *t*. Three rules match the sensory input. Since the state transition is not deterministic, selection of the same rules does not always lead to the same state. The branching arcs indicate such cases. We term a part of the state transition diagram around one sensory input as a *conflict structure*. Figure 1 is an example of it.

Figure 2a) is an environment consisting of three sensory inputs, x, y and z, denoted by circles. Two actions, a and b, can be selected in each sensory input. An arrow means a state transition by execution of an action described on the arrow.

We term the rule sequence selected between two rewards, or an initial sensory input and a reward, as an *episode*. For example, when the agent selects xb, xa, ya, za, yb, xa, za and yb in Figure2a), there exist two episodes (xb·xa·ya·za·yb) and (xa·za·yb), as shown in Figure 2b). We term the subsequent episode as a *detour* when the sensory input of the first selection rule and the sensory output of the last selection rule are the same although both rules are different. For example, the episode (xb·xa·ya·za·yb) has two de-

tours (xb) and (ya·za), as shown in Figure 2b). The rules on a detour may not contribute to obtain a reward. We term a rule as *irrational* if and only if it always exists on a detour. Otherwise, a rule is termed as *rational*. An irrational rule should not be selected when they conflict with a rational rule.

PS reinforces all the weights of rules on an episode at the same time when the agent obtains a reward. We term the number of rules on an episode as a *reinforcement interval*. Also, we term the function that shares a reward among the weights of rules on an episode as a *reinforcement function*. f_i denotes the reinforcement value for the weight of the rule selected at step *i* before a reward is obtained. We assume that the weight of each rule is reinforced by $\omega_{r_i} = \omega_{r_i} + f_i$ for the episode ($r_W \cdots r_i \cdots r_2 \cdot r_1$), where *W* denotes the reinforcement interval of the episode.

Properties of the Target Environments

We focus on the POMDPs environments where the number of types of a reward is one. In POMDPs, the agent may sense different states on an environment as the same sensory input. We call the sensory input a *Partially Observable* (PO) sensory input.

We recognize that the learning in POMDPs must overcome two deceptive problems (Miyazaki et al., 1998). We term the indistinguishable of *state values*, that are assigned for each state on the environment in POMDPs, as a *type 1 confusion*. Figure 3a) is an example of the type 1 confusion. In this example, the state value (*v*) is estimated by the minimum number of steps required to obtain a reward[1]. The values for the states 1a and 1b are 2 and 8, respectively. Although the state 1a and 1b are different states, the agent senses them as the same sensory input 1 hatched in Figure 3. If the agent experiences the state 1a

Figure 2. a) An environment consisting of three sensory inputs and two actions. b) An example of an episode and a detour

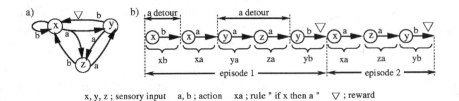

x, y, z ; sensory input   a, b ; action   xa ; rule " if x then a "   ▽ ; reward

Figure 3. Examples of type 1(a) and type 2(b) confusions

Figure 4. Three classes of the target environments

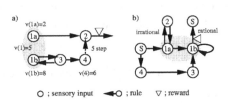

○ ; sensory input   ◄●○ ; rule   ▽ ; reward

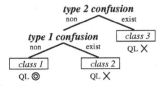

and 1b equally likely, the value of the sensory input 1 becomes 5 ($=\dfrac{2+8}{2}$). Therefore, the value is higher than that of the state 4 (that is 6). If the agent uses the state values for its learning, it would prefer to move *left* in the sensory input ₃. On the other hand, the agent has to move *right* in the sensory input 1. This implies that the agent has learned the irrational policy, where it only transits between the states 1b and 3.

We term the indistinguishable of rational and irrational rules as a *type 2 confusion*. Figure 3b) is an example of the type 2 confusion. Although the action of moving *up* in the state 1a is irrational, it is rational in the state 1b. Since the agent senses the states 1a and 1b as the same sensory input 1, the action of moving *up* in the sensory input 1 is regarded as rational. If the agent learns the action of moving *right* in the sensory input S,

it will fall into an irrational policy where the agent only transits between the states 1a and 2.

In general, if there is a type 2 confusion in some sensory input, there is a type 1 confusion in it. On account of these confusions, we can classify the target environments into three classes, as shown in Figure 4. MDPs belong to class 1. Q-learning (QL) (Watkins and Dayan, 1992), that is a representative RL system, is deceived by a type 1 confusion since it uses state values to formulate a policy. As PS does not use state values, it is not deceived by the confusion. On the other hand, RL systems that use the weight (including QL and PS) might be deceived by a type 2 confusion. In the next section, we show the rationality theorem of PS that guarantees the acquisition of a rational policy in the POMDPs where there is no type 2 confusion and the number of types of a reward is one.

RATIONALITY OF PROFIT SHARING

Rationality Theorem of Profit Sharing

In this section, we consider the rationality of PS. We will introduce the following theorem called *the Rationality Theorem of PS*.

Theorem 1 (The Rationality Theorem of PS)

PS can learn a rational policy in the POMDPs where there is no type 2 confusion and the number of types of a reward is one if and only if

$$\forall i = 1, 2, \ldots W. \qquad L\sum_{j=i}^{W} f_j < f_{i-1}, \qquad (1)$$

where L is the upper bound of the number of conflicting rational rules, and W is the upper bound of the reinforcement interval. ∎

We term Equation(1) as *suppression conditions*. The proof is presented in Appendix A.

Theorem 1 guarantees the *rationality* that can learn a rational policy. We cannot determine the number L in general. However, in practice, we can set $L=M-1$, where M is the number of actions.

There are several functions that satisfy the theorem. We present an example of those functions in Figure 5c) and d). It should be noted that conventional reinforcement functions, such as a constant function (Grefenstette, 1988) (Figure 5a), Func-a) or some decreasing one (Liepins et al., 1989) (Figure 5b), Func-b), do not satisfy the theorem.

The following geometrically decreasing function satisfies the theorem (Figure 5c), Func-c).

$$\text{Func-c} \quad : \quad f_n = \frac{1}{S} f_{n-1}$$
$$where \qquad S \geq L+1$$
$$n = 1, 2, \cdots, W-1$$

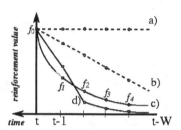

We show another function that satisfies the theorem (Figure 5d), Func-d).

$$\text{Func-d:} \quad : \quad g_0 = f_0, g_1 = f_1 + h, g_2 = f_2 - h, g_3 = f_3$$
$$g_4 + \cdots + g_W < \frac{f_2}{2} - h$$
$$and \quad g_n = \frac{1}{2} g_{n-1}, n = 5, \cdots, W$$
$$where \quad f_n = \frac{1}{2} f_{n-1}$$
$$0 < h < \frac{f_2}{2}$$

A Multi-Start Method for Profit Sharing

For the class where there is a type 2 confusion, we cannot always obtain a rational policy by the rationality theorem of PS. In this case, we should experiment with a *multi-start* method, where each memory is initialized to formulate a new policy.

For example, if no reward is obtained within *2k* steps, where *k* is the number of steps from initial step to the final reward obtaining step, we can establish that there is a type 2 confusion in the environment. Subsequently, we use the multi-start method to guarantee the rationality in the environment.

In *Expansion of PS to POMDPS*, we show another method to approach to a type 2 confusion.

Figure 6. The problem used in the experiment

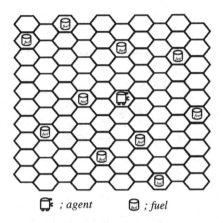

☐ ; agent   ⬒ ; fuel

Application to a Maze-like Environment

Setting

We compare PS with QL in the environment that consists of 10 x 10 honeycomb cells as shown in Figure 6. There are always 10 fuel pots; the capacity of each is 20 liters. The agent consumes one liter of fuel per an action. The agent obtains a reward when it takes a fuel pot. A fuel pot disappears when it is taken by the agent, and instantly appears at a random position on the same vertical axis. In this environment, there is no type 2 confusion. The agent can sense three sensory inputs as follows: 1) **SF**: there is something in the forward direction, 2) **NF**: there is nothing in the forward direction, and 3) **ON**: there is something at the position where there is the agent.

The agent can perform four actions as follows; 1) **MOVE**: move forward, 2) **TURN**: turn by 60 degrees clockwise, 3) **RIGHT**: take a fuel pot in right hand, and 4) **LEFT**: take a fuel pot in the left hand. LEFT fails with a probability of 50%.

Figure 7 shows the state transition diagram of this example. There exist two rational policies. One is the policy 1: {NF → TURN, SF → MOVE, ON → RIGHT}, the other is the policy

Figure 7. The state transition diagram used in the example

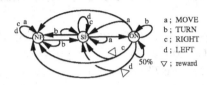

Figure 8. The rates of rational policies

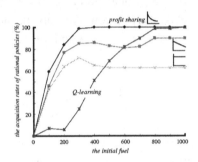

2: {NF → TURN, SF → MOVE, ON → LEFT}. The policy 1 is the optimal policy.

Results and Discussion

In this environment, we change the initial amount of fuel that the agent has taken. We carried out 100 trials with different random seeds. Figure 8 shows the rates of rational policies acquired until the agent has consumed the initial fuel.

Three types of PS are tested in this environment. Two types of PS ("■" and "X" marks in Figure 8), that do not satisfy theorem 1, cannot always learn a rational policy. On the other hand, the PS ("■" marks in Figure 8), that satisfies theorem 1, guarantees the acquisition of a rational policy. In this environment, if the agent takes a fuel pot once, it can learn a rational policy. Therefore, the PS that satisfies theorem 1 can learn a rational policy with a high ratio in the case of a small amount of the initial fuel.

Figure 9. A rule sequence when 3 agents select an action in some order

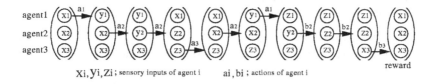

Figure 10. Three episodes and one detour in Figure 9

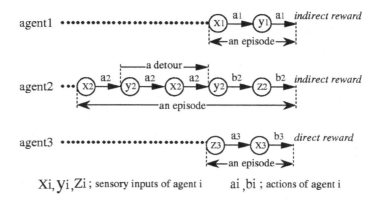

The learning rate of QL is 0.02 and its discount rate is 0.8. Though QL guarantees the acquisition of the optimal policy, it requires numerous trials. In this example, QL requires more initial fuel than the PS to acquire a rational policy. Furthermore, though it is not shown in Figure 8, QL requires 5000 liters to guarantee the acquisition of the optimal policy.

PROFIT SHARING IN MULTI-AGENT ENVIRONMENTS

Approaches to Multi-agent Environments

In this section, we consider n ($n>1$) agents. At each discrete time step, an *agent i* ($i=1,2,...n$) is selected from n agents based on the selection probabilities $P_i (P_i > 0, \sum_{i=1}^{n} P_i = 1$, and it

senses the environment and performs an action. The agent senses a set of discrete attribute-value pairs and performs an action in M discrete varieties. We denote *agent i*'s sensory inputs as $x_i, y_i, \cdots$ and its actions as $a_i, b_i, \cdots$.

When the n'th agent ($0<n'\leq n$) has a *special sensory input* on condition that (n'-1) agents have special sensory inputs at some time step, the n'th agent obtains a *direct reward R* ($R>0$) and the other (n-1) agents obtain an *indirect reward μR* ($\mu \geq 0$). We call the n'th agent the *direct-reward agent* and *the* other (n−1) agents *indirect-reward agents*. We do not have any information about the n' and the special sensory input. Furthermore, nobody knows whether (n-1) agents except for the n'th agent are important or not. A set of n' agents that are necessary for obtaining a direct reward is termed the *goal-agent set*. In order to preserve the rationality in the multi-agent environments all agents in a goal-agent set must learn a rational policy. When a reward is given to an agent, PS reinforces rules on the episode. In our multi-agent environments, the episode is interpreted by each agent. For example, when 3 agents select the rule sequence $(x_1a_1, x_2a_2, y_2a_2, z_3a_3, x_2a_2, y_1a_1, y_2b_2, z_2b_2$ and $x_3b_3)$ (Figure 9) and have *special sensory inputs* (for obtaining a reward), it contains the episode $(x_1a_1 \cdot y_1a_1), (x_2a_2 \cdot y_2a_2 \cdot x_2a_2 \cdot y_2b_2 \cdot z_2b_2)$ and $(z_3a_3 \cdot x_3b_3)$ for agent 1, 2 and 3, respectively (Figure 10). In this case, agent 3 obtains a direct reward and the other agent obtain an indirect reward. We assume that the initial sensory input on the episode for a direct-reward agent is the same.

When the agent obtains a reward, we use the following reinforcement function that satisfies the rationality theorem of PS (theorem 1),

$$f_n = \frac{1}{M} f_{n-1}, \quad n = 1, 2, \ldots W_a - 1. \qquad (2)$$

In Equation (2), (f_0, W_a) is (R, W) for the direct-reward agent and $(\mu R, W_0 (W_0 \leq W))$ for indirect-reward agents, where W and W_0 are reinforcement intervals for direct and indirect-reward agents, respectively. For example, in Figure 10, the weight of x_1a_1 and y_1a_1 are reinforced by $\omega_{x_1a_1} = \omega_{x_1a_1} + (\frac{1}{M})^2 \mu R$ and $\omega_{y_1a_1} = \omega_{y_1a_1} + \frac{1}{M} \mu R$, respectively.

Rationality Theorem of PS in the Multi-agent Environments

In order to preserve the rationality in the multi-agent environments discussed in the previous section, all irrational rules in a goal-agent set must be suppressed. On the other hand, if a goal-agent set is constructed by the agents that all irrational rules have been suppressed, we can preserve the rationality. Therefore, we can derive the necessary and sufficient condition about the range of μ to suppress all irrational rules in some goal-agent set.

Theorem 2 (The Rationality Theorem of PS in the Multi-agent Environments)

In the POMDPs where there is no type 2 confusion and the number of types of a reward is one, any irrational rule in some goal-agent set can be suppressed if and only if

$$\mu < \frac{M - 1}{M^W (1 - (\frac{1}{M})^{W_0})(n - 1)L}, \qquad (3)$$

where M is the upper bound of conflicting rules in the same sensory input, L is the upper bound of conflicting rational rules, n is the number of agents, and W and W_0 are the upper bounds of the reinforcement intervals for a direct and indirect-reward agents, respectively. ∎

The proof is presented in Appendix B.

Theorem 2 is derived by avoiding the least desirable situation where expected reward per an action is zero. Therefore, if we use this theorem, we can expect multiple efficient aspects of indirect rewards including improvement of learning speeds and qualities.

We cannot know the number of L in general. However, in practice, we can set $L=M-1$. We cannot know the number of W in general. However, in practice, we can set $\mu=0$, if a reinforcement interval is larger than some number that is determined in advance, or the assumption of the initial sensory input for a goal-direct agent has been broken. If we set $L=M-1$ and $W_0=W$ where indirect-reward agents have the same reinforcement interval of direct-reward agent, Equation (3) is simplified as follows;

$$\mu = \frac{1}{(M^W - 1)(n - 1)}. \qquad (4)$$

Figure 11. Roulette-like environments

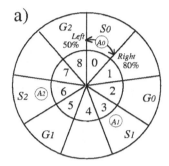 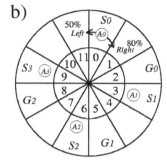

Application to Roulette-like Environments

Setting

Consider the roulette-like environments in Figure 11. There are 3 and 4 learning agents in the roulette a) and b), respectively. The initial position of an agent $i(A_i)$ is S_i. The number shown in the center of both the roulettes (from 0 to 8 or 11) is given to each agent as a sensory input. There are two actions for each agent; move right (20% failure) or move left (50% failure). If an action fails, the agent cannot move. There is no situation where another agent gets the same sensory input. At each discrete time step, A_i is selected based on the selection probabilities $P_i(P_i > 0, \sum_{i=1}^{n} P_i = 1)$. (P_0, P_1, P_2) is $(0.9, 0.05, 0.05)$ for the roulette a), and (P_0, P_1, P_2, P_3) is $(0.72, 0.04, 0.04, 0.2)$ for the roulette b). When A_i reaches the goal $i(G_i)$, P_i sets 0.0 and P_j $(j \neq i)$ are modified proportionally.

When A_r reaches G_r on condition that A_i $(i \neq R)$ have reached G_i, the direct reward $R(=100.0)$ is given to A_R and the indirect rewards μR are given to the other agents. When some agent obtains the direct reward or Ai reaches Gj $(j \neq i)$, all agents return to the initial position shown in Figure 11. The initial weights for all rules are 100.0.

If all agents learn the policy '*move right in any sensory input*', it is the optimal. If at least two agents learn the policy '*move left in the initial position*' or '*move right in the initial position, and move left in the right side of the initial position*', it is an irrational. When the optimal policy does not have been destroyed in 100 episodes, the learning is judged to be successful. We will stop the learning if agent 0, 1 and 2 learn the policy '*move left in the initial position*' or the number of actions is larger than 10 thousand. Initially, we set $W=3$. If the length of an episode is larger than 3, we set $\mu=0$. From Equation (4), we set $\mu<0.0714...$ for the roulette a) and $\mu<0.0333...$ for the roulette b) to preserve theorem 2.

Results and Discussion

We show *the quality*, that is evaluated by acquiring times of an irrational or the optimal policies in a thousand different trials where random seeds are changing, and *the speed*, that is evaluated by total action numbers to learn a thousand the optimal policies in Figure 12. Figure 12a) and 1b) are the results of the roulette a) and b), respectively. Figure 13a) and b) are details of the speeds in the roulette a) and b), respectively.

Though theorem 2 satisfies the rationality, it does not guarantee the optimality. However, in both the roulettes, the optimal policy always has been learned beyond the range of theorem 2.

Figure 12. The results of the roulette-like environments in Figure 11

a)

μ	learning qualities		learning speeds	
	irrational	optimal	Ave.	S.D.
0.0	0	1000	1201.1	273.0
10^{-6}	0	1000	1031.2	119.4
0.07	0	1000	946.7	107.0
0.3	**0**	**1000**	**900.7**	**172.8**
0.4	1	999	910.3	221.3
1.0	4	939	1120.0	794.3

b)

μ	learning qualities		learning speeds	
	irrational	optimal	Ave.	S.D.
0.0	0	0	–	–
10^{-6}	0	1000	2690.2	263.8
0.03	0	1000	2570.6	265.2
0.2	**0**	**1000**	**2474.9**	**402.6**
0.4	1	998	2671.8	945.2
1.0	13	909	3103.8	1561.6

Figure 13. Details of the learning speeds in the roulette a) and b)

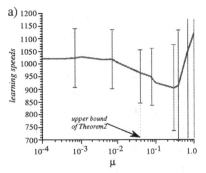
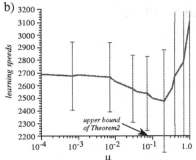

In the roulette a), $\mu=0.3$ makes the learning speed the best (Figure 12a)), Figure 13a)). On the other hand, if we set $\mu \geq 0.4$, there is a case that irrational policies have been learned. For example, consider the case that A_0, A_1 and A_2 in the roulette a) get three rule sequences in Figure 14. In this case, if we set $\mu=1.0$, A_0, A_1 and A_2 approach to G_2, G_0 and G_1, respectively. If we set $\mu<0.0714...$, such irrational policies do not have been learned. Furthermore, we have improved the learning speeds. Though it is possible to improve the learning speeds beyond the range of theorem 2, we should preserve it to guarantee the rationality in any environment.

In the roulette b), A_3 cannot learn anything because there is no G_3. Therefore, if we set $\mu=0$, the optimal policy does not have been learned (Figure 12b)). In this case, we should use the indirect reward. Figure 12b) and Figure 13b) show that $\mu=0.2$ makes the learning speed the best. On

the other hand, if we set $\mu \geq 0.3$, there is a case that irrational policies have been learned. It is an important property of the indirect reward that the learning qualities exceed those of the case of $\mu=0$.

Though theorem 2 only guarantees the rationality, numerical examples show that it is possible to improve the learning speeds and qualities.

EXPANSION OF PS TO POMDPS

Approaches to POMDPs

The traditional approach to POMDPs is the *memory-based approach* (Chrisman, 1992; Ma-Callum, 1995; Boutilier et al., 1996) that uses the history of sensor-action pairs or a model to identify the environmental states corresponding a partially observable (PO) sensory input. Although the memory-based approach can attain the opti-

Figure 14. An example of rule sequences in the roulette a)

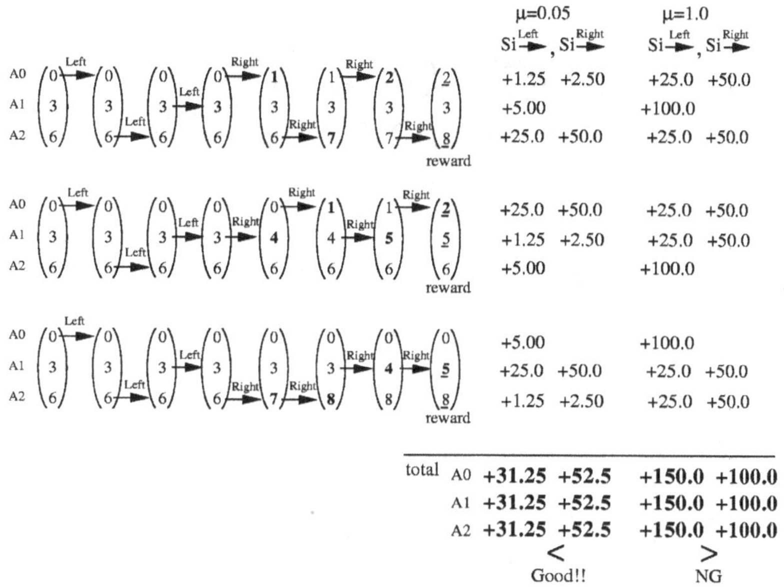

	$\mu=0.05$		$\mu=1.0$	
	$Si\xrightarrow{\text{Left}}, Si\xrightarrow{\text{Right}}$		$Si\xrightarrow{\text{Left}}, Si\xrightarrow{\text{Right}}$	
A0	+1.25	+2.50	+25.0	+50.0
A1	+5.00		+100.0	
A2	+25.0	+50.0	+25.0	+50.0
A0	+25.0	+50.0	+25.0	+50.0
A1	+1.25	+2.50	+25.0	+50.0
A2	+5.00		+100.0	
A0	+5.00		+100.0	
A1	+25.0	+50.0	+25.0	+50.0
A2	+1.25	+2.50	+25.0	+50.0

total				
A0	+31.25	+52.5	+150.0	+100.0
A1	+31.25	+52.5	+150.0	+100.0
A2	+31.25	+52.5	+150.0	+100.0
	<		>	
	Good!!		NG	

mality, it is hardware intensive since it requires a huge memory.

To resolve the problem using the memory-based approach, a *stochastic policy* (Singh et al., 1994) is proposed, where the agent selects an action based on the non-zero probability of every action in every sensory input in order to escape the PO sensory inputs. The simplest stochastic policy is the *random walk* (RA) that assigns the same probability to every action. On the other hand, the existing RL systems of learning a stochastic policy (Williams, 1992; Jaakkola et al., 1994; Kimura et al., 1995; Baird et al., 1999; Konda et al., 2000; Sutton et al., 2000; Aberdeen and Baxter, 2002; Perkins, 2002) are types of hill-climbing methods. They are often used in POMDPs since they can attain a *local optimum*. However, they cannot always improve RA. Furthermore, we know of a case where they change for a policy worse than RA.

For example, in Figure 15, the average number of steps required obtain a reward by RA is 4.

If the agent selects action-a in S0 and S1, it can obtain a reward in 3 steps, that is the minimum number of steps required to obtain a reward. On the other hand, if the agent selects the action-b in S0, it requires 4 steps to obtain a reward that is the same as RA. When we improved RA using the *Stochastic Gradient Ascent* (SGA) (Kimura et al., 1995), that is a type of hill-climbing methods, the average number of steps required to obtain a reward was 3.78 in 100 trials. Although 25 trials in 100 trials were able to improve RA, 73 trials were the same as RA and the other 2 trials resulted in a deteriorated RA.[2]

The hill-climbing methods that were used previous RL systems resulted in the stochastic policy function being learned well in many cases. However, once they converged to a local optimum i.e., a policy worse than RA, it is not possible for them to improve it. This implies that they change for a policy worse than RA. To avoid the fault, we focus on PS that belongs in XoL and is not a hill-climbing approach.

Figure 15. The environment demonstrating the limit of SGA

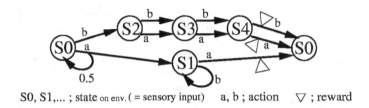

S0, S1,... ; state on env. (= sensory input)   a, b ; action   ▽ ; reward

Figure 16. The PS-r algorithm

```
procedure  PS-r ;
  begin
    Initializes the 1st memory.
    All rules are regarded as irrational rules and initialized by r .
  do
    Senses the environment as sensory input s.
    Selects an an action a based on RA.
    Describes a in the current sensory input on the 1st memory.
    if gets a reward then
       the points r are updated to 1 in all rules on the 1st memory.
    while
  end;
```

The PS-r Algorithm

Proposition of PS-r

The rationality theorem of PS guarantees rationality in which there is no type 2 confusion. In the first, we propose *PS-r* to analyze the behavior of PS in which there is a type 2 confusion.

The principal objective of the rationality theorem is determining the rational rules. This can be implemented by the *1st memory* that is an array whose length is the number of sensory inputs. After an action is selected, the action is described in the current sensory input on the 1st memory. If the agent obtains a reward, the contents of the 1st memory are rational rules. PS-r is an algorithm to learn a rational policy by the 1st memory.

The PS-r algorithm is shown in Figure 16. At the beginning of learning, all rules are regarded as irrational and are initialized by r ($0 < r < 1$)

points. If a rule is regarded as rational, r of the rule is updated to 1, which means that is a rational rule. Once r is set to 1, this value is maintained throughout. While learning is in progress, an action is selected based on RA. Therefore, in practice, PS-r can determine all rational rules. When we have evaluated or utilized a policy that is learned by PS-r termed *Policy(PS-r)*, an action is selected based on the *roulette selection* in proportion to r.

The rationality theorem of PS is always reinforced by a rational rule as opposed to an irrational rule in the same sensory input. On the other hand, PS-r is an abstract algorithm of PS wherein the values of all rules are evaluated by r or 1.

Properties of PS-r

If there is no type 2 confusion in an environment, it is evident that the average number of steps required to obtain a reward by *Policy(PS-r)* is not larger than that of RA. On the other hand, we can derive the following theorem in the POMDPs where there is a type 2 confusion and the number of types of a reward is one.

Theorem 3 (Comparison between PS-r and RA in the POMDPs)

The maximum value of the average number of steps to obtain a reward by Policy(PS-r) divided by that of RA in the POMDPs where the number of types of a reward is one is given by

$$r \frac{(1 + \frac{M-1}{r})^n}{M^n} \tag{5}$$

where n is the upper bound of the number of different environmental states that are sensed as the same sensory input. ∎

The proof is presented in Appendix C.

Theorem 3 is derived from the worst case, where an environment is constructed by the most difficult environmental structure termed the structure Ω (see Figure 24 in Appendix C) only. Therefore, if there is no structure Ω in an environment, the behavior of *Policy(PS-r)* will be better than that estimated using Equation (5). Furthermore, when there is no type 2 confusion in some part of an environment and there is an irrational rule in it, its behavior will increasingly improve.

From PS-r to PS-r*

Improvement of PS-r to fit on POMDPs

If we do not identify the environmental states that correspond to a PO sensory input, we should use a stochastic policy to escape from the PO sensory inputs. Although the simplest stochastic policy is RA, the existing RL systems to learn a stochastic policy cannot always improve RA;

however, it is also possible for them to deteriorate it. On the other hand, we do not have to select an irrational rule in the non-PO (¬ PO) sensory inputs. Therefore, we only select a rational rule in them and follow a stochastic policy in the other sensory inputs. In particular, we use RA as the stochastic policy to avoid change for a policy worse than RA.

To implement the above idea, it is important to judge whether sensory inputs are PO. If a sensory input is ¬ PO, each transition probability to one of the following sensory inputs by a rule that has been selected on the sensory input converges to some constant value, even if an action is selected based on different policies. On the other hand, if a sensory input is PO, the transition probability will be changed depending on the policy that is used to reach the sensory input. We aim at judging whether sensory inputs are PO, by comparing them with the transition probabilities between RA and the other policy.

The comparison is executed using the χ^2-goodness-of-fit test. It only requires transition probabilities to the following sensory inputs by all rules. Therefore, it requires a memory of $O(MN^2)$, where M and N are the numbers of actions and sensory inputs, that is less than previous memory-based approaches. After the test of a rule based on each RA, and provided the other policy enough sampling, if a transition probability to one of the following sensory inputs is not coincident between RA and the other policy, we can regard the sensory input in which the rule can be selected as PO. Otherwise, if all the transition probabilities are coincident, the sensory input can be regarded as ¬ PO. It should be noted that although it is possible not to determine a part of the PO sensory inputs, it could be resolved by changing the policy that has been compared with RA.

In general, we require several actions in order to execute a correct χ^2-goodness-of-fit test. When we set the *significant level* and *detection power* as α and $1 - \beta$, respectively, the number of actions (n) required to achieve the correct test regarding a transition by a rule can be statistically estimated by the following criteria;

$$n = \frac{1}{2}\left(\frac{u(\alpha) + u(2\beta)}{sin^{-1}\sqrt{\pi_1} - sin^{-1}\sqrt{\pi_2}} \right)^2 \qquad (6)$$

where π_1 and π_2 are transition probabilities of the following sensory input by the rule when RA and the other policy are used, respectively. In addition, $u()$ is derived using a *normal distribution table*, for example, by setting $\alpha = 0.05$ and $\beta = 0.10$, $u(\alpha) = 1.960$ and $u(2\beta)=1.282$.

The PS-r* Algorithm

We propose PS-r* to implement the above idea. We show the algorithm in Figure 17. It is mainly divided into the learning mode and the test mode. It requires the following five types of memory: the 1st memory to determine rational rules that are the same as PS-r, $\neg$ PO-judge flags to judge whether sensory inputs, whose length are the number of a rule, are PO, PO flags to store the result of the judgment (PO/$\neg$ PO/unknown) regarding the sensory inputs that are unknown during initialization, the number of ways of selecting each rule in two modes (*NofR(learning)* and *NofR(test)*) that requires a memory of $O(MN)$, and the number of transitions to each following sensory input by each rule in two modes (*NofT(learning)* and *NofT(test)*) that requires a memory of $O(MN^2)$.

PS-r* starts from the learning mode. In the learning mode, the agent selects an action based on RA to determine all rational rules. Rational rules are determined using the same algorithm as PS-r. If all *NofR(learning)* are larger than *CV* that is calculated by the upper bound of Equation (6), the mode is changed to the test one. If we set $\alpha = 0.05$, $\beta = 0.10$, and $max|\pi_1\_\pi_2|=0.05$, which means that the maximum error of estimation of a transition probability is 0.05, *CV* is 2059.09.

In the test mode, if the agent senses an unknown sensory input, it returns to the learning mode. Otherwise, an action is selected based on *Policy(PS-r*(test))* that is the policy learned in the learning mode. Usually, an action is selected by the roulette selection in proportion to *r* in *Policy(PS-r*(test))*. However, we can use another policy such that some action is not selected to improve the accuracy of the χ^2-goodness-of-fit test.

If *NofR(test)* for a rule that can be selected by the sensory inputs, where they do not decide whether PO or $\neg$ PO is larger than *CV*, the χ^2-goodness-of-fit test between *NofT(learning)* and *NofT(test)* for the rule is executed. If the result of the test indicates that they are not coincident, the sensory input that can select the rule is PO. Otherwise, a $\neg$ PO-judge flag for the rule will be raised. Subsequently, if all the $\neg$ PO-judge flags for rules that can be selected on the same sensory input are raised, the sensory input is $\neg$ PO. PS-r* is stopped when all *NofR(test)* are larger than *CV*.

When we have evaluated or utilized a policy that is learned by PS-r* termed *Policy(PS-r)*, an action is selected based on the roulette selection in proportion to r, where *r* is set to 0 if it is less than 1 in the $\neg$ PO sensory inputs. Therefore, *Policy(PS-r*)* is coincident to RA if all the sensory inputs are PO. On the other hand, if there exist several irrational rules in the $\neg$ PO sensory inputs, *Policy(PS-r*)* increasingly better than RA.

Features of PS-r*

Policy(PS-r)* is coincident to RA in the PO sensory inputs. Therefore, if all sensory inputs are PO, it is the most difficult task for PS-r*.

On the other hand, *Policy(PS-r*)* selects a rational rule in the $\neg$ PO sensory inputs. Therefore, if there exist many irrational rules in the $\neg$ PO sensory inputs, *Policy(PS-r*)* is increasingly better than RA. They are very important properties of PS-r* that are not guaranteed in PS-r and the

Figure 17. The PS-r algorithm*

```
procedure  PS-r * ;
  begin
    Initiallizes the 1st memory, ¬PO-judge flags, PO flags,
      NofR(learning/test) and NofT(learning/test).
    Sets max|π_r−π_a|, and calculates CV by Eq. (6).
    All rules are regarded as irrational and initialized by r.
    Senses the environment as sensory input s.
    call(learnning mode).
  end;

procedure  learning mode ;
  begin
    do
      Selects an action a based on RA.
      Updates NofR(learning).
      Describes a in the current sensory input on the 1st memory.
      if gets a reward then
        the points r are updated to 1 for all rules on the 1st memory.
      if all of NofR(learning) > CV then
        call(test mode) with Policy(PS-r*(test)).
      Senses the environment as sensory input s.
      Updates NofT(learning).
    while
  end;

procedure  test mode ;
  begin
    Senses the environment as sensory input s.
    do
      if s is an unknown sensory input then call(learning mode).
      else selects an action a based on Policy(PS-r*(test)).
      Updates NofR(test).
      if NofR(test) for a rule that can be selected on the sensory
        input where they are not decided to PO or ¬PO > CV
        then executes χ²-goodness-of-fit test about between
          NofT(learning) and NofT(test) for the rule.
      if the result of the test says that they are not concident
        then the sensory input that can select the rule is PO.
      else raises a ¬PO-judge flag for the rule
        if all the ¬PO-judge flags for rules that can be selected
          on the same sensory input are raised
          then the sensory input is ¬PO.
      if all of NofR(test) > CV then break;
      Senses the environment as sensory input s.
      Updates NofT(test).
    while
    Outputs Policy(PS-r*).
  end;
```

existing RL systems when learning a stochastic policy.

PS-r* requires a memory of $O(MN^2)$. It is larger than PS-r, that only requires a memory of $O(MN)$. However, this value is much smaller than those of previous memory-based approaches.

Application to the Most Difficult Environmental Structure

Setting

We compare PS-r* with RA, PS-r, and SGA in the environment that is shown in Figure 18. In this environment, the different environmental states $Z_a, Z_b, Z_c,$ and Z_d are sensed as the same sensory input Z. In addition, it adopts the structure Ω, that is the most difficult environmental structure, in the sensory input Z. After the agent selects action-a in sensory input X, it moves the sensory inputs $S1$ and X with p and $1-p$ probabilities, respectively. States from S_1 to S_n are sensed as n different sensory inputs. If we adjust n and p, the average number of steps required to obtain a reward can be changed.

Figure 18. The environment in which PS-r is compared with RA, PS-r, and SGA*

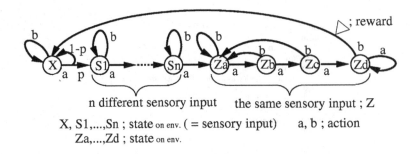

n different sensory input   the same sensory input ; Z

X, S1,...,Sn ; state on env. (= sensory input)   a, b ; action
Za,...,Zd ; state on env.

Comparison between RA and PS-r

We can calculate the average number of steps required to obtain a reward by RA and *Policy(PS-r)* as $\left[\left(\frac{1}{p}+1\right)+(2n)+(16)\right]$ and

$$\left[\left(\frac{1}{p}+r\right)+((1+r)n)+\left(\frac{(1+r)^4}{r^3}\right)\right],$$ respectively. Therefore, if we set $\left[n>\frac{1}{1-r}\left(\frac{(1+r)^4}{r^3}+(r-17)\right)\right]$, *Policy(PS-r)* is better than RA.

Policy(PS-r)* coincides with RA when the χ^2-goodness-of-fit test is not executed. If the test is completed, the average number of steps required to obtain a reward for *Policy(PS-r*)* is $\left[\frac{1}{p}+n+16\right]$, that is better than both RA and PS-r.

Comparison with SGA

Next, we compare PS-r* with SGA empirically. The learning parameter and discounted rate of SGA are 0.1 and 0.99, respectively. The initial policy given to SGA is RA. In PS-r*, since we set $\alpha = 0.05$, $\beta = 0.10$, and $max|\pi_1\_\pi_2|=0.05$, CV is 2095.09.

We list *the quality*, that is the average number of steps required to obtain a reward and its standard deviation of *Policy(PS-r*)* and SGA in the left side of Table 1. We have set $n=7$, $n=14$, and $n=21$. Furthermore, we list *the speed*, that is the average number of steps required to reach the quality and its standard deviation in the right side of Table 1.

PS-r* is better than SGA with regards to the quality, though it is worse than SGA in the speed. If n is larger, the quality of *Policy(PS-r*)* is much better than SGA. This is because SGA selects an irrational rule at the sensory input S_i that does not need a stochastic policy. On the other hand, an irrational rule is never selected by PS-r* at the sensory inputs. It is a very important property of PS-r* in comparison with the previous RL systems to learn a stochastic policy.

CONCLUSION

In this chapter, we have proposed *Exploitation-oriented Learning (XoL)* that is a new approach to goal-directed learning from interaction. XoL does not require a sophisticated design of the value of a reward signal and aims to pursue the rationality that can obtain a reward very quickly. We have focused on the *Partially Observed Markov Decision Processes* (POMDPs) environments and classified these difficulties into a type 1 and type 2 confusions. We have defined that a type 2 confusion is the most difficult property in POMDPs.

Table 1. The results of the comparison with PS-r* and SGA in Figure 18

	n	The Quality		The Speed	
		Ave.	S.D.	Ave.	S.D.
PS-r*	7	24.2	0.218	2.38×10^4	4.10×10^4
	14	31.2	0.237	1.73×10^5	2.48×10^4
	21	38.2	0.237	2.19×10^5	6.77×10^4
SGA	7	26.3	6.83	2.21×10^3	2.91×10^3
	14	38.0	10.9	4.27×10^3	4.67×10^3
	21	50.8	9.47	4.42×10^3	5.47×10^3

For the POMDPs environments where there is no type 2 confusion and the number of types of a reward is one, we have proved the *Rationality Theorem of Profit Sharing (PS)*. Next, we have proved the *Rationality Theorem of PS in multi-agent environments*. Last, we have analyzed the behavior of *PS-r*, that is an abstract algorithm of PS, in the POMDPs. Furthermore, we have proposed *PS-r**, that is an extended algorithm of PS-r, to fit on the POMDPs environments where there is a type 2 confusion and the number of types of a reward is one.

We have shown several numerical examples to support how to use these XoL methods. Also, we have shown that the performance of PS-r* is not less than that of *random walk* (RA) and it exhibits exceptional potential to improve RA using a lower memory than the previous memory-based approaches.

Our future projects include: improving RA in PS-r*, extending XoL to multi-dimensional reward and penalty environments and discovering efficient real-world applications, and so on.

REFERENCES

Abbeel, P., & Ng, A. Y. (2005). Exploration and apprenticeship learning in reinforcement learning. In *Proceedings of the Twentyfirst International Conference on Machine Learning* (pp. 1-8).

Aberdeen, D., & Baxter, J. (2002). Scalable Internal-State Policy-Gradient Methods for POMDPs. In *Proceedings of the Nineteenth International Conference on Machine Learning* (pp. 3-10).

Baird, L., & Poole, D. (1999). Gradient Descent for General Reinforcement Learning. *Advances in Neural Information Processing Systems, 11*, 968–974.

Boutilier, C., & Poole, D. (1996). Computing Optimal Policies for Partially Observable Decision Processes using Compact Representations. In *Proceedings of the Thirteenth National Conference on Artificial Intelligence* (pp. 1168-1175).

Chrisman, L. (1992). Reinforcement Learning with Perceptual Aliasing: The Perceptual Distinctions Approach. In *Proceedings of the Tenth National Conference on Artificial Intelligence* (pp. 183-188).

Gosavi, A. (2004). A Reinforcement Learning Algorithm Based on Policy Iteration for Average Reward: Empirical Results with Yield Management and Convergence Analysis. *Machine Learning, 55*, 5–29. doi:10.1023/B:MACH.0000019802.64038.6c

Grefenstette, J. J. (1988). Credit Assignment in Rule Discovery Systems Based on Genetic Algorithms. *Machine Learning, 3*, 225–245. doi:10.1007/BF00113898

Jaakkola, T., Singh, S. P., & Jordan, M. I. (1994). Reinforcement Learning Algorithm for Partially Observable Markov Decision Problems. *Advances in Neural Information Processing Systems, 7*, 345–352.

Kimura, H., Yamamura, M., & Kobayashi, S. (1995). Reinforcement Learning by Stochastic Hill Climbing on Discounted Reward. In *Proceedings of the Twelfth International Conference on Machine Learning* (pp. 295-303).

Konda, V. R., & Tsitsiklis, J. N. (2000). Actor-Critic Algorithms. *Advances in Neural Information Processing Systems, 12*, 1008–1014.

Liepins, G. E., Hilliard, M. R., Palmer, M., & Rangarajan, G. (1989). Alternatives for Classifier System Credit Assignment. In *Proceedings of the Eleventh International Joint Conference on Artificial Intelligent* (pp. 756-761).

McCallum, R. A. (1995). Instance-Based Utile Distinctions for Reinforcement Learning with Hidden State. In *Proceedings of the Twelfth International Conference on Machine Learning* (pp. 387-395).

Merrick, K., & Maher, M. L. (2007). Motivated Reinforcement Learning for Adaptive Characters in Open-Ended Simulation Games. In *Proceedings of the International Conference on Advanced in Computer Entertainment Technology* (pp. 127-134).

Miyazaki, K., & Kobayashi, S. (1998). Learning Deterministic Policies in Partially Observable Markov Decision Processes. In *Proceedings of the Fifth International Conference on Intelligent Autonomous System* (pp. 250-257).

Miyazaki, K., & Kobayashi, S. (2000). Reinforcement Learning for Penalty Avoiding Policy Making. In *Proceedings of the 2000 IEEE International Conference on Systems, Man and Cybernetics* (pp. 206-211).

Miyazaki, K., & Kobayashi, S. (2001). Rationality of Reward Sharing in Multi-agent Reinforcement Learning. *New Generation Computing, 91*, 157–172. doi:10.1007/BF03037252

Miyazaki, K., & Kobayashi, S. (2003). An Extension of Profit Sharing to Partially Observable Markov Decision Processes: Proposition of PS-r* and its Evaluation. [in Japanese]. *Journal of the Japanese Society for Artificial Intelligence, 18*(5), 286–296. doi:10.1527/tjsai.18.286

Miyazaki, K., Yamaumra, M., & Kobayashi, S. (1994). On the Rationality of Profit Sharing in Reinforcement Learning. In *Proceedings of the Third International Conference on Fuzzy Logic, Neural Nets and Soft Computing* (pp. 285-288).

Ng, A. Y., Harada, D., & Russell, S. J. (1999). Policy Invariance Under Reward Transformations: Theory and Application to Reward Shaping. In *Proceedings of the Sixteenth International Conference on Machine Learning* (pp. 278-287).

Ng, A. Y., & Russell, S. J. (2000). Algorithms for Inverse Reinforcement Learning. In *Proceedings of the Seventeenth International Conference on Machine Learning* (pp. 663-670).

Perkins, T. J. (2002). Reinforcement Learning for POMDPs based on Action Values and Stochastic Optimization. In *Proceedings of the Eighteenth National Conference on Artificial Intelligence* (pp. 199-204).

Singh, S. P., Jaakkola, T., & Jordan, M. I. (1994). Learning Without State-Estimation in Partially Observable Markovian Decision Processes. In *Proceedings of the Eleventh International Conference on Machine Learning* (pp. 284-292).

Sutton, R. S. (1988). Learning to Predict by the Methods of Temporal Differences. *Machine Learning, 3*, 9–44. doi:10.1007/BF00115009

Sutton, R. S., & Barto, A. (1998). *Reinforcement Learning: An Introduction.* Cambridge, MA: MIT Press.

Sutton, R. S., McAllester, D., Singh, S. P., & Mansour, Y. (2000). Policy Gradient Methods for Reinforcement Learning with Function Approximation. *Advances in Neural Information Processing Systems, 12,* 1057–1063.

Watkins, C. J. H., & Dayan, P. (1992). Technical note: Q-learning . *Machine Learning, 8,* 55–68. doi:10.1023/A:1022676722315

Williams, R. J. (1992). Simple Statistical Gradient Following Algorithms for Connectionist Reinforcement Learning. *Machine Learning, 8,* 229–256. doi:10.1007/BF00992696

ENDNOTES

[1] Remark that the highest state value is 1 that is assigned in state 2.

[2] Theoretically, the performance of SGA is not less than that of RA; however, in practice, the property does not always hold true since some assumption that is required by the theory has been broken.

APPENDIX A: PROOF OF THEOREM 1

We derive the necessary and sufficient condition to suppress irrational rules in the POMDPs where there is no type 2 confusion and the number of types of a reward is one. In the first, we consider the local rationality that can suppress any irrational rule (Theorem A.1). It is derived by two lemmas (Lemma A.1 and A.2). We characterize a conflict structure where it is the most difficult to suppress irrational rules (Lemma A.1). For two conflict structures A and B, we say A is more difficult than B when the class of reinforcement functions that can suppress any irrational rule of A is included in that of B. Then, we derive the necessary and sufficient condition to suppress any irrational rule for the most difficult conflict structure (Lemma A.2).

Lemma A.1 (The Most Difficult Conflict Structure)

The most difficult conflict structure has only one irrational rule with a self-loop.

Figure 19 show the most difficult conflict structure where only one irrational rule with a self-loop conflicts with L rational rules.

Proof of lemma A.1

Although we set $L=1$, we can easily extend it to any number. Reinforcement of an irrational rule makes it difficult to learn a rational policy under any reinforcement function. Therefore, the difficulty of a conflict structure varies monotonically with the number of reinforcements for irrational rules. We enumerate conflict structures according to the *branching factor b*, that is the number of state transitions in the same sensory input, the *conflict factor c*, that is the number of conflicting rules in it, and examine the number of reinforcements for irrational rules.

$b=1$: It is clearly not difficult since there are no conflicts (Figure 20a).

$b=2$: When there are no conflicts (Figure 20b), it is the same as $b=1$. We divide the structures of $c=2$ into two subclasses. One contains a self-loop (Figure 20c), and the other does not it (Figure 20d). In the case given in Figure 20c, there is a possibility that the self-loop rule is repeatedly selected, while the non-self-loop rule is selected at most once. Therefore, if the self-loop rule is irrational, it will be reinforced more than the irrational rule of Figure 20d.

$b \geq 3$: When there are no conflicts (Figure 20e), it is the same as $b=1$. Consider the structure of $c=2$ (Figure 20f). Although the most difficult case is that the conflict structure has an irrational rule as a self-loop, even such a structure is less difficult than Figure 20c. Considering the structure of $c=3$ (Figure 20g), two of the conflict rules are irrational. Therefore, the expected number of reinforcement for one irrational rule is less than that of Figure 20f.

Similarly, conflict structures of $b>3$ are less difficult than Figure 20c.

From the above discussion, it is concluded that the most difficult conflict structure is expressed in Figure 20c. Q.E.D.

Lemma A.2 (Suppressing Only One Irrational Rule with a Self-Loop)

Only one irrational rule with a self-loop can be suppressed if and only if suppression conditions hold.

Figure 19. The most difficult conflict structure

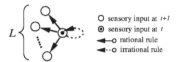

Figure 20. Conflict structures used in the proof

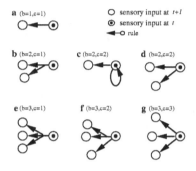

Proof of lemma A.2

Although we set $L=1$, we can easily extend it to any number. If the rational rule of the most difficult conflict structure is reinforced by f_N, the value of the reinforcement for the conflicting irrational rule becomes maximal when it has been selected W-N times before the selection of the rational rule. Subsequently, the weight of the irrational rule is increased by $f_{N+1} + \cdots + f_W$, and that of the rational rule by f_N. From a viewpoint of rationality, the increased weight of the rational rule must be larger than that of the irrational rule. Such a condition must hold for any later part of the reinforcement interval. Therefore, suppression conditions are necessary. The sufficiency is evident. Q.E.D.

Using *the law of transitivity*, the following theorem is directly derived from these lemmas.

Theorem A.1 (Irrational Rules Suppression Theorem)

Any irrational rule can be suppressed if and only if suppression conditions hold.

Theorem A.1 guarantees that the local rationality in reinforcement learning will suppress any irrational rule. A policy that is constructed by rational rules satisfies local rationality. However, we can construct an example such that the policy is irrational. Next, we discuss the global rationality that can learn a rational policy.

In the global rationality, the necessity of suppression conditions is evident. We investigate the sufficiency (Lemma A.3).

Figure 21. The environment used in the proof

Lemma A.3 (Sufficiency of Suppression Conditions)

If PS satisfies suppression conditions, it can learn a rational policy in the POMDPs where there is no type 2 confusion and the number of types of a reward is one.

Proof of lemma A.3

Although we set $L=2$, we can easily extend it to any number. If a policy is irrational, the policy has to contain a *rewardless loop* that does not include any rewards. We need at least two episodes to construct a rewardless loop. Furthermore, the following inequalities must be maintained in sensory inputs x and y, that are the exits in the loop of Figure 21,

$$\Delta\omega_{xo} < \Delta\omega_{xi}, \Delta\omega_{yo} < \Delta\omega_{yi} \tag{7}$$

where xo and yo are rules that exit the loop, xi and yi are rules that enter the loop, and Δ represents the total reinforcement to be accumulated in a rule.

If the episode with xi contains xo and the irrational rule suppression theorem is satisfied, from theorem A.1, $\Delta\omega_{xo} > \Delta\omega_{xi}$. Therefore, the episode with xi requires the rule that is not xo to exit the loop. This also applied to yi. The following inequalities are derived using theorem A.1.

$$\Delta\omega_{yo} > \Delta\omega_{xi} + \Delta\omega_{yi}, \quad \Delta\omega_{xo} > \Delta\omega_{yi} + \Delta\omega_{xi} \tag{8}$$

The following inequality is derived using Equation (7) and (8).

$$\Delta\omega_{xi} + \Delta\omega_{yi} > \Delta\omega_{xo} + \Delta\omega_{yo} > 2(\Delta\omega_{xi} + \Delta\omega_{yi}) \tag{9}$$

There is no solution to satisfy the inequality as $\Delta\omega > 0$. Then, we cannot construct a rewardless loop. Therefore, if we use reinforcement functions that satisfy suppression conditions, we can always obtain a rational policy. Q.E.D.

Theorem 1 is directly derived from this lemma. Q.E.D.

APPENDIX B: PROOF OF THEOREM 2

First, we derive the necessary and sufficient condition to suppress any irrational rule for the most difficult conflict structure. If it can be derived, we can extend it to any conflict structure by using *the law of transitivity*.

From lemma A.1, the most difficult conflict structure has only one irrational rule with a self-loop. In this structure, we can derive the following lemma.

Lemma B.1 (Suppressing Only One Irrational Rule with a Self-Loop)

Only one irrational rule with a self-loop in some goal-agent set can be suppressed if and only if Equation (3)

Proof of lemma B.1

For any reinforcement interval k ($k=0,1,\dots,W$-1) in some goal-agent set, we show that there is j ($j=1,2,\dots$,L) satisfying the following condition,

$$\omega_{ij}^k > \omega_{i0}^k \tag{10}$$

where ω_{ij}^k and ω_{i0}^k are weights of jth rational rule (r_{ij}^k) in agent i ($i=0,1,\dots,n$-1) and only one irrational rule with a self-loop (r_{i0}^k) in the agent, respectively (Figure 22).

First, we consider the ratio of the selection number of (r_{ij}^k) to (r_{i0}^k). When $n'=1$ (the number of the goal-agent set is one) and L rational rules for each agent are selected by all agents in turn, the minimum of the ratio is maximized (Figure 23). In this case, the following ratio holds (Figure 24),

$$r_{ij}^k : r_{i0}^k = 1 : (n-1)L \tag{11}$$

Second, we consider weights given to (r_{ij}^k) and (r_{i0}^k). When the agent that obtains the direct reward senses no similar sensory input in W, the weight given to (r_{ij}^k) is minimized. It is $\dfrac{R}{M^{W-1}}$ in $k=W$. On the other hand, when agents that obtain the indirect reward sense the same sensory input in W, the weight given to r_{i0}^k is maximized. It is $\mu R \dfrac{M}{M-1}\left(1-\left(\dfrac{1}{M}\right)^{W_0}\right)$ in $W \geq W_0$.

Figure 22. The most difficult multi-agent structure

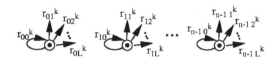

Figure 23. The rule sequence:

$$
\begin{aligned}
&\big(r_{01}^k, r_{10}^k, \cdots, r_{n-1\ 0}^k, r_{02}^k, r_{10}^k, \cdots, r_{n-1\ 0}^k, \cdots, r_{0L}^k, r_{10}^k, \cdots, r_{n-1\ 0}^k, r_{00}^k, r_{11}^k, \cdots, r_{n-1\ 0}^k, r_{00}^k, r_{12}^k, \cdots, r_{n-1\ 0}^k, \\
&\cdots, r_{00}^k, r_{1L}^k, \cdots, r_{n-1\ 0}^k, \cdots, r_{00}^k, r_{10}^k, \cdots, r_{n-1\ 1}^k, r_{00}^k, r_{10}^k, \cdots, r_{n-1\ 2}^k, \cdots, r_{00}^k, r_{10}^k, \cdots, r_{n-1\ L}^k\big)
\end{aligned}
$$

or all agents on some k. For example, if 'x changes to O1' or 'O2 changes to O1' in this figure, the learning in the agent that can select the changing rule occurs more easily

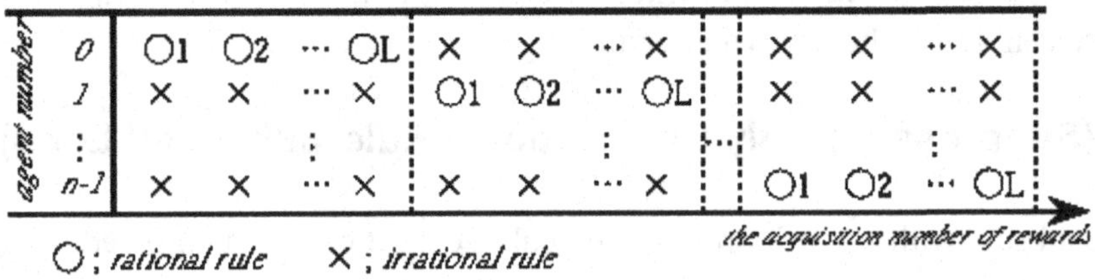

Figure 24. Sample rule sequences at n=3 and L=3. Though the sequence 1 has some partial selection of rules, the sequence 2 does not have it. The sequence 2 corresponds to Figure 23. Sequence 1 is more easily learned than sequence 2 as discussed on Figure 23.

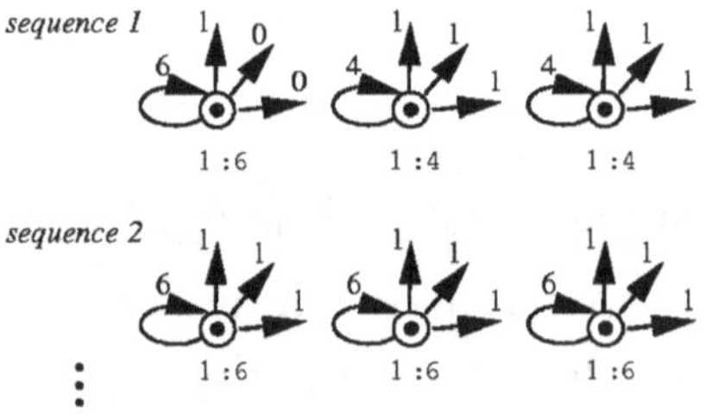

Therefore, it is necessary for satisfying condition Equation (10) to hold the following condition,

$$
\frac{R}{M^{W-1}} > \mu R \frac{M}{M-1}\left[1 - \left(\frac{1}{M}\right)^{W_0}(n-1)L\right], \tag{12}
$$

that is,

$$
\mu < \frac{M-1}{M^W\left[1 - \left(\dfrac{1}{M}\right)^{W_0}(n-1)L\right]}. \tag{13}
$$

It is clearly the sufficient condition. Q.E.D.
Theorem 2 is directly derived from this lemma. Q.E.D.

APPENDIX C: PROOF OF THEOREM 3

First, we show the *most difficult environmental structure*, where the average number of steps required to obtain a reward by *Policy(PS-r)* is the worst in comparison with RA (Lemma C.1). Next, we analyze its behavior in the structure (Lemma C.2). If it can be derived, we can extend it to all the classes of the POMDPs where the number of types of a reward is one.

Lemma C.1 (The Most Difficult Environmental Structure)

The most difficult environmental structure is an environment that is shown in Figure 25 where there exist M−1 actions that are the same as action-b. We term it as structure Ω.

Proof of lemma C.1

If all rules are rational, there is no difference between PS-r and RA. Therefore, we treat the case as one where there is an irrational rule.

(i)   *M*=2

In structure Ω, the rules that are constructed by action-a and b are regarded as irrational and rational, respectively. If we select action-a in the state that is not state B, we can approach a reward. On the other hand, if we select action-b in its states, we move to state A, that is the furthest from a reward. Therefore, in structure Ω at *M*=2, if we select a rational rule, the number of steps required to obtain a reward will be larger. In a structure that has a lesser effect than the case of structure Ω, the difference between PS-r and RA reduces. Therefore, in the case of M=2, structure Ω requires the largest number of steps to obtain a reward by PS-r in comparison with that of RA.

(ii)   *M*>2

Figure 25. The most difficult environmental structure: structure Ω

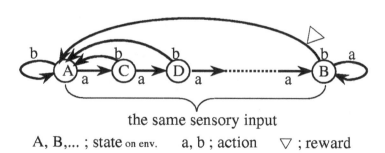

the same sensory input

A, B,... ; state on env.   a, b ; action $\triangledown$; reward

At first, we consider the case where the other irrational rules are added to structure Ω at M=2. If the selection probabilities of their rules are not zero, the average number of steps required to obtain a reward should be larger than that of structure Ω at M=2. In RA, the selection probability of the rule is $\frac{1}{M}$. On the other hand, in PS-r, it is the less than $\frac{1}{M}$. Therefore, if we compare the same structure, when the other irrational rules are added, the difference between RA and PS-r reduces.

Next, we consider the case where the other rational rules are added to structure Ω at M=2. In this case, when all the other rules are the same as the rule that is constructed by action-b, that is the largest average number of steps required to obtain a reward by PS-r in comparison with RA.

Therefore, structure Ω requires the largest number of steps to obtain a reward by PS-r in comparison with RA. Q.E.D.

Lemma C.2 (Comparison Between PS-r and RA in Structure Ω)

In structure Ω, the maximum value of the average number of steps required to obtain a reward by Policy(PS-r) divided by that of RA is given by Equation (5)

Proof of lemma C.2

In structure Ω, the average number of steps required to obtain a reward (V_a) is $V_a = \dfrac{1}{s(1-s)^{n-1}}$ where s is the selection probability of action-b where there exist the same M-1 actions. We can calculate $s = \dfrac{M-1}{M}$ and $\dfrac{M-1}{(M-1)+r}$ for RA and PS-r, respectively. Therefore, calculating V_a] for each s, we can get the rate $r \dfrac{\left(1 + \dfrac{M-1}{r}\right)^n}{M^n}$. Q.E.D.

Theorem 3 is directly derived from these lemmas. Q.E.D.

Section 7
Miscellaneous

Chapter 16
Pheromone–Style Communication for Swarm Intelligence

Hidenori Kawamura
Hokkaido University, Japan

Keiji Suzuki
Hokkaido University, Japan

ABSTRACT

Pheromones are the important chemical substances for social insects to realize cooperative collective behavior. The most famous example of pheromone-based behavior is foraging. Real ants use pheromone trail to inform each other where food source exists and they effectively reach and forage the food. This sophisticated but simple communication method is useful to design artificial multiagent systems. In this chapter, the evolutionary pheromone communication is proposed on a competitive ant environment model, and we show two patterns of pheromone communication emerged through co-evolutionary process by genetic algorithm. In addition, such communication patterns are investigated with Shannon's entropy.

INTRODUCTION

Swarm intelligence is a type of artificial intelligence based on the collective behavior of decentralized, self-organized systems (Dorigo & Theraulaz, 1999). The knowledge and information processors in swarm intelligence are widely decentralized in parts of the system, which are called agents. All agents basically have a decision-making mechanism obtained from local knowledge, by local-information processing, and from communication channels with cooperative agents.

Swarm behavior is the aggregation of such local interactions. The usability of multi-agent systems and swarm intelligence is well-known in various applications in robotics, optimization problems, distributed computing, Web services, and mobile technologies. The main topic of such applications has always been how to design local agents that will emerge to demonstrate sophisticated global behavior. We can only design local behavior by implementing agents, although we need sophisticated global behavior. To obtain good designs, we must understand the relationship between local design and emerging results.

DOI: 10.4018/978-1-60566-898-7.ch016

One good way of introducing such relationships is provided by nature. Real ants and bees are called social insects. Their colonies consist of many members who attend to various jobs to preserve the life of each colony (Sheely, 1995). The sizes of colonies are much too large for members, even queens, to comprehend all activities and information. In other words, although individual abilities to assess the condition of each colony are limited, they can still do their work based only on this limited information. The total activities of colonies, e.g., defense against enemies, repairing nests, childcare, and foraging, emerge due to the aggregation of such individual behaviors. Moreover, colonies must balance out the total work appropriately according to changes in the situation to optimize their operating costs. It is easy to see that communication between members is very important to attend to the many matters that each colony requires.

Many species of social insects not only have direct- but also indirect-communication channels that are equally important attained by using special chemical substances, which are called "pheromones." (Agosta, 1992) A pheromone is a chemical that triggers a natural behavioral response in another member of the same species. When one member of an ant colony senses a particular internal condition or external stimulus, it responds to such a situation and it releases a corresponding kind of pheromone into the environment. The pheromone is diffused through the environment by natural characteristics, e.g., evaporation from the ground, diffusion in the air, and physical contact between the members or the members and their enemies. By using the effect of such translation from the sender to the environment, the pheromone signal sends not only a message from the sender but also information about the current environment. When the receiver senses such a pheromone, it causes a particular reaction in the receiver due to the natural characteristics of the species. One kind of honeybee handles over thirty types of pheromones, which include alarm pheromones for warnings about enemy attacks,

Nasanov pheromones for gathering their mates, and queen pheromones as a signal to indicate the queen is alive.

A good example enabling the relationship between pheromone communication and swarm intelligence to be understood is the foraging behavior of real ants. In the first stage of typical ant-foraging behavior, scouting worker ants individually begin searching for routes from the nest in random directions. When a scouting worker discovers a food source along the route, it picks up the food and brings it back to the nest while laying down a pheromone. Consequently, it releases the first pheromone trail on the ground from the food source to the nest. The pheromone trail plays an important role in collective foraging behavior. If other workers around the nest find the pheromone trail, they try to follow it to arrive at the food source. These workers also discover the food source and then return to the nest while reinforcing the intensity of the pheromone trail. The intensity of the pheromone trail is successively reinforced by the large numbers of workers who continue to march to the food source until all the food is consumed. No ants reinforce the pheromone trail after they have removed all the food. The pheromone gradually evaporates from the ground and the trail automatically dissipates into the air.

This type of sophisticated collective behavior can emerge due to the complex effect of local decision-making, pheromone-communication channels, and natural characteristics of the environment. The mechanism based on such complex effects is called "stigmergy" in the research area of ethology. In the pheromone mechanism of stigmergy, a member releases a pheromone into the environment, which interferences with its propagation, and the other members detect the pheromone from the environment. This communication channel enables the entire colony to organize all of its members and achieve high-level tasks that require coordination and decentralization between them (Dorigo & Theraulaz, 1999).

To investigate what effect stigmergy has had with artificial pheromones, some researchers have tried to create models of swarming behavior. Collins et al. and Bennett studied the evolutionary design of foraging behavior with neural networks and genetic algorithms (Collins & Jefferson, 1991; Bennett III, 1996). Nakamura et al. reported the relationship between global effects and local behavior in a foraging model with artificial ants (Nakamura & Kurumatani, 1997). Suzuki et al. demonstrated the possibility of pheromones solving a deadlocked situation with food-scrambling agents (Suzuki & Ohuchi, 1997). These researchers revealed the possibility that artificial stigmergy based on pheromone models could be used to replicate swarm intelligence.

Some good applications of artificial stigmergy to help solve these problems have been proposed. One of the most successful examples is the ant colony optimization (ACO) algorithm proposed by Dorigo et al. (Dorigo, Maniezzo & Colorni, 1991; Colorni, Dorigo & Maniezzo, 1991) ACO is a probabilistic technique for solving computational problems that can be reduced to finding good paths through the use of graphs (Dorigo & Stutzle, 2004). In ACO, artificial pheromones are introduced to each path to represent how a corresponding path is useful for constructing a solution to a given computational problem. The series of ACOs perform well in solving various computational problems. In other research, Sauter et al. proposed the use of digital pheromones for controlling and coordinating swarms of unmanned vehicles, and demonstrated the effectiveness of these pheromone algorithms for surveillance, target acquisition, and tracking (Sauter, Matthews, Parunak & Brueckner, 2005). Mamei et al. proposed a simple low-cost and general-purpose implementation of a pheromone-based interaction mechanism for pervasive environments with RFID tags (Mamei & Zambonelli, 2007). Sole et al. discussed that behavioral rules at the individual level with a pheromone model could produce optimal colony-level patterns (Sole, Bonabeau, Delgado,

Fernandez & Marin, 2000). Ando et al. succeeded in predicting the future density of traffic congestion by using an artificial pheromone model on a road map (Ando, Masutani, Honiden, Fukazawa & Iwasaki, 2006). These researchers revealed that artificial stigmergy with pheromone models are useful for applications to multi-agent systems.

This chapter focuses on the design of multi-agent systems to successfully complete given tasks with collective behavior and artificial stigmergy. We particularly selected the evolutionary design in pheromone communication. If an agent has a specific advance rule to output a pheromone, another agent should optimize its reaction to the pheromone to establish communication. However, if an agent has a specific advance reaction to a pheromone, another agent should optimize the situation as to when and where it outputs the pheromone. The possibility of designing evolutionary-pheromone communication is not so easy in these cases because it is necessary for the communication to simultaneously specify the reaction to the pheromone and the situation to output it.

Section 2 proposes an ant war as a competitive environment for a task requiring effective agent communication. Section 3 explains the evolutionary design process for the agent system. Section 4 presents the setting for computer simulation, and Section 5 concludes the chapter by clarifying the evolutionary process to form artificial pheromone communication from the viewpoint of information theory.

ANT WAR AS COMPETITIVE ENVIRONMENT

An ant war is a competitive environment for two teams of ant-like agents (Kawamura, Yamamoto, Suzuki & Ohuchi, 1999; Kawamura, Yamamoto & Ohuchi, 2001). The environment is constructed on 44X80 grids, and these two are for a blue-ant and a red-ant team. Each team consists of 80

Figure 1. The outline of ant war competition. This figure has been reproduced by permission from Kawamura, H., Yamamoto, M. & Ohuchi, A. (2001).: Investigation of Evolutionary Pheromone Communication Based on External Measurement and Emergence of Swarm Intelligence, Japanese Journal of the Society of Instrument and Control Engineers, 37(5), 455–464.

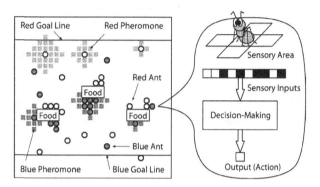

homogeneous agents initially placed in random positions. Some food packs are initially placed on the center line of the environment. The purpose of each team is to carry the food packs to its own goal line and to collect more food packs than the opposing team. The competition in the ant war is outlined in Figure 1.

The pheromone field for both teams in the environment is defined with two variables $T(x, y, t)$ and $P(x, y, t)$. $T(x, y, t)$ is the intensity of the pheromone on the grid, (x, y), at time t, and $P(x, y, t)$ is the aerial density of the pheromone on grid (x, y) at time t. Between times t and $t + 1$ these pheromone field variables are updated as

$$T\left(x,\ y,\ t+1\right) = \left(1 - \gamma_{eva}\right) \cdot T\left(x,\ y,\ t\right) + \sum_{k} \Delta T^{k}(x,y,t),$$

$$\Delta T^{k}\left(x,y,t\right) = \begin{cases} Q \\ 0 \end{cases},$$

If agent k lays off the pheromone on grid (x,y) at time t otherwise:

$$P\left(x,y,t+1\right) = P\left(x,y,t\right) + \gamma_{dif} \cdot \begin{pmatrix} P\left(x-1,y,t\right) + P\left(x+1,y,t\right) + \\ P\left(x,y-1,t\right) + P\left(x,y+1,t\right) - \\ 5P\left(x,y,t\right) \end{pmatrix} + \gamma_{eva} \cdot T(x,y,t),$$

where γ_{eva} is the pheromone evaporation rate from the ground and γ_{dif} is the pheromone diffusion rate into the air. Q is the amount of pheromone released by one ant agent. According to these equations the pheromone released by the ant agent gradually evaporates from the ground and diffuses into the space.

Each agent has a limited 10-bit sensor, denoted as $i_1 i_2, \ldots, i_{10}$, that can obtain information about its neighboring grids. First, six bits, $i_1 i_2, \ldots, i_6$, are determined according to the six rules.

- Whether the agent touches a food pack or not.
- Whether the agent contacts a mate who tries to carry a food pack or not.
- Whether the agent contacts a mate in front of it or not.

Figure 2. The firing probability of pheromone sensory inputs. This figure has been reproduced by permission from Kawamura, H., Yamamoto, M. & Ohuchi, A. (2001).: Investigation of Evolutionary Pheromone Communication Based on External Measurement and Emergence of Swarm Intelligence, Japanese Journal of the Society of Instrument and Control Engineers, 37(5), 455–464.

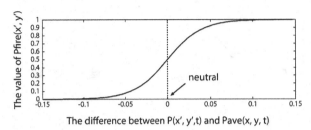

The difference between P(x′, y′,t) and Pave(x, y, t)

- Whether the agent contacts a mate to its right or not.
- Whether the agent contacts a mate to its left or not.
- Whether the agent contacts a mate behind it or not.

An additional four bits, $i_7 i_8 i_9 i_{10}$, stochastically respond to the aerial densities of the pheromone on four neighboring grids. Here, let (x, y) be the position of agent k and

$$(x', y') \in \left\{ \begin{array}{l} (x-1, y), (x+1, y), \\ (x, y-1), (x, y+1) \end{array} \right\}$$ be the

sensing position of agent k. The firing probability of each pheromone bit is defined as

$$P_{fire}(x', y') = \frac{1}{\left[1 + \exp\left(-\frac{P(x', y', t) - P_{ave}(x, y, t)}{T} \right) \right]}$$

$$P_{ave}(x, y, t) = \left\{ \begin{array}{l} P(x-1, y, t) + \\ P(x+1, y, t) + \\ P(x, y-1, t) + \\ P(x, y+1, t) \end{array} \right\} / 4,$$

where T is the parameter of sensitivity. The firing probability takes a higher value in responding to relatively high-density pheromones (see Figure 2). Although the pheromone sensor only has binary values, agent k can react according to the density of pheromones. Since the agent in this model must determine its actions according to sensory information only about the neighborhood, effective communication through pheromone channels is important to win in this competition.

At time t, each individual agent can select one of seven actions, going forward, backward, left, right, standing by in the current position, pulling a food pack, or laying a pheromone in the current position. If the agent wants to pull a food pack and actually apply force to the target food, it has to satisfy a condition where it touches the target food pack or a mate who is actually pulling the target food pack. The food pack is moved in a tug of war. More agents can move the food pack more quickly. Then, the food pack crossing the goal line is removed from the environment and the team is awarded a score.

The competition is finished when time t expires or all the food packs have crossed the goal line. The winner is the team who has collected more food packs than its opponent.

DECISION-MAKING

Each time every agent receives 10 bits of sensory information and selects one action from seven choices. A simple two-layer neural network for individual decision-making is determined, whose network has a simple structure and is flexible to represent various decision-making functions. All the agents in each team have an identical neural network with the same set of weight parameters. Let i_j be the j-th input bit and O_k be the probability to select the k-th action from seven actions. The neural network maps from the input vector to the output vector. The weight vector of the neural network is represented as w_{jk}, and the mapping is done as follows.

$$o_k = \left(1 + \exp\left(-\sum_{j=1}^{11} w_{jk} \bullet i_j\right)\right)^{-1} \Bigg/ \sum_{h=1}^{7}\left(1 + \exp\left(-\sum_{j=1}^{11} w_{jh} \bullet i_j\right)\right)^{-1},$$

where i_{11} is the additional input and -1 is always set as the bias for the neural network.

EVOLUTIONARY PROCESS

The main topic of this chapter is to explain how multi-agent systems organize artificial pheromone communication with evolutionary computations. The chromosomes in the computations are constructed with a set of weights, w_{jk}. At the initial stage of evolution, all w_{jk} are initialized with a random value from [-0.5, 0.5]. The evolutionary computation has N chromosomes and these chromosomes evolve through a five-step procedure (see Figure 3).

Figure 3. The outline of evolutionary computation. This figure has been reproduced by permission from Kawamura, H., Yamamoto, M. & Ohuchi, A. (2001).: Investigation of Evolutionary Pheromone Communication Based on External Measurement and Emergence of Swarm Intelligence, Japanese Journal of the Society of Instrument and Control Engineers, 37(5), 455–464.

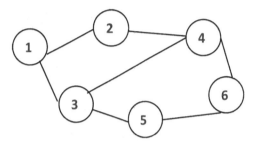

- Step 1: Two chromosomes are randomly selected from the population.
- Step 2: Ant-agent teams with the selected chromosomes compete in the ant war. The loser is removed from the population.
- Step 3: The chromosomes of the winner are copied to two prototypes and the chromosomes of these two are varied by cross-over and mutation operations.
- Step 4: Two new chromosomes are returned to the population.
- Step 5: Go back to Step 1 until the final iteration.

Crossover operation exchanges the weight values at the same locus of two chromosomes with the probability, P_c. Mutation operation adds noise from [-0.5, 0.5] to each weight with the probability, P_m.

EXPERIMENT

A computer experiment was carried out with 10 trials. The maximum simulation time in each

Figure 4. An example distribution of obtained pheromones. This figure has been reproduced by permission from Kawamura, H., Yamamoto, M. & Ohuchi, A. (2001).: Investigation of Evolutionary Pheromone Communication Based on External Measurement and Emergence of Swarm Intelligence, Japanese Journal of the Society of Instrument and Control Engineers, 37(5), 455–464.

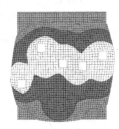 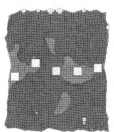

(A)Attractive Pheromone   (B)Repulsive Pheromone

ant-war environment was 2000. The parameters of the artificial pheromones, Q, γ_{eva}, and γ_{dif} were set to correspond to 50, 0.1, and 0.1. There were 20 chromosomes, and the final generation of the evolutionary process was 50000. The parameters for evolutionary operations, P_c and P_m, were set to correspond to 0.04 and 0.08. These settings were determined through various preliminary experiments.

RESULTS

Two types of pheromone communications were self-organized in the final generation through 10 trials. We called these two attractive and repulsive pheromones. There were seven attractive pheromones, and three repulsive pheromones.

Attractive pheromone: An ant based on this type of pheromone basically walks randomly and goes about exploring the environment when not sensing a special stimulus. Once sensing contact with a food pack, the ant tries to carry it and releases the pheromone on the ground. The pheromone released by such an ant diffuses near the food pack, and its intensity in the environment is based on the food pack. The other ants who have sensed the pheromone try to effectively reach the food pack based on the pheromone's level of intensity. This type of pheromone attracts their mates and we called it an "attractive pheromone." There is an example distribution of attractive pheromones in Figure 4 (A).

Repulsive pheromone: An ant using this type of pheromone has a tendency to dislike it. When exploring, this ant scatters the pheromone on the ground. Once the ant finds a food pack, it stops to release the pheromone. As a result of such behavior, the pheromone field leaves the food pack and enters the environment. This means that ants mark space with the pheromone that has already been explored, which is unnecessary to re-explore, and they therefore save time in effectively finding the food pack. As this type of pheromone repulses their mates, we called it a "repulsive pheromone." There is an example distribution of repulsive pheromones in Figure 4 (B).

Although an artificial evolutionary mechanism where winners survive and losers disappear generated these two types of ant behaviors, it is not clear whether these are sufficiently dominant in the evolutionary process. To evaluate the evolution of ant strategies, we depicted the winning percentage of 100 competitions between successive generations versus the final generation. The evolutionary

Figure 5. The evolutionary transition of the winning percentage in the two types. The opponent of the competition is the final generations. This figure has been reproduced by permission from Kawamura, H., Yamamoto, M. & Ohuchi, A. (2001).: Investigation of Evolutionary Pheromone Communication Based on External Measurement and Emergence of Swarm Intelligence, Japanese Journal of the Society of Instrument and Control Engineers, 37(5), 455–464.

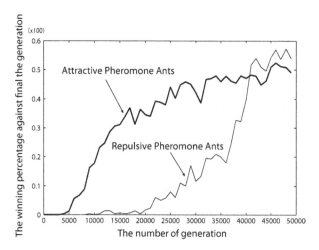

transition of the winning percentage in the two types is shown in Figure 5. The X-axis indicates the number of generations, and the Y-axis indicates the winning percentage. The percentages of both types increase as each generation progresses. This graph plots the evolutionary process for dominant behavior to emerge.

The tendencies of both lines differ in terms of evolution speed, i.e., the attractive pheromone evolved quicker than the repulsive. The difference in evolutionary speed may have been caused by the strength of the relationship between the outcome of competition and pheromone communication. The use of the attractive pheromone was simple and the ant climbing up the pheromone field contributed effectively to the competition. However, the use of the repulsive pheromone was a little complicated because the ant was going down the pheromone field and did not always reach the food pack. The relationship between the outcome and communication was weaker than for the attractive pheromone.

We next measured the evolutionary process of communication by quantifying the effectiveness of pheromone sensor inputs and the uniqueness of the situation where the ant released the pheromone. Shannon's entropy and mutual information from information theory were selected to measure the degree of communication (Shannon & Weaver, 1964). Formally, the entropy of a discrete variable, X, is defined as:

$$H(X) = -\sum_{x \in X} p(x) \log p(x),$$

where p(x) is the marginal probability-distribution function of X.

The mutual information of two discrete variables X and Y is defined as:

$$I(X;Y) = H(X) - H(X \mid Y)$$

Figure 6. The evolutionary transition of entropy and mutual information. This figure has been reproduced by permission from Kawamura, H., Yamamoto, M. & Ohuchi, A. (2001).: Investigation of Evolutionary Pheromone Communication Based on External Measurement and Emergence of Swarm Intelligence, Japanese Journal of the Society of Instrument and Control Engineers, 37(5), 455–464.

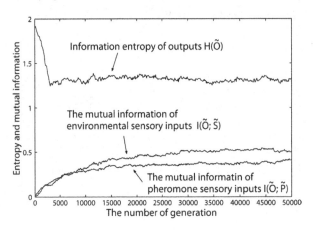

$$H\left(X|Y\right) = -\sum_{x \in X, y \in Y} p\left(x, y\right) \log p(x \mid y),$$

$$\tilde{P} = \left\{\left(i_7, i_8, \ldots, i_{10}\right); \; i_7, i_8, \ldots, i_{10} \in \left\{0,1\right\}\right\},$$

where H(X|Y) represents the conditional entropies, p(x,y) is the joint-probability-distribution function of X and Y, and p(x | y) is the conditional-distribution function of X given Y.

Entropy H(X) represents the uncertainty of X, and mutual information I(X;Y) derives how much knowing one of these variables reduces our uncertainty about the other.

To measure the effect of sensory inputs for decision-making, three discrete variables, $\tilde{O}$, $\tilde{S}$, and $\tilde{P}$, are introduced. $\tilde{O}$ is the variable of output, $\tilde{S}$ is the set of sensory inputs without the pheromone, and $\tilde{P}$ is the set of the pheromone sensory inputs. The formal description of these variables is:

$\tilde{O}$={forward, backward, left, right, standby, pull a food pack, drop off pheromone,}

$$\tilde{S} = \left\{\left(i_1, i_2, \ldots, i_6\right); \; i_1, i_2, \ldots, i_6 \in \left\{0,1\right\}\right\},$$

where $\tilde{O}$ has seven elements, $\tilde{S}$ has 2^6=64 elements, and $\tilde{P}$ has 2^4=16 elements. The probability of each element is denoted by p(•), and each p(•) is calculated by using the results of a competition. Using these probabilities, the entropy of output, H($\tilde{O}$), mutual information on sensory inputs without pheromones, $I\left(\tilde{O}, \tilde{S}\right)$, and mutual information on pheromone sensory inputs, $I\left(\tilde{O}, \tilde{P}\right)$, are measured.

Figures 6 and 7 show the evolutionary transitions of H($\tilde{O}$), $I\left(\tilde{O}, \tilde{S}\right)$, and $I\left(\tilde{O}, \tilde{P}\right)$ for attractive and repulsive pheromones. The X-axis indicates the number of generations and the Y-axis indicates the degrees of entropy and mutual information.

The H($\tilde{O}$), is falling to around 1.3 at the 2,500th generation for the attractive pheromone and is almost stable after that. This means that the ant self-organized the most important reaction in the early stage of evolution, and that the reaction was

Figure 7. The evolutionary transition of entropy and mutual information. This figure has been reproduced by permission from Kawamura, H., Yamamoto, M. & Ohuchi, A. (2001).: Investigation of Evolutionary Pheromone Communication Based on External Measurement and Emergence of Swarm Intelligence, Japanese Journal of the Society of Instrument and Control Engineers, 37(5), 455–464.

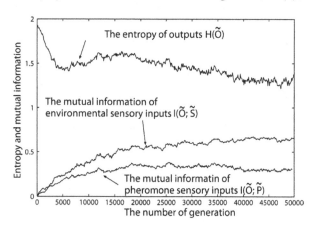

pulling a food pack after finding it. The $I\left(\tilde{O}, \tilde{S}\right)$ and $I\left(\tilde{O}, \tilde{P}\right)$ are gradually increasing from the first generation to around the 25,000th, and saturate after that. From the viewpoint of information theory, this measured transition means that the ants were self-organizing the effectiveness of the environmental inputs and pheromone inputs for decision-making. The importance of the environmental inputs and pheromone inputs in this evolutionary path are almost same to determine action.

The $H(\tilde{O})$ has fallen once to around 1.4 at the 5,000th generation for the repulsive pheromone, then has gradually increased and decreased, before finally reaching around 1.3. The $I\left(\tilde{O}, \tilde{S}\right)$ and $I\left(\tilde{O}, \tilde{P}\right)$ are gradually increasing according to the progress of evolution. Here, the value of environmental inputs is larger than that of pheromone inputs. This means that the environmental sensors play a more important role than the pheromone sensors to determine action. The field of repulsive pheromones is unstable compared with that of attractive pheromones.

To measure the uniqueness of the pheromone-releasing action in sensory inputs, two additional

discrete variables, $\tilde{I}$ and $\tilde{A}$, are introduced. $\tilde{I}$ is the set of whole sensory inputs, and $\tilde{A}$ is a discrete variable that represents whether the ant releases pheromones or not. The formal description of these variables is:

$$\tilde{I} = \left\{\left(i_1, i_2, \ldots, i_{10}\right); \ i_1, i_2, \ldots, i_{10} \in \{0,1\}\right\},$$

$\tilde{A}=\{$drop off pheromone, otherwise$\}$.

The $\tilde{I}$ has $2^{10} = 1024$ elements, and $\tilde{A}$ only has two elements. The $H\left(\tilde{I}\right)$ is the entropy of the set of sensory inputs, and $I\left(\tilde{I}, \ \tilde{A}\right)$ is the mutual information of $\tilde{A}$.

Figures 8 and 9 show the evolutionary transitions of $H\left(\tilde{I}\right)$ and $I\left(\tilde{I}, \ \tilde{A}\right)$ for attractive and repulsive pheromones. In both cases, $H\left(\tilde{I}\right)$ takes a lower value in the early generations and gradually increases as evolution progresses. The lower value of $H\left(\tilde{I}\right)$ in the early generations was caused by inequality in action-selecting probabilities. The randomly generated initial-neural networks were not equal in action-selecting probabilities. Consequently, almost all ants gathered in a corner of the environment and their sensory inputs were

Figure 8. The evolutionary transition of entropy and mutual information. This figure has been reproduced by permission from Kawamura, H., Yamamoto, M. & Ohuchi, A. (2001).: Investigation of Evolutionary Pheromone Communication Based on External Measurement and Emergence of Swarm Intelligence, Japanese Journal of the Society of Instrument and Control Engineers, 37(5), 455–464.

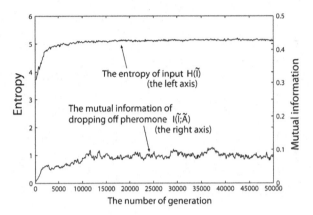

Figure 9. The evolutionary transition of entropy and mutual information. This figure has been reproduced by permission from Kawamura, H., Yamamoto, M. & Ohuchi, A. (2001).: Investigation of Evolutionary Pheromone Communication Based on External Measurement and Emergence of Swarm Intelligence, Japanese Journal of the Society of Instrument and Control Engineers, 37(5), 455–464.

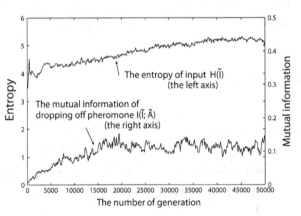

biased in specific patterns. After evolution, the ants scattered here and there to effectively search for a food pack and the variations in sensory inputs were wider than those in the early generations.

To investigate the evolutionary process of $I(\tilde{O},\tilde{P})$ and $I(\tilde{I},\tilde{A})$, we plotted scatter diagrams with pairs observed with these values. The scatter diagrams for attractive-pheromone evolution and repulsive-pheromone evolution are in Figures 10

and 11. Here again, $I(\tilde{O},\tilde{P})$ corresponds to the effectiveness of pheromone sensory inputs on decision-making, and $I(\tilde{I},\tilde{A})$ corresponds to the uniqueness of sensory-input patterns for selecting pheromone-release actions. The graphs indicate that both values of $I(\tilde{O},\tilde{P})$ and $I(\tilde{I},\tilde{A})$ have a distinct positive correlation and the values of the pair are increasing together step by step. This suggests that the situation's uniqueness in sending

Figure 10. The scatter diagrams of mutual information for attractive-pheromone evolution. This figure has been reproduced by permission from Kawamura, H., Yamamoto, M. & Ohuchi, A. (2001).: Investigation of Evolutionary Pheromone Communication Based on External Measurement and Emergence of Swarm Intelligence, Japanese Journal of the Society of Instrument and Control Engineers, 37(5), 455–464.

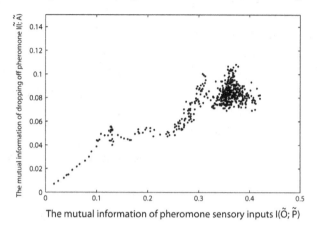

Figure 11. The scatter diagrams of mutual information for repulsive-pheromone evolution. This figure has been reproduced by permission from Kawamura, H., Yamamoto, M. & Ohuchi, A. (2001).: Investigation of Evolutionary Pheromone Communication Based on External Measurement and Emergence of Swarm Intelligence, Japanese Journal of the Society of Instrument and Control Engineers, 37(5), 455–464.

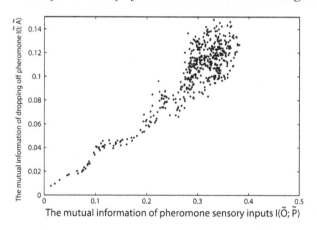

a signal and the reaction's uniqueness in receiving a signal are two sides of the same coin in evolutionary agent design and artificial-pheromone communication is gradually formed while the same pace is maintained.

CONCLUSION

This chapter proposed an ant war as a competitive environment enabling studies of artificial stigmergy and the origins of pheromone-style communication in evolutionary agent design. Two types of collective intelligence were acquired through computer simulations, i.e., ants with attractive and repulsive pheromones. Both types demonstrated rational strategies to win the competition and these strategies effectively utilized the characteristics of artificial pheromones and the environment. We introduced Shannon's entropy and mutual information on artificial pheromones to measure the situation's uniqueness in sending pheromones and the reaction's uniqueness in receiving pheromones. Such uniqueness represented two sides of the same coin and artificial-pheromone communication was gradually formed while the same pace was maintained.

REFERENCES

Agosta, W. (1992). *Chemical Communication – The Language of Pheromone*. W. H. Freeman and Company.

Ando, Y., Masutani, O., Honiden, S., Fukazawa, Y., & Iwasaki, H. (2006). Performance of Pheromone Model for Predicting Traffic Congestion. In . *Proceedings of AAMAS, 2006*, 73–80.

Bennett, F., III. (1996). Emergence of a Multi-Agent Architecture and New Tactics for the Ant Colony Food Foraging Problem Using Genetic Programming. In *From Animals to Animats 4, Proceedings of the Fourth International Conference on Simulations of Adaptive Behavior* (pp. 430–439).

Bonabeau, E., Dorigo, M., & Theraulaz, G. (1999). *Swarm Intelligence from Natural to Artificial Systems*. Oxford University Press.

Collins, R., & Jeffersion, D. (1991). AntFarm: Towards Simulated Evolution. In *Artificial Life II, Proceedings of the Second International Conference on Artificial Life* (pp. 159–168).

Colorni, A., Dorigo, M., & Maniezzo, V. (1991). Distributed Optimization by Ant Colonies. In . *Proceedings, ECAL91*, 134–142.

Dorigo, M., Maniezzo, V., & Colorni, A. (1991). *Positive Feedback as a Search Strategy* (Technical Report No. 91-016). Politecnico di Milano.

Dorigo, M., & Stutzle, T. (2004). *Ant Colony Optimization*. Cambridge, MA: The MIT Press.

Kawamura, H., & Yamamoto, M. Suzuki & Ohuchi, A. (1999). Ants War with Evolutive Pheromone Style Communication. In *Advances in Artificial Life, ECAL'99* (LNAI 1674, pp. 639-643).

Kawamura, H., Yamamoto, M., & Ohuchi, A. (2001). (in Japanese). Investigation of Evolutionary Pheromone Communication Based on External Measurement and Emergence of Swarm Intelligence. *Japanese Journal of the Society of Instrument and Control Engineers*, 37(5), 455–464.

Mamei, M. & Zambonelli, F. (2007). Pervasive pheromone-based interaction with RFID tags. *ACM Transactions on Autonomous and Adaptive Systems (TAAS) archive*, 2(2).

Nakamura, M., & Kurumatani, K. (1997). Formation Mechanism of Pheromone Pattern and Control of Foraging Behavior in an Ant Colony Model. In *Proceedings of the Fifth International Workshop on the Synthesis and Simulation of Living Systems* (pp. 67 -74).

Sauter, J., Matthews, R., Parunak, H., & Brueckner, S. (2005). Performance of digital pheromones for swarming vehicle control. In *Proceedings of the fourth international joint conference on Autonomous agents and multiagent systems* (pp. 903-910).

Shannon, C., & Weaver, W. (1964). *The Mathematical Theory of Communication*. The University of Illinois Press.

Sheely, T. (1995). *The Wisdom of the Hive: The Social Physiology of Honey Bee Colonies*. Harvard University Press.

Sole, R., Bonabeau, E., Delgado, J., Fernandez, P., & Marin, J. (2000). Pattern Formation and Optimization in Army Raids. [The MIT Press.]. *Artificial Life*, 6(3), 219–226. doi:10.1162/106454600568843

Suzuki, K., & Ohuchi, A. (1997). Reorganization of Agents with Pheromone Style Communication in Mulltiple Monkey Banana Problem. In . *Proceedings of Intelligent Autonomous Systems*, 5, 615–622.

308

Chapter 17
Evolutionary Search for Cellular Automata with Self-Organizing Properties toward Controlling Decentralized Pervasive Systems

Yusuke Iwase
Nagoya University, Japan

Reiji Suzuki
Nagoya University, Japan

Takaya Arita
Nagoya University, Japan

ABSTRACT

Cellular Automata (CAs) have been investigated extensively as abstract models of the decentralized systems composed of autonomous entities characterized by local interactions. However, it is poorly understood how CAs can interact with their external environment, which would be useful for implementing decentralized pervasive systems that consist of billions of components (nodes, sensors, etc.) distributed in our everyday environments. This chapter focuses on the emergent properties of CAs induced by external perturbations toward controlling decentralized pervasive systems. We assumed a minimum task in which a CA has to change its global state drastically after every occurrence of a perturbation period. In the perturbation period, each cell state is modified by using an external rule with a small probability. By conducting evolutionary searches for rules of CAs, we obtained interesting behaviors of CAs in which their global state cyclically transited among different stable states in either ascending or descending order. The self-organizing behaviors are due to the clusters of cell states that dynamically grow through occurrences of perturbation periods. These results imply that we can dynamically control the global behaviors of decentralized systems by states of randomly selected components only.

DOI: 10.4018/978-1-60566-898-7.ch017

INTRODUCTION

A cellular automaton (CA) is a discrete model consisting of a regular grid of finite automata called cells. The next state of each cell is completely decided by the current states of its neighbors including itself. CAs are well-suited for investigating the global behavior emerging from local interactions among component parts. A CA can be interpreted as an abstract model of multi-agent systems (MAS) if we regard each cell in a CA as an agent in a MAS, because the multiple autonomous agents in a MAS locally interact with each other in general, as shown as Figure 1. MAS can be constructed as a large-scale space by using CA and it is also easy to visualize. Therefore, there are several applications for MAS which make use of computational and emergent properties of CAs, including traffic simulations (Sakai, Nishinari & Iida, 2006), ecological modeling (Nagata, Morita, Yoshimura, Nitta & Tainaka, 2008), controlling decentralized pervasive systems (Mamei, Roli & Zambonelli, 2005).

There have been a number of studies on basic characteristic of CAs under the assumption of strong restrictions such as a regular arrangement of cells and synchronous update of the cell states.

It is well known that Wolfram suggested that elementary (one-dimensional two-state three-neighbor) CAs fall into four classes. In particular, he pointed out that CAs in class four exhibit complex behaviors, and some of them achieve computational universality (Wolfram, 2002).

On the one hand, there are studies on CAs with relaxed restrictions so as to investigate the behaviors of the decentralized systems in realistic situations from a variety of viewpoints. Ingerson & Buvel (1984) pointed out that the synchronous updating rule is unnatural if we regard them as abstractions of decentralized systems in a real world, and analyzed elementary CAs with asynchronous updating rules. They demonstrated that the global pattern of the asynchronous CA self-organized into a regular array, and argued that asynchronous CA might be useful for understanding self-organization in a real world.

Also, there are several discussions on the effects of influences on the system from outside such as boundary conditions and perturbations. Ninagawa, Yoneda & Hirose (1997) focused on influences of the differences, and compared the effects of the dissipative boundary conditions (in which each cell state is randomly decided at each step) on the global behaviors of CAs with those of the standard periodic boundary conditions. They showed that the CAs with the former condition can eliminate the size effects of the grid on the global behaviors which are prominent in CAs with the latter condition. Marr & Hütt (2006) showed that based on comparison of space-time diagrams of configurations and some of indices, the modifications of the neighboring structures on the regular grid can be functionally equivalent to the introduction of stochastic noises on the states of the cells in CAs that basically exhibit chaotic behaviors.

Figure 1. A cellular automaton (CA) as an abstract model of multi-agent systems (MAS)

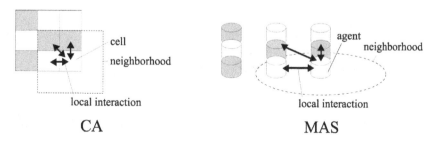

The self-organizing behaviors of CAs can be dynamically affected by the external influences. Several models of CAs which have such a property have been proposed for controlling decentralized pervasive systems (Figure 2) (Mamei *et al.*, 2005 and Kwak, Baryshnikov & Coffman, 2008). Decentralized pervasive systems consist of distributed components (nodes, sensors, etc.) in our everyday environments and the systems perform global tasks using their whole components. Mamei *et al.* (2005) focused on the influences of the external world on decentralized pervasive systems, and constructed asynchronous CA with external perturbations termed ``dissipative cellular automata'' (Roli & Zambonelli, 2002) in that the external environment can somehow inject energy to dynamically influence the evolution of the automata. They regarded the asynchronous CA as a group of autonomous individuals which locally interact with each other, and introduced continuously occurring perturbations on the states of the cells into their model. The perturbations correspond to the influences caused by the external world on the group. The CAs presented regular patterns if and only if they are induced by the perturbations, such as a stripe pattern. Moreover, they argued about applications of such self-organized features for controlling decentralized pervasive systems such as sensor networks. For example, based on the experiments of dissipative CAs, they showed possible scenarios that the global state of the CA can be changed to the desired pattern only by filling up the limited area of the CA with a fixed pattern of the states. However, these discussions were based on several hand-coded rules for CAs. Thus, it is still an open question how CAs can show emergent properties through the interactions with an external world.

However, in general, there are complex relationships between the global behaviors of CAs and the local behaviors of cells. It is difficult to design the rules for CAs by hand-coding which exhibit the desired emergent behaviors. Thus, there have been various studies based on evolutionary searches for rules of CAs that can exhibit emergent behaviors (Mitchell, Crutchfield & Hraber, 1994). Rocha (2004) evolved CA rules that can solve the density task, and discussed about the nature of the memory-like interactions among the particles of the cell states emerged for storing and manipulating information. Ninagawa (2005) also evolved CAs which generate $1/f$ noise where the power is inversely proportional to the frequency, and obtained a CA of which behavior is similar to the Game of Life. As above, an evolutionary search for rules of CAs will be also useful for understanding the characteristics of systems that exhibit self-organizing properties caused by interactions with an external environment.

This chapter focuses on the emergent properties of CAs induced by external perturbations toward controlling decentralized pervasive systems. We assumed a minimum task in which CAs have to change its global state after every occurrence of perturbation period, and searched the rules for

Figure 2. CAs with external perturbations aimed at controlling decentralized pervasive systems

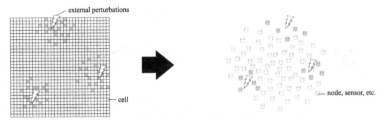

The self-organizing behaviors of CAs   Controlling decentralized pervasive systems

CAs which can solve the task by using a genetic algorithm (GA) (Iwase, Suzuki & Arita, 2007). We obtained the rules for CAs in which global state of evolved CAs cyclically transited among different stable states of which the number is more than that of distinct cell states, and looked into the self-organizing properties that a drastic change in its global state occurs every two successive occurrences of perturbation periods.

TASK

We constructed a task that self-organizing behaviors induced by external perturbations are required to solve, and we searched the rules of CAs for solving the task by using GA. In an evaluation process of a CA, there are the fixed number of perturbation periods in which each cell state is modified by using an external rule with a small probability. The CA has to change its configuration represented by the distribution ratio of cell states after every occurrence of a perturbation period. This is non-trivial if the number of the perturbation periods is larger than the number of possible cell states because the global behavior of CAs must stably exhibit different configurations composed of intermediate distribution ratios. Thus the CAs should need some kind of emergent properties that utilize the occurrences of perturbations effectively.

Cellular Automata and Perturbations

We adopt a two-dimensional ($N \times N$) M-state nine-neighbor (Moore neighborhood) cellular automata with periodic boundary condition as an abstract model of the distributed systems composed of autonomous entities characterized by local interactions.

A cell (i, j) have a state $q^t_{i,j} \in \{0, \dots, M-1\}$ at time step t. At each time step t, each cell state is asynchronously updated with a probability P_a by

$$q_{i,j}^{t+1} = \begin{cases} \delta(S_{i,j}^t) & (P_a) \\ q_{i,j}^t & (1 - P_a), \end{cases} \tag{1}$$

where δ is a local transition rule which maps a configuration of cell states in a neighborhood (3×3 cells around the focal cell (i, j)) $S^t_{i,j}$ into a state.

Also we introduce an external perturbation ε which changes a cell state independently of δ. ε expresses a simple transition rule that increments the value of the cell state as defined by

$$\varepsilon : q_{i,j}^{t+1} := \left(q_{i,j}^{t+1} + 1 \right) \bmod M. \tag{2}$$

ε is applied to each cell with a probability P_e every after transitions of cell states by Equation 1. We expect that some of relationship between the global behavior of CAs and external perturbations can occur by introducing Equation 2. It is because that the actual effect of a perturbation on the CA (a change in the cell state) is deterministic although it occurs probabilistically.

Transition and Evaluation

Figure 3 is a time diagram of an evaluation process for a rule (δ) of a CA described above. Starting from the initial condition in which each cell state is randomly assigned, the transitions without perturbations ($P_e = 0.0$) occur for L_{pre} steps so that effects of the initial condition are eliminated. Next, a perturbation period of L_d steps occurs every after a normal period of approximately L_{int} steps. For each cell, a perturbation occurs with a probability $P_e = \beta$ during perturbation periods, and it does not occur during normal periods ($P_e = 0$). The evaluation stops when the normal periods have occurred for $D + 1$ times. Note that the actual time step at which each perturbation period starts fluctuates randomly within a specific range ($\pm L_{fluct}$ steps) as shown in Figure 3.

A density distribution of cell states ρ^t at time t is a vector consisting of the ratios $\rho^t(s)$ of the cell state s among all cells, which is defined by

$$\rho^t(s) = \frac{1}{N \times N} \sum_{(i,j) \in N \times N} \begin{cases} 1 & \text{if } q_{i,j}^t = s \\ 0 & \text{otherwise,} \end{cases}$$
$$\hat{A}^t = \left\{ \rho^t(0), \rho^t(1), \ldots, \rho^t(M-1), \right\}. \tag{3}$$

Also, in order to stress the existence of a small amount of each cell state, we define a scaled density distribution $\rho^t_{\theta,\varphi}$ by

$$SF_{\theta,\varphi}(x) = \frac{1}{1 + e^{-(x-\theta) \times \varphi}},$$
$$\rho^t_{\theta,\varphi}(s) = SF_{\theta,\varphi}(\rho^t(s)),$$
$$\hat{A}^t_{\theta,\varphi} = \left\{ \rho^t_{\theta,\varphi}(0), \rho^t_{\theta,\varphi}(1), \ldots, \rho^t_{\theta,\varphi}(M-1), \right\}, \tag{4}$$

where $SF_{\theta,\varphi}$ is a sigmoid function which scales $\rho^t(s)$ with a threshold θ, and φ is a parameter for a degree of scaling. Equation 4 means that if $\rho^t(s)$ is more than θ, $\rho^t_{\theta,\varphi}$ becomes close to 1. Otherwise, it becomes close to 0.

Here, we take $\rho^t_{\theta,\varphi}$ as the global configuration of the CA, and define the fitness by using $\rho^t_{\theta,\varphi}$ at the last steps of normal periods ($\rho^{a(0)}_{\theta,\varphi}, \rho^{a(1)}_{\theta,\varphi}, \ldots,$ $\rho^{a(D)}_{\theta,\varphi}$ in Figure 3) as follows:

$$\left| \hat{A}^i_{\theta,\varphi} - \hat{A}^j_{\theta,\varphi} \right| = \left| \rho^i_{\theta,\varphi}(0) - \rho^j_{\theta,\varphi}(0) \right| +$$
$$\left| \rho^i_{\theta,\varphi}(1) - \rho^j_{\theta,\varphi}(1) \right| + \cdots$$

$$+ \left| \rho^i_{\theta,\varphi}(M-1) - \rho^j_{\theta,\varphi}(M-1) \right|, \tag{5}$$

$$f_{\theta,\varphi} = \sum_{i \neq j \in \{a(0), a(1), \ldots, a(D)\}} \left| \hat{A}^i_{\theta,\varphi} - \hat{A}^j_{\theta,\varphi} \right|. \tag{6}$$

Equation (5) defines the difference between the scaled density distributions as the sum of the absolute differences between the corresponding elements of the distributions. The fitness of δ is the sum of the differences over all possible pairs of the scaled density distributions at the last steps of the normal periods.

Thus, the CAs have to use an occurrence of a perturbation period as a trigger to change their own global configuration dynamically.

EVOLUTIONARY SEARCH BY GENETIC ALGORITHM

Transition Rule

We optimize the rules for CAs to maximize the fitness defined above by using GA. We adopt a transition rule based on the number of each cell state in the neighborhood, expecting emergence of interesting behaviors of the CA and reduction of the search domain for GA.

Figure 4 illustrates an example of the transition rules in the case of $M = 3$. This rule is an extended version of outer-totalistic rules. The pattern of the neighborhood configuration $S'_{i,j}$ at the cell (i, j) is given by

Figure 3. Repeated occurrences of external perturbations

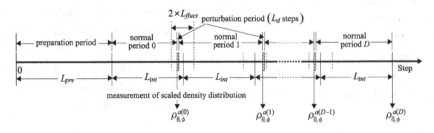

$$n_{i,j}^t(s) \;=\; \sum_{|k|\leq 1, |l|\leq 1, |k|+|l|\neq 0} \begin{cases} 1 & \text{if } q_{i+k,j+l}^t = s \\ 0 & \text{otherwise,} \end{cases}$$

$$S_{i,j}^t \;=\; \left\{ q_{i,j}^t, n_{i,j}^t(0), n_{i,j}^t(1), \ldots, n_{i,j}^t(M-1) \right\}.$$

$$(7)$$

$n_{i,j}^t(s)$ is the number of the cell state s in the neighborhood except for the cell (i, j) itself. $S_{i,j}^t$ is a set of the focal cell state and the number of each cell state as illustrated in Figure 4 - A.

Since the changes of the CAs' global state are evaluated with the densities of cell states only, there would be existing rules which can be applied to some situations if the cell states on the rules are uniformly changed by something method. Therefore, in concert with the external perturbations, we further introduced a transitivity into the transition rule δ defined as follows:

$$\delta_q(n(0), n(1), \ldots, n(M-1)) = \delta(q, n(0), n(1), \ldots, n(M-1)),$$

$$\delta_q(n(0), n(1), \ldots, n(M-1)) = \qquad (8)$$

$$\begin{pmatrix} \delta_0(n(cs(q,0)), n(cs(q,1)), \ldots, \\ n(cs(q, M-1))) + q \end{pmatrix} \bmod M$$

$$\big(cs(q, x) = (q+x)\bmod M\big),$$

where δ_q is a map from a neighboring configuration in which the focal cell state is q to a state. Equation 8 means that δ_q ($q = 1, \ldots, M-1$) can be obtained by incrementing each value in the equation of δ_0 as shown in Figure 4 - B.

The above description of the rule largely reduces the search domain for GA. In case of employing two-dimensional M-state nine-neighbor CAs, the maximum number of the transition rules is M^{M^9} because there are M^9 distinct neighborhood configurations at a maximum. On the contrary, if we adopt rules described above, the possible number comes down to $M^{8+(M-1)CM-1}$.

Genetic Information and Evolutionary Operations

We used a form of GA to evolve CAs to perform the task described above. Each individual in GA has a string consists of genes g_l. Each g_l represents the cell state into which the corresponding pattern of the neighborhood is mapped in δ_0 (Figure 4 - C). The population of I individuals is evolved over G generations as follows:

1. All g_l s in the initial population is initialized with 0.

2. The string of genes of each individual is translated to the transition rule, and then its fitness is calculated by using the evaluation process described above.

3. The top E individuals are copied without modification to the next generation. The remaining I - E individuals for the next generation are selected with the probability equal to its relative fitness within the whole population, and they pair with each other as parents.

Figure 4. An example of applying the transition rules (M = 3)

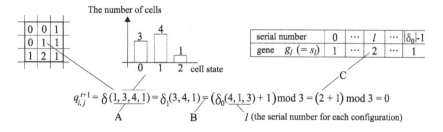

313

4. Two offsprings are generated from the pair based on a two-point crossover with a probability $P_{crossover}$ and a mutation for each gene with a probability $P_{mutation}$. A mutation changes g_l to a random value ($0 \leq g_l < M$) except for the current value.

5. The E elites and $I - E$ offsprings form the population of the new generation, and the process goes back to the step 2 until the generation reaches G.

EXPERIMENTAL RESULTS AND ANALYSES

Course of Evolution

We adopted the settings of parameters as follows: $N = 64$, $M = 3$, $P_a = 0.2$, $\beta = 0.01$, $D=5$, $L_{pre} = 2048$, $L_{int} = 1024$, $L_{fluct} = 128$, $L_d = 8$, $\theta = 0.1$, $\varphi = 100$, $I = 32$, $E = 8$, $G = 256$, $P_{crossover} = 0.75$ and $P_{mutation} = 0.05$.

We conducted 12 runs, and it turned out that the fitness reached approximately 27 in 10 runs. The fitness of remaining 2 runs went up to about 21 or 25. Here, we focus on the run in which the fitness reached the best value among them.

Figure 5 shows the best, average and worst fitness at each generation. The best fitness was about zero in the initial population and it rapidly went up to approximately 19 until the 10th generation, then eventually converged to approximately 27 around the 50th generation. Also, we see the average fitness tended to be a half of the best

fitness, and the worst fitness was almost zero through the experiment.

As defined in Equation (6), the fitness is the sum of the differences between the scaled density distributions. So as to grasp the main reason for the increase in fitness through the experiment, we plotted the all differences measured during the evaluation processes of all individuals at each generation in Figure 6. After approximately the 30th generation, we see that the difference often became the maximum value 3.0 which was obtained when the difference between the mutually opposite distributions were measured. It clearly shows that the evolved CAs successfully exhibited the self-organizing behaviors that their global configurations drastically changed after the occurrences of external perturbations.

Emergence of State Transition Cycles Induced by External Perturbations

In the previous section, we observed that the population successfully evolved, and individuals were expected to exhibit self-organizing behaviors induced by perturbations. Next, we analyze the behavior of the adaptive individuals in detail, and discuss their self-organizing properties.

Figure 7 illustrates the transitions of several indices during an evaluation of a typical individual

Figure 6. The difference between the scaled density distributions. The dots represent all values of Equation (5) measured when all individuals were evaluated in each generation.

Figure 5. The fitness of the population

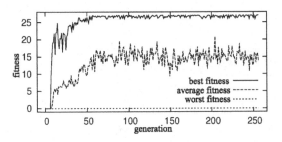

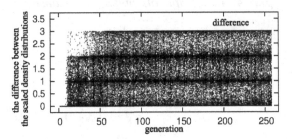

in the last generation of the same run as in Figure 5. The string of genes is ``020110010011110120010 1110101210120100000000000''[2]. The lower graph shows the transitions of the elements of density distribution and the entropy[3] during the evaluation. The above images also illustrate the configuration of cell states at the end of each period.

Through the preparation period, the configuration of the CA gradually changed from the random initial configuration to the stable one which was characterized by the decrease in the entropy. At the end of the preparation period, there were small clusters of the state 1 in a sea of the state 0 (Figure 7 - 1). The configuration did not change through the first normal period (2).

Because the most of the cell states were 0, the occurrences of the first perturbation period increased the density of the state 1 (3). Then, the clusters of 1 gradually expanded their size and finally occupied the whole configuration through the subsequent normal period (4).

In contrast, the effect of the second perturbation period was not strong. Although the density of the state 2 was increased (5), the global configuration did not change any further (6). However the effect of the third perturbation period (7) caused the significant change in the global configuration (8). The clusters of the state 2 appeared, expanded their size, and finally occupied the whole configuration. As explained, we observed similar changes in the dominant cell state (12) every two occurrences of perturbation periods (9 - 11). The detailed analyses on the effects of perturbations on the local interac-

tions among cells showed that if the number of the subsequent dominant cell state exceeds a certain threshold by perturbations, it begins to increase during the subsequent normal period.

Figure 8 is the trajectory of the density distribution during this evaluation. We see that the density distribution showed a triangle cycle on this space as a result of the emergent dynamics explained above. This cycle is *ascending* in that the value of the dominant cell state increases as the global configuration changes.

The global configuration of the CA in Figure 7 and Figure 8 showed cyclic transitions between 6 different configurations occupied by almost one cell state (see Figure 7 - 4) or two cell states (*i. e.* a number of state 2 in a sea of state 1, (6)). Because the scaling function increases the small density of the cell state, the actual differences in these scaled density distributions become large, and as a result, the differences between the several pairs (*i. e.* (6) and (12)) become the highest. Also, in each normal period, the global configuration completely converges before the end of the period as shown in Figure 7. Thus, we can say that the cyclic behavior emerged through the course of evolution because of these adaptive and stable properties.

On the other hand, we also observed another interesting rule at the last generation in other runs. The string of genes is ``112211000100211 0012121122221202222200000000000''. Its typical behavior is illustrated in Figure 9 and Figure 10. As we can see from this figure, the density

Figure 7. The behavior of cellular automata (ascending cycle)

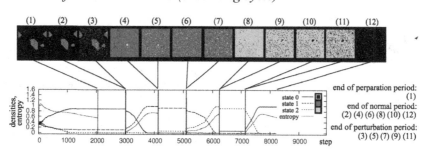

Figure 8. The trajectory of density distribution (ascending cycle)

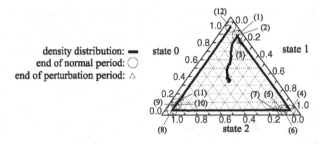

distribution exhibited a reverse (*descending*) cycle in comparison with the previous one. This is an unexpected phenomenon because the global configuration changes in a descending order despite that perturbations increment the values of cell states. The detailed analyses also showed that the perturbations worked like a catalyst in that the perturbed cells can decrement the values of cell states in neighbors. For example, if the configuration consists of a sea of the state 0, the perturbed cells of 1 can change their neighboring cell states of 0 to 2.

Among the successful runs, the rules of CA that exhibited the ascending cycles were clearly observed in 5 runs, and those exhibited the descending cycles were obtained in 3 runs. We can say that there were two opposite solutions to solve the task, but both rules cyclically changed the global configuration of CAs by using their self-organizing properties that a drastic change in its global state occurs every if the number of the subsequent dominant cells goes beyond a certain threshold.

Stability of Emerged Cycles

We obtained the emergent behaviors of CAs with cyclic trajectories of the density distribution triggered by perturbations. However, it needs more consideration to discuss the stability of the cyclic behaviors of CAs during their long-term transitions because the number of the perturbation periods in each evaluation process was merely 5. Consequently, we conducted additional evaluations on the two typical rules explained in the previous section, in which the settings of the parameters were the same as the previous experiments except for the number of perturbation periods $D = 17$.

In order to understand the stability of transitions between configurations, we calculated the transition probability between configurations at the steps when the global configurations were used for fitness evaluation ($a(0), a(1), ..., a(D)$ in Figure 3). Specifically, all global configurations were divided into $2^3 = 8$ classes depending on whether each cell state exists or not on the global configuration. The existence of a cell state on the

Figure 9. The behavior of cellular automata (descending cycle)

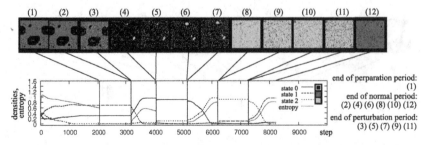

Figure 10. The trajectory of density distribution (descending cycle)

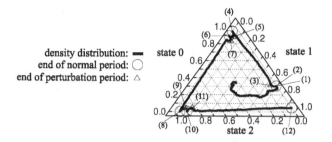

global configuration was decided by whether the density of the cell state can be increased by the scaling function Equation 4 or not[4]. For example, if a density distribution is {0.80, 0.15, 0.05} and the scaled density distribution is {1.00, 0.99, 0.01}, then it is regarded that the cell state 0 and 1 exist on the configuration. Then, we measured the transition probabilities between the classes during the evaluation process.

The table in Figure 11 displays the average transition probabilities between configurations over 100 evaluations of the individual which showed the ascending cycle in the previous experiment. Each set of cell states in row and column is the set of existent cell states in the corresponding class, and each value is the specific transition probability from the column to the row class. The transition diagram on the right also visualizes the same distribution probabilities, in which the line types of arrows correspond to different ranges of the value. As shown from the table and diagram, all the transition probabilities corresponding to the ascending cycle {0} → {0, 1} → {1} → {1, 2} → {2} → {2, 0} → ..., were greater than 0.65. As above, the ascending cycle with 6 different configurations is stable through the long-term evaluation.

Figure 11. The transition table and diagram for global configuration (ascending cycle)

Transition probability between configurations.

	{0}	{0, 1}	{1}	{1, 2}	{2}	{2, 0}	{0, 1, 2}
{0}	0.137	0.687	0.176	0.000	0.000	0.000	0.000
{0, 1}	0.000	0.016	0.959	0.024	0.000	0.000	0.000
{1}	0.000	0.000	0.147	0.668	0.185	0.000	0.000
{1, 2}	0.000	0.000	0.000	0.029	0.955	0.016	0.000
{2}	0.185	0.000	0.000	0.000	0.152	0.663	0.000
{2, 0}	0.979	0.013	0.000	0.000	0.000	0.009	0.000
{0, 1, 2}	0.000	0.313	0.000	0.313	0.000	0.188	0.188

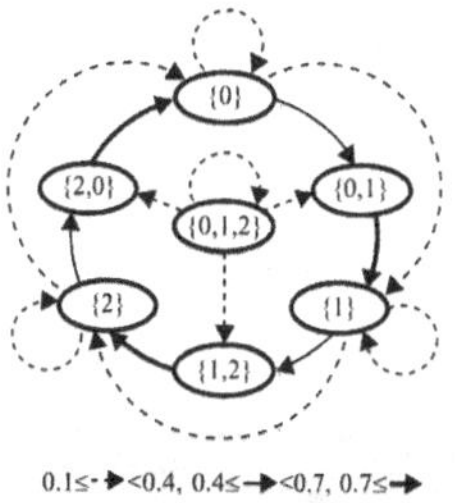

$0.1 \leq \cdots < 0.4, \ 0.4 \leq \twoheadrightarrow < 0.7, \ 0.7 \leq \Rightarrow$

Figure 12. The transition table and diagram for global configuration (descending cycle)

Transition probability between configurations.

	{0}	{0, 1}	{1}	{1, 2}	{2}	{2, 0}	{0, 1, 2}
{0}	0.000	0.000	0.000	0.000	0.038	0.962	0.000
{0, 1}	0.721	0.208	0.000	0.000	0.000	0.071	0.000
{1}	0.048	0.952	0.000	0.000	0.000	0.000	0.000
{1, 2}	0.000	0.059	0.648	0.293	0.000	0.000	0.000
{2}	0.000	0.000	0.017	0.983	0.000	0.000	0.000
{2, 0}	0.000	0.000	0.000	0.086	0.676	0.239	0.000
{0, 1, 2}	0.000	0.333	0.000	0.111	0.000	0.111	0.444

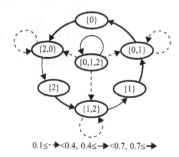

$0.1 \leq \cdots < 0.4, \ 0.4 \leq \twoheadrightarrow < 0.7, \ 0.7 \leq \Rightarrow$

The table in Figure 12 displays the transition probability between configurations of the individual which showed the descending cycle in the previous experiments. The transition probabilities corresponding to the descending cycle $\{0\} \rightarrow \{2, 0\} \rightarrow \{2\} \rightarrow \{1, 2\} \rightarrow \{1\} \rightarrow \{0, 1\} \rightarrow \dots$, were greater than 0.65 approximately, which are similar to the transition probabilities of the ascending cycle. The transition diagram clearly shows that the cycle is also stable while it is reversed compared to the previous one.

CONCLUSION

We have investigated emergent properties of CAs induced by external perturbations. We introduced the transitivity into an extended version of outer-totalistic rules of CAs, and adopted the scaled density distribution of cell states as the global

configuration, expecting emergences of the self-organizing behaviors and the reduction of the search space for a GA. We assumed a minimal task in which a CA has to change its global state every perturbation, and then searched the rules for CAs which can solve the task by using a GA. We obtained the rules for the CA in which the global configuration cyclically transited among different stable configurations, and these stable configurations composed of not only homogeneous but also heterogeneous cell states. As a result, the number of stable configurations became twice as that of possible cell states. These interesting results were obtained only when we introduced the transitivity (Equation (8)) into the rule of CAs. It should be emphasized that we found both ascending and descending cycles of global configurations even though a perturbation always increments the value of a cell state. Detailed analyses showed that the ascending cycle was due to the self-organizing

Figure 13. The emergent behaviors of the CAs in which the cells performed random walk on a two dimensional space. The neighborhood of a cell is defined by all those cells that are within a specific distance. Each cell can be regarded as a unit of decentralized mobile robots. We adopted the same rule as that of the CA which exhibited an ascending cycle in Figure 7. The center graph shows the transitions of the elements of density distribution during the evaluation. Each image also illustrates the configuration of cell states at the end of each period. Each circle denotes a cell, and its color is assigned to its state. The occurrence of the second perturbation period increased the density of the state 0 (2, 3). Then, the clusters of 0 gradually expanded their size and finally occupied the whole configuration through the subsequent normal period (4).

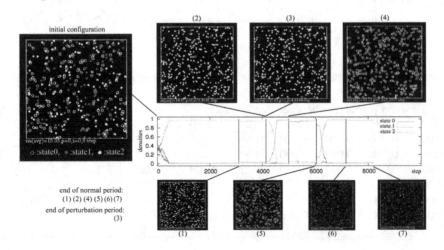

feature that a drastic change in its global state occurs every when the accumulation of the perturbed cells goes beyond a certain threshold. Also, the descending cycle was due to the catalytic effect of perturbed cells which change cell states in their neighborhood into the dominant cell state in the subsequent stable configuration.

Mamei *et al.* (2005) argued about applications of self-organizing features in the dissipative CAs for controlling decentralized pervasive systems. In our model, we can regard the external perturbations as the signals for controlling the global behavior of decentralized systems. Our results imply that we can dynamically control the global behaviors of decentralized systems by changing several states of randomly selected components only.

In real systems such as swarm robots or decentralized pervasive systems, the nodes are not regularly distributed in general. Thus, one might assume that the dynamics of CAs could not be applied to these systems. However, Mamei *et al.* (2005) conducted additional experiments with CAs in which the cells were randomly distributed and their neighborhood of a cell is defined by all those cells that are within a specific distance, and showed that their schemes could be also adapted to these CAs. As shown in Figure 13, we confirmed that the self-organizing behaviors described above also appeared on a CA in which the cells performed random walk on a two dimensional space.

Future work includes investigating our model with different several parameters, obtaining more complex behaviors induced by different kinds of perturbations, such as conditional branches of the behavior trajectories based on the kinds of perturbations, and designing multi-agent systems which make use of the self-organizing properties of the CAs.

ACKNOWLEDGMENT

This work was supported in part by a Grant-in-Aid for 21st Century COE ``Frontiers of Computational Science''.

REFERENCES

Ingerson, T. E., & Buvel, R. L. (1984). Structure in Asynchronous Cellular Automata. *Physica D. Nonlinear Phenomena, 10*, 59–68. doi:10.1016/0167-2789(84)90249-5

Iwase, Y., Suzuki, R., & Arita, T. (2007). Evolutionary Search for Cellular Automata that Exhibit Self-Organizing Properties Induced by External Perturbations. In *Proc. 2007 IEEE Congress on Evolutionary Computation* (CEC2007) (pp. 759-765).

Kwak, K. J., Baryshnikov, Y. M., & Coffman, E. G. (2008). Self-Organizing Sleep-Wake Sensor Systems. In *Proc. the 2nd IEEE International Conference on Self-Adaptive and Self-Organizing Systems* (SASO2008) (pp. 393-402).

Mamei, M., Roli, A., & Zambonelli, F. (2005). Emergence and Control of Macro-Spatial Structures in Perturbed Cellular Automata, and Implications for Pervasive Computing Systems. *IEEE Trans. Systems, Man, and Cybernetics . Part A: Systems and Humans, 35*(3), 337–348. doi:10.1109/TSMCA.2005.846379

Marr, C., & Hütt, M. T. (2006). Similar Impact of Topological and Dynamic Noise on Complex Patterns. *Physics Letters. [Part A], 349*, 302–305. doi:10.1016/j.physleta.2005.08.096

Mitchell, M., Crutchfield, J. P., & Hraber, P. T. (1994). Evolving Cellular Automata to Perform Computations: Mechanisms and Impediments. *Physica D. Nonlinear Phenomena, 75*, 361–391. doi:10.1016/0167-2789(94)90293-3

Nagata, H., Morita, S., Yoshimura, J., Nitta, T., & Tainaka, K. (2008). Perturbation Experiments and Fluctuation Enhancement in Finite Size of Lattice Ecosystems: Uncertainty in Top-Predator Conservation. *Ecological Informatics, 3*(2), 191–201. doi:10.1016/j.ecoinf.2008.01.005

Ninagawa, S. (2005). Evolving Cellular Automata by 1/f Noise. In *Proc. the 8th European Conference on Artificial Life* (ECAL2005) (pp. 453-460).

Ninagawa, S., Yoneda, M., & Hirose, S. (1997). Cellular Automata in Dissipative Boundary Conditions [in Japanese]. *Transactions of Information Processing Society of Japan, 38*(4), 927–930.

Rocha, L. M. (2004). Evolving Memory: Logical Tasks for Cellular Automata. In *Proc. the 9th International Conference on the Simulation and Synthesis of Living Systems* (ALIFE9) (pp. 256-261).

Roli, A., & Zambonelli, F. (2002). Emergence of Macro Spatial Structures in Dissipative Cellular Automata. In *Proc. the 5th International Conference on Cellular Automata for Research and Industry* (ACRI2002) (pp. 144-155).

Sakai, S., Nishinari, K., & Iida, S. (2006). A New Stochastic Cellular Automaton Model on Traffic Flow and Its Jamming Phase Transition. *Journal of Physics. A, Mathematical and General, 39*(50),15327–15339. doi:10.1088/0305-4470/39/50/002

Wolfram, S. (2002). *A New Kind of Science.* Wolfram Media Inc.

ENDNOTES

[1] In the case of $M = 3$, the every possible number of the transition rule in our model comes down from $M^{M9} = 3^{39} \approx 1.505 \times 10^{9391}$ to $M^{8 + (M-1)CM-1} = 3^{10C2} \approx 2.954 \times 10^{21}$.

[2] Each value in the string represents the value in δ_0 as follows:
"$\delta_0(0,0,8)\, \delta_0(0,1,7) \ldots \delta_0(1,0,7)\, \delta_0(1,1,6) \ldots \ldots \delta_0(7,0,1)\, \delta_0(7,1,0)\, \delta_0(8,0,0)$".

[3] The entropy of the global configuration **H** is defined by

$$\mathbf{H} = \frac{1}{N \times N} \sum_{(i,j) \in N \times N} H_{i,j},$$

$$H_{i,j} = -\sum_{s=0}^{M-1} P_{i,j}(s) \log_2 P_{i,j}(s),$$

where $H_{i,j}$ is the entropy of the cell (i,j), and $P_{i,j}(s)$ is the probability of the occurrence of the cell state s at (i,j) during each period.

[4] Actually, we defined that the density can be increased if the density is larger than the x-value (approximately 0.075) at the intersection of $y = SF_{0.1,\,100}(x)$ with $y = x$ around $\theta = 0.1$.

Compilation of References

Abbeel, P., & Ng, A. Y. (2005). Exploration and apprenticeship learning in reinforcement learning. In *Proceedings of the Twentyfirst International Conference on Machine Learning* (pp. 1-8).

Abbott, A., Doering, C., Caves, C., Lidar, D., Brandt, H., & Hamilton, A. (2003). Dreams versus Reality: Plenary Debate Session on Quantum Computing. *Quantum Information Processing, 2*(6), 449–472. doi:10.1023/B:QINP.0000042203.24782.9a

Aberdeen, D., & Baxter, J. (2002). Scalable Internal-State Policy-Gradient Methods for POMDPs. In *Proceedings of the Nineteenth International Conference on Machine Learning* (pp. 3-10).

Acerbi, A., et al. (2007). Social Facilitation on the Development of Foraging Behaviors in a Population of Autonomous Robots. In *Proceedings of the 9th European Conference in Artificial Life* (pp. 625-634).

Agogino, A. K., & Tumer, K. (2004). Unifying Temporal and Structural Credit Assignment Problems. In *Proceedings of the Third International Joint Conference on Autonomous Agents and Multi-Agent Systems* (pp. 980-987).

Agosta, W. (1992). *Chemical Communication – The Language of Pheromone*. W. H. Freeman and Company.

Alfarano, S., Wagner, F., & Lux, T. (2004). *Estimation of Agent-Based Models: the case of an asymmetric herding model.*

Ambler, S. (2008). Scaling Scrum – Meeting Real World Development Needs. *Dr. Dobbs Journal*. Retrieved April 23, 2008 from http://www.drdobbsonline.net/architect/207100381.

Ando, Y., Masutani, O., Honiden, S., Fukazawa, Y., & Iwasaki, H. (2006). Performance of Pheromone Model for Predicting Traffic Congestion. In . *Proceedings of AAMAS, 2006*, 73–80.

Angeline, P. J., Sauders, G. M., & Pollack, J. B. (1994). An evolutionary algorithms that constructs recurrent neural networks. *IEEE Transactions on Neural Networks, 5*, 54–65. doi:10.1109/72.265960

Appleton-Young, L. (2008). 2008 real estate market forecast. California Association of Realtors. Retrieved December 2008, from http://bayareahousingreview.com/wp-content/uploads/2008/02/ leslie_appleton_young _preso _read-only1.pdf.

Arai, S. & Tanaka, N. (2006). Experimental Analysis of Reward Design for Continuing Task in Multiagent Domains. *Journal of Japanese Society for Artificial Intelligence, in Japanese, 13*(5), 537-546.

Aranha, C., & Iba, H. (2007). Portfolio Management by Genetic Algorithms with Error Modeling. In *JCIS Online Proceedings of International Conference on Computational Intelligence in Economics & Finance.*

Arthur, W. B. (1993). On designing economic agents that behave like human agents. *Journal of Evolutionary Economics, 3*, 1–22. doi:10.1007/BF01199986

Arthur, W. B., Holland, J. H., LeBaron, B., Palmer, R. G., & Taylor, P. (1997). Asset Pricing under Endogenous Expectations in an Artificial Stock Market. [Addison-Wesley.]. *The Economy as an Evolving Complex System, II*, 15–44.

Atanassov, K. T. (1999). *Intuitionistic Fuzzy Sets, Physica Verlag*. Heidelberg: Springer.

Axelrod, R. (1997). *The Complexity of Cooperation -Agent-Based Model of Competition and Collaboration*. Princeton University Press.

Axtell, R. (2000). Why Agents? On the Varied Motivation For Agent Computing In the Social Sciences. *The Brookings Institution Center on Social and Economic Dynamics Working Paper*, November, No.17.

Bäck, T. (1996). *Evolutionary Algorithms in Theory and Practice: Evolution Strategies, Evolutionary Programming, Genetic Algorithms*. Oxford University Press.

Bagnall, A. J., & Smith, G. D. (2005). A Multi agent Model of UK Market in Electricity Generation. *IEEE Transactions on Evolutionary Computation*, 522–536. doi:10.1109/TEVC.2005.850264

Baird, L., & Poole, D. (1999). Gradient Descent for General Reinforcement Learning. *Advances in Neural Information Processing Systems*, *11*, 968–974.

Baki, B., Bouzid, M., Ligęza, A., & Mouaddib, A. (2006). A centralized planning technique with temporal constraints and uncertainty for multi-agent systems. *Journal of Experimental & Theoretical Artificial Intelligence*, *18*(3), 331–364. doi:10.1080/09528130600906340

Baldassarre, G., Nolfi, S., & Parisi, D. (2003). Evolving Mobile Robots Able to Display Collective Behaviours . *Artificial Life*, *9*(3), 255–267. doi:10.1162/106454603322392460

Baldassarre, G. (2007, June). *Research on brain and behaviour, and agent-based modelling, will deeply impact investigations on well-being (and theoretical economics)*. Paper presented at International Conference on Policies for Happiness, Certosa di Pontignano, Siena, Italy.

Barto, A. (1996). Muti-agent reinforcement learning and adaptive neural networks. Retrieved December 2008, from http://stinet.dtic.mil/cgi-bin/GetTRDoc?AD=ADA315266&Location=U2&doc =GetTRDoc.pdf.

Bazerman, M. (1998). *Judgment in Managerial Decision Making*. John Wiley & Sons.

Becker, M., & Szczerbicka, H. (2005). Parameters Influencing the Performance of Ant Algorithm Applied to Optimisation of Buffer Size in Manufacturing. *Industrial Engineering and Management Systems*, *4*(2), 184–191.

Beer, R. D. (1996). Toward the Evolution of Dynamical Neural Networks for Minimally Cognitive. In *From Animals to Animats 4: Proceedings of the Fourth International Conference on Simulation of Adaptive Behavior* (pp. 421-429).

Benenti, G. (2004). *Principles of Quantum Computation and Information* (*Vol. 1*). New Jersey: World Scientific.

Beni, G., & Wang, J. (1989). Swarm Intelligence in Cellular Robotic Systems. In *Proceed. NATO Advanced Workshop on Robots and Biological Systems*, Tuscany, Italy, June 26–30

Benjamin, D., Brown, S., & Shapiro, J. (2006). *Who is 'behavioral'? Cognitive ability and anomalous preferences*. Levine's Working Paper Archive 122247000000001334, UCLA Department of Economics.

Bennett, F., III. (1996). Emergence of a Multi-Agent Architecture and New Tactics for the Ant Colony Food Foraging Problem Using Genetic Programming. In *From Animals to Animats 4, Proceedings of the Fourth International Conference on Simulations of Adaptive Behavior* (pp. 430–439).

Binder, W. J., Hulaas, G., & Villazon, A. (2001). Portable Resource Control in the J-SEAL2 Mobile Agent System. In *Proceedings of International Conference on Autonomous Agents* (pp. 222-223).

Black, F., & Litterman, R. (1992, Sept/Oct). Global Portfolio Optimization. *Financial Analysts Journal*, 28–43. doi:10.2469/faj.v48.n5.28

Blynel, J., & Floreano, D. (2003). Exploring the T-Maze: Evolving Learning-Like Robot Behaviors using CTRNNs. In *Proceedings of the 2nd European Workshop on Evolutionary Robotics (EvoRob'2003)* (LNCS).

Boehm, B., & Turner, R. (2004). *Balancing Agility and discipline: A Guide for the Perplexed*. Addison-Wesley Press.

Bonabeau, E., Dorigo, M., & Theraulaz, G. (1999). *Swarm Intelligence from Natural to Artificial Systems*. Oxford University Press.

Bornholdt, S. (2001). Expectation bubbles in a spin model of markets. *International Journal of Modern Physics C*, *12*(5), 667–674. doi:10.1142/S0129183101001845

Bossaerts, P., Beierholm, U., Anen, C., Tzieropoulos, H., Quartz, S., de Peralta, R., & Gonzalez, S. (2008, September). *Neurobiological foundations for "dual system"' theory in decision making under uncertainty: fMRI and EEG evidence*. Paper presented at Annual Conference on Neuroeconomics, Park City, Utah.

Bostonbubble.com. (2007). S&P/Case-Shiller Boston snapshot Q3 2007. Retrieved December 2008, from http://www.bostonbubble.com/forums/viewtopic.php?t=598.

Boswijk H. P., Hommes C. H, & Manzan, S. (2004). *Behavioral Heterogeneity in Stock Prices.*

Boutilier, C., & Poole, D. (1996). Computing Optimal Policies for Partially Observable Decision Processes using Compact Representations. In *Proceedings of the Thirteenth National Conference on Artificial Intelligence* (pp. 1168-1175).

Brocas, I., & Carrillo, J. (2008a). The brain as a hierarchical organization. *The American Economic Review*, *98*(4), 1312–1346. doi:10.1257/aer.98.4.1312

Brocas, I., & Carrillo, J. (2008b). Theories of the mind. *American Economic Review: Papers\& Proceedings, 98*(2), 175-180.

Brunnermeier, M. K. (2001). *Asset Pricing under Asymmetric Information*. Oxford University Press. doi:10.1093/0198296983.001.0001

Budhraja, V. S. (2001). California's electricity crisis. *IEEE Power Engineering Society Summer Meeting.*

C.A.R. (2008). U.S. economic outlook: 2008. Retrieved December 2008, from http://rodomino.realtor.org/Research.nsf/files/ currentforecast.pdf/$FILE/current-forecast.pdf.

Callebaut, W., & Rasskin-Gutman, D. (Eds.). (2005). *Modularity: Understanding the development and evolution of natural complex systems*. MA: MIT Press.

Caplin, A., & Dean, M. (2008). Economic insights from ``neuroeconomic'' data. *The American Economic Review, 98*(2), 169–174. doi:10.1257/aer.98.2.169

Carnap, R., & Jeffrey, R. (1971). *Studies in Inductive Logics and Probability* (*Vol. 1*, pp. 35–165). Berkeley, CA: University of California Press.

Casari, M. (2004). Can genetic algorithms explain experimental anomalies? An application to common property resources. *Computational Economics, 24*, 257–275. doi:10.1007/s10614-004-4197-5

Case, K., Glaeser, E., & Parker, J. (2000). Real estate and the macroeconomy. *Brookings Papers on Economic Activity, 2*, 119–162. doi:.doi:10.1353/eca.2000.0011

Case, K., & Shiller, R. (1989). The efficiency of the market for single-family homes. *The American Economic Review, 79*, 125–137.

Case, K., & Shiller, R. (1990). Forecasting prices and excess returns in the housing market. *American Real Estate and Urban Economics Association Journal, 18*, 263–273. doi:.doi:10.1111/1540-6229.00521

Case, K., & Shiller, R. (2003). Is there a bubble in the housing market? *Brookings Papers on Economic Activity, 1*, 299–342. doi:.doi:10.1353/eca.2004.0004

Chalkiadakis, G., & Boutilier, C. (2008). Sequential Decision Making in Repeated Coalition Formation under Uncertainty, In: Proc. of 7th Int. Conf. on Autonomous Agents and Multi-agent Systems (AA-MAS 2008), Padgham, Parkes, Müller and Parsons (eds.), May, 12-16, 2008, Estoril, Portugal, http://eprints.ecs.soton.ac.uk/15174/1/BayesRLCF08.pdf

Chan, N. T., LeBaron, B., Lo, A. W., & Poggio, T. (2008). Agent-based models of financial markets: A comparison with experimental markets. MIT Artificial Markets Project, Paper No. 124, September. Retrieved January 1, 2008, from http://citeseer.ist.psu.edu/chan99agentbased.html.

Chang, T. J., Meade, N., Beasley, J. E., & Sharaiha, Y. M. (2000). Heuristics for Cardinality Constrained Portfolio Optimization . *Computers & Operations Research*, *27*, 1271–1302. doi:10.1016/S0305-0548(99)00074-X

Charness, G., & Levin, D. (2005). When optimal choices feel wrong: A laboratory study of Bayesian updating, complexity, and affect. *The American Economic Review*, *95*(4), 1300–1309. doi:10.1257/0002828054825583

Chattoe, E. (1998). Just how (un)realistic are evolutionary algorithms as representations of social processes? *Journal of Artificial Societies and Social Simulation*, 1.

Chen, S. H., & Yeh, C. H. (2002). On the Emergent Properties of Artificial Stock Markets: The Efficient Market Hypothesis and the Rational Expectations Hypothesis. *Journal of Behavior & Organization*, *49*, 217–239. doi:10.1016/S0167-2681(02)00068-9

Chen, S.-H. (2008). Software-agent designs in economics: An interdisciplinary framework. *IEEE Computational Intelligence Magazine*, *3*(4), 18–22. doi:10.1109/MCI.2008.929844

Chen, S.-H., & Chie, B.-T. (2004). Agent-based economic modeling of the evolution of technology: The relevance of functional modularity and genetic programming. *International Journal of Modern Physics B*, *18*(17-19), 2376–2386. doi:10.1142/S0217979204025403

Chen, S.-H., & Huang, Y.-C. (2008). Risk preference, forecasting accuracy and survival dynamics: Simulations based on a multi-asset agent-based artificial stock market. *Journal of Economic Behavior & Organization*, *67*(3), 702–717. doi:10.1016/j.jebo.2006.11.006

Chen, S.-H., & Tai, C.-C. (2003). Trading restrictions, price dynamics and allocative efficiency in double auction markets: an analysis based on agent-based modeling and simulations. *Advances in Complex Systems*, *6*(3), 283–302. doi:10.1142/S021952590300089X

Chen, S.-H., Zeng, R.-J., & Yu, T. (2009a). Analysis of Micro-Behavior and Bounded Rationality in Double Auction Markets Using Co-evolutionary GP . In *Proceedings of World Summit on Genetic and Evolutionary Computation*. ACM.

Chen, S., & Yeh, C. (1996). Genetic programming learning and the cobweb model . In Angeline, P. (Ed.), *Advances in Genetic Programming* (*Vol. 2*, pp. 443–466). Cambridge, MA: MIT Press.

Chen, S.-H., Zeng, R.-J., & Yu, T. (2009). Co-evolving trading strategies to analyze bounded rationality in double auction markets . In Riolo, R., Soule, T., & Worzel, B. (Eds.), *Genetic Programming: Theory and Practice VI* (pp. 195–213). Springer. doi:10.1007/978-0-387-87623-8_13

Chen, S.-H., Zeng, R.-J., & Yu, T. (2008). Co-evolving trading strategies to analyze bounded rationality in double auction markets . In Riolo, R., Soule, T., & Worzel, B. (Eds.), *Genetic Programming Theory and Practice VI* (pp. 195–213). Springer.

Chen, S.-H., & Chie, B.-T. (2007). Modularity, product innovation, and consumer satisfaction: An agent-based approach . In Yin, H., Tino, P., Corchado, E., Byrne, W., & Yao, X. (Eds.), *Intelligent Data Engineering and Automated Learning* (pp. 1053–1062). Heidelberg, Germany: Springer. doi:10.1007/978-3-540-77226-2_105

Chen, L., Xu, X., & Chen, Y. (2004). An adaptive ant colony clustering algorithm. In *Proceedings of the Third IEEE International Conference on Machine Learning and Cybernetics* (pp. 1387-1392).

Chen, S., & Yeh, C. (1995). Predicting stock returns with genetic programming: Do the short-run nonlinear regularities exist? In D. Fisher (Ed.), *Proceedings of the Fifth International Workshop on Artificial Intelligence and Statistics* (pp. 95-101). Ft. Lauderdale, FL.

Chen, S.-H., Chie, B.-T., & Tai, C.-C. (2001). Evolving bargaining strategies with genetic programming: An overview of AIE-DA Ver. 2, Part 2. In B. Verma & A. Ohuchi (Eds.), *Proceedings of Fourth International Conference on Computational Intelligence and Multimedia Applications (ICCIMA 2001)* (pp. 55–60). IEEE Computer Society Press.

Chen, X., & Tokinaga, S. (2006). Analysis of price fluctuation in double auction markets consisting of multi-agents using the genetic programming for learning. Retrieved from https://qir.kyushuu.ac.jp/dspace/bitstream /2324/8706/ 1/ p147-167.pdf.

China Bystanders. (2008). Bank profits trimmed by subprime losses. Retrieved from http://chinabystander. wordpress.com /2008/03/25/bank-profits-trimmed-by-subprime-losses/.

Chrisman, L. (1992). Reinforcement Learning with Perceptual Aliasing: The Perceptual Distinctions Approach. In *Proceedings of the Tenth National Conference on Artificial Intelligence* (pp. 183-188).

Cincotti, S., Focardi, S., Marchesi, M., & Raberto, M. (2003). Who wins? Study of long-run trader survival in an artificial stock market. *Physica A, 324*, 227–233. doi:10.1016/S0378-4371(02)01902-7

Cliff, D., Harvey, I., & Husbands, P. (1993). Explorations in Evolutionary Robotics. *Adaptive Behavior, 2*(1), 71–104. doi:10.1177/105971239300200104

Cliff, D., & Bruten, J. (1997). *Zero is not enough: On the lower limit of agent intelligence for continuous double auction markets* (Technical Report no. HPL-97-141). Hewlett-Packard Laboratories. Retrieved January 1, 2008, from http://citeseer.ist.psu.edu/cliff97zero.html

CME 2007. (n.d.). Retrieved December 2008, from http:// www.cme.com/trading/prd/re/housing.html.

Collins, R., & Jefferson, D. (1991). AntFarm: Towards Simulated Evolution. In *Artificial Life II, Proceedings of the Second International Conference on Artificial Life* (pp. 159–168).

Colorni, A., Dorigo, M., & Maniezzo, V. (1991). Distributed Optimization by Ant Colonies. In . *Proceedings, ECAL91*, 134–142.

Colyvan, M. (2004). The Philosophical Significance of Cox's Theorem. *International Journal of Approximate Reasoning, 37*(1), 71–85. doi:10.1016/j.ijar.2003.11.001

Colyvan, M. (2008). Is Probability the Only Coherent Approach to Uncertainty? *Risk Analysis, 28*, 645–652. doi:10.1111/j.1539-6924.2008.01058.x

Commonweal of Australia. (2001). Economic Outlook. Retrieved December 2008, from http://www.budget.gov. au/2000-01/papers/ bp1/html/bs2.htm.

Cont, R., & Bouchaud, J.-P. (2000). Herd behavior and aggregate fluctuations in financial markets. *Macroeconomics Dynamics, 4*, 170–196.

Covaci, S. (1999). Autonomous Agent Technology. In *Proceedings of the 4th international symposium on Autonomous Decentralized Systems*. Washington, DC: IEEE Computer Science Society.

Croley, T., & Lewis, C. (2006).. . *Journal of Great Lakes Research, 32*, 852–869. doi:10.3394/0380-1330(2006)32[852:WADCTM]2.0.CO;2

D'Espagnat, B. (1999). *Conceptual Foundation of Quantum mechanics* (2nd ed.). Perseus Books.

d'Acremont, M., & Bossaerts, P. (2008, September). *Grasping the fundamental difference between expected utility and mean-variance theories.* Paper presented at Annual Conference on Neuroeconomics, Park City, Utah.

Das, R., Hanson, J. E., Kephart, J. O., & Tesauro, G. (2001). Agent-human interactions in the continuous double auction. In *Proceedings of the 17th International Joint Conference on Artificial Intelligence (IJCAI)*, San Francisco. CA: Morgan-Kaufmann.

De Long, J. B., Shleifer, A. L., Summers, H., & Waldmann, R. J. (1991). The survival of noise traders in financial markets. *The Journal of Business, 64*(1), 1–19. doi:10.1086/296523

Deneuburg, J., Goss, S., Franks, N., Sendova-Franks, A., Detrain, C., & Chretien, L. (1991). The Dynamics of Collective Sorting: Robot-Like Ant and Ant-Like Robot. In *Proceedings of First Conference on Simulation of Adaptive Behavior: From Animals to Animats* (pp. 356-363). Cambridge: MIT Press.

Detterman, D. K., & Daniel, M. H. (1989). Correlations of mental tests with each other and with cognitive variables are highest for low-IQ groups. *Intelligence, 13*, 349–359. doi:10.1016/S0160-2896(89)80007-8

Devetag, G., & Warglien, M. (2003). Games and phone numbers: Do short-term memory bounds affect strategic behavior? *Journal of Economic Psychology, 24*, 189–202. doi:10.1016/S0167-4870(02)00202-7

Devetag, G., & Warglien, M. (2008). Playing the wrong game: An experimental analysis of relational complexity and strategic misrepresentation. *Games and Economic Behavior, 62*, 364–382. doi:10.1016/j.geb.2007.05.007

Dimeas, A. L., & Hatziargyriou, N. D. (2007). Agent based control of Virtual Power Plants. *International Conference on Intelligent Systems Applications to Power Systems.*

DiVincenzo, D. (1995). Quantum Computation. *Science, 270*(5234), 255–261. doi:10.1126/science.270.5234.255

DiVincenzo, D. (2000). The Physical Implementation of Quantum Computation. *Experimental Proposals for Quantum Computation.* arXiv:quant-ph/0002077

Dorigo, M., & Gambardella, L. M. (1996). Ant Colony System: a Cooperative Learning Approach to the Traveling Salesman. *IEEE Transactions on Evolutionary Computation, 1*(1), 53–66. doi:10.1109/4235.585892

Dorigo, M., & Stutzle, T. (2004). *Ant Colony Optimization.* Cambridge, MA: The MIT Press.

Dorigo, M., Maniezzo, V., & Colorni, A. (1991). *Positive Feedback as a Search Strategy* (Technical Report No. 91-016). Politecnico di Milano.

Duffy, J. (2006). Agent-based models and human subject experiments . In Tesfatsion, L., & Judd, K. (Eds.), *Handbook of Computational Economics* (*Vol. 2*). North Holland.

Durlauf, S. N., & Young, H. P. (2001). *Social Dynamics.* Brookings Institution Press.

Easley, D., & Ledyard, J. (1993). Theories of price formation and exchange in double oral auction . In Friedman, D., & Rust, J. (Eds.), *The Double Auction Market-Institutions, Theories, and Evidence.* Addison-Wesley.

Economist.com. (2007). The world economy: Rocky terrain ahead. Retrieved December 2008, from http://www.economist.com/ daily/news/displaystory. cfm?storyid=9725432&top_story=1.

Edmonds, B. (2002). Review of Reasoning about Rational Agents by Michael Wooldridge. *Journal of Artificial Societies and Social Simulation, 5*(1). Retrieved from http://jasss.soc.surrey.ac.uk/5/1/reviews/edmonds.html.

Edmonds, B. (1998). Modelling socially intelligent agents. *Applied Artificial Intelligence, 12*, 677–699. doi:10.1080/088395198117587

Ellis, C., Kenyon, I., & Spence, M. (1990). Occasional Publication of the London Chapter . *OAS, 5*, 65–124.

Elton, E., Gruber, G., & Blake, C. (1996). Survivorship Bias and Mutual Fund Performance. *Review of Financial Studies, 9*, 1097–1120. doi:10.1093/rfs/9.4.1097

Epstein, J. M., & Axtell, R. (1996). *Growing Artificial Societies Social Science From the The Bottom Up.* MIT Press.

Evolution Robotics Ltd. Homepage (2008). Retrieved from http://www.evolution.com/

Fagin, R., & Halpern, J. (1994). Reasoning about Knowledge and Probability. *Journal of the ACM, 41*(2), 340–367. doi:10.1145/174652.174658

Fair, R., & Jaffee, D. (1972). Methods of estimation for markets in disequilibrium. *Econometrica, 40*, 497–514. doi:.doi:10.2307/1913181

Fama, E. (1970). Efficient Capital Markets: A Review of Theory and Empirical Work. *The Journal of Finance, 25*, 383–417. doi:10.2307/2325486

Feldman, J. (1962). Computer simulation of cognitive processes . In Broko, H. (Ed.), *Computer applications in the behavioral sciences*. Upper Saddle River, NJ: Prentice Hall.

Ferber, J. (1999). *Multi Agent Systems*. Addison Wesley.

Feynman, R. (1982). Simulating physics with computers. *International Journal of Theoretical Physics*, *21*, 467. doi:10.1007/BF02650179

Figner, B., Johnson, E., Lai, G., Krosch, A., Steffener, J., & Weber, E. (2008, September). *Asymmetries in intertemporal discounting: Neural systems and the directional evaluation of immediate vs future rewards*. Paper presented at Annual Conference on Neuroeconomics, Park City, Utah.

Fischhoff, B. (1991). Value elicitation: Is there anything in there? *The American Psychologist*, *46*, 835–847. doi:10.1037/0003-066X.46.8.835

Flament, C. (1963). *Applications of graphs theory to group structure*. London: Prentice Hall.

Freddie Mac. (2008a). CMHPI data. Retrieved December 2008, from http://www.freddiemac.com/finance/cmhpi/#old.

Freddie Mac. (2008b). 30-year fixed rate historical Tables. Historical PMMS® Data. Retrieved December 2008, from http://www.freddiemac.com/pmms/pmms30.htm.

Frederick, S., Loewenstein, G., & O'Donoghue, T. (2002). Time discounting and time preference: A critical review. *Journal of Economic Literature*, *XL*, 351–401. doi:10.1257/002205102320161311

Friedman, D. (1991). A simple testable model of double auction markets. *Journal of Economic Behavior & Organization*, *15*, 47–70. doi:10.1016/0167-2681(91)90004-H

Friedman, M. (1953). *Essays in Positive Economics*. University of Chicago Press.

Fudenberg, D., & Levine, D. (2006). A dual-self model of impulse control. *The American Economic Review*, *96*(5), 1449–1476. doi:10.1257/aer.96.5.1449

Gigerenzer, G., & Selten, R. (2002). *Bounded Rationality*. Cambridge: The MIT Press.

Gjerstad, S., & Dickhaut, J. (1998). Price formation in double auctions. *Games and Economic Behavior*, *22*, 1–29. doi:10.1006/game.1997.0576

Gode, D. K., & Sunder, S. (1993). Allocative efficiency of markets with zero-intelligence traders: markets as a partial substitute for individual rationality. *The Journal of Political Economy*, *101*, 119–137. doi:10.1086/261868

Gode, D., & Sunder, S. (1993). Allocative efficiency of markets with zero-intelligence traders: Market as a partial substitute for individual rationality. *The Journal of Political Economy*, *101*, 119–137. doi:10.1086/261868

Goldberg, D. E. (1989). *Genetic Algorithms in Search, Optimization and Machine Learning*. Addison-Wesley.

Gomez, F. J. and Miikkulainen, R. (1999). Solving Non-Markovian Control Tasks with Neuroevolution, In *Proceedings of the International Joint Conference on Artificial Intelligence* (pp. 1356-1361).

Gosavi, A. (2004). A Reinforcement Learning Algorithm Based on Policy Iteration for Average Reward: Empirical Results with Yield Management and Convergence Analysis. *Machine Learning*, *55*, 5–29. doi:10.1023/B:MACH.0000019802.64038.6c

Gottfredson, L. S. (1997). Mainstream science on intelligence: An editorial with 52 signatories, history, and bibliography. *Intelligence*, *24*(1), 13–23. doi:10.1016/S0160-2896(97)90011-8

Grefenstette, J. J. (1988). Credit Assignment in Rule Discovery Systems Based on Genetic Algorithms. *Machine Learning*, *3*, 225–245. doi:10.1007/BF00113898

Gregg, L., & Simon, H. (1979). Process models and stochastic theories of simple concept formation. In H. Simon, *Models of Thought* (Vol. I). New Haven, CT: Yale Uniersity Press.

Grossklags, J., & Schmidt, C. (2006). Software agents and market (in)efficiency—a human trader experiment. *IEEE Transactions on System, Man, and Cybernetics: Part C*. *Special Issue on Game-theoretic Analysis & Simulation of Negotiation Agents, 36*(1), 56–67.

Group, C. M. E. (2007). S&P/Case-Shiller Price Index: Futures and options. Retrieved December 2008, from http://housingderivatives. typepad.com/housing_derivatives/files/cme_housing _fact_sheet.pdf.

Gruber, M. J. (1996). Another Puzzle: The Growth in Actively Managed Mutual Funds. *The Journal of Finance, 51*(3), 783–810. doi:10.2307/2329222

Haji, K. (2007). Subprime mortgage crisis casts a global shadow – medium-term economic forecast (FY 2007~2017). Retrieved December 2008, from http://www.nli-research.co.jp/english/economics/2007/ eco071228. pdf.

Halpern, J. (2005). *Reasoning about uncertainty*. MIT Press.

Hanaki, N. (2005). Individual and social learning. *Computational Economics, 26*, 213–232. doi:10.1007/s10614-005-9003-5

Harmanec, D., Resconi, G., Klir, G. J., & Pan, Y. (1995). On the computation of uncertainty measure in Dempster-Shafer theory. *International Journal of General Systems, 25*(2), 153–163. doi:10.1080/03081079608945140

Harvey, I., Di Paolo, E., Wood, A., & Quinn, R., M., & Tuci, E. (2005). Evolutionary Robotics: A New Scientific Tool for Studying Cognition. *Artificial Life, 11*(3/4), 79–98. doi:10.1162/1064546053278991

Harvey, I., Husbands, P., Cliff, D., Thompson, A., & Jakobi, N. (1997). Evolutionary robotics: The sussex approach. *Robotics and Autonomous Systems, 20*, 205–224. doi:10.1016/S0921-8890(96)00067-X

Hatziargyriou, N. D., Dimeas, A., Tsikalakis, A. G., Lopes, J. A. P., Kariniotakis, G., & Oyarzabal, J. (2005). Management of Microgrids in Market Environment. *International Conference on Future Power Systems*.

Hiroshi, I., & Masahito, H. (2006). *Quantum Computation and Information*. Berlin: Springer.

Hisdal, E. (1998). *Logical Structures for Representation of Knowledge and Uncertainty*. Springer.

Holland, J. H. (1975). *Adaptation in Natural and Artificial Systems*. University of Michigan Press.

Hough, J. (1958). Geology of the Great Lakes. [Univ. of Illinois Press.]. *Urbana (Caracas, Venezuela)*, IL.

Housing Predictor. (2008). Independent real estate housing forecast. Retrieved December 2008, from http://www.housingpredictor.com/ california.html.

Hunt, E. (1995). The role of intelligence in modern society. *American Scientist*, (July/August): 356–368.

Huynh, T. D., Jennings, N. R., & Shadbolt, N. R. (2006). An integrated trust and reputation model for open multi-agent systems. *Journal of Autonomous agents and multi agent systems*.

Iacono, T. (2008). Case-Shiller® Home Price Index forecasts: Exclusive house-price forecasts based on Fiserv's leading Case-Shiller Home Price Indexes. Retrieved December 2008, from http://www.economy.com/home/products/ case_shiller_indexes.asp.

Ingerson, T. E., & Buvel, R. L. (1984). Structure in Asynchronous Cellular Automata. *Physica D. Nonlinear Phenomena, 10*, 59–68. doi:10.1016/0167-2789(84)90249-5

Iwase, Y., Suzuki, R., & Arita, T. (2007). Evolutionary Search for Cellular Automata that Exhibit Self-Organizing Properties Induced by External Perturbations. In *Proc. 2007 IEEE Congress on Evolutionary Computation (CEC2007)* (pp. 759-765).

Iyengar, S., & Lepper, M. (2000). When choice is demotivating: Can one desire too much of a good thing? *Journal of Personality and Social Psychology, 79*(6), 995–1006. doi:10.1037/0022-3514.79.6.995

Jaakkola, T., Singh, S. P., & Jordan, M. I. (1994). Reinforcement Learning Algorithm for Partially Observable Markov Decision Problems. *Advances in Neural Information Processing Systems, 7*, 345–352.

Jaeger, G. (2006). *Quantum Information: An Overview*. Berlin: Springer.

Jamison, J., Saxton, K., Aungle, P., & Francis, D. (2008). *The development of preferences in rat pups*. Paper presented at Annual Conference on Neuroeconomics, Park City, Utah.

Jayantilal, A., Cheung, K. W., Shamsollahi, P., & Bresler, F. S. (2001). Market Based Regulation for the PJM Electricity Market. *IEEE International Conference on Innovative Computing for Power Electric Energy Meets the Markets* (pp. 155-160).

Jevons, W. (1879). *The Theory of Political Economy, 2nd Edtion. Edited and introduced by R. Black (1970)*. Harmondsworth: Penguin.

Johnson, E., Haeubl, G., & Keinan, A. (2007). Aspects of endowment: A query theory account of loss aversion for simple objects. *Journal of Experimental Psychology. Learning, Memory, and Cognition, 33*, 461–474. doi:10.1037/0278-7393.33.3.461

Johnson, N., Jeffries, P., & Hui, P. M. (2003). *Financial Market Complexity*. Oxford.

Jurca, R., & Faltings, B. (2003). Towards Incentive-Compatible Reputation Management. *Trust, Reputation and Security: Theories and Practice* (LNAI 2631, pp. 138-147).

Kaboudan, M. (2001). Genetically evolved models and normality of their residuals. *Journal of Economic Dynamics & Control, 25*, 1719–1749. doi:.doi:10.1016/S0165-1889(00)00004-X

Kaboudan, M. (2004). TSGP: A time series genetic programming software. Retrieved December 2008, from http://bulldog2.redlands.edu/ fac/mak_kaboudan/tsgp.

Kagan, H. (2006). *The Psychological Immune System: A New Look at Protection and Survival*. Bloomington, IN: AuthorHouse.

Kagel, J. (1995). Auction: A survey of experimental research . In Kagel, J., & Roth, A. (Eds.), *The Handbook of Experimental Economics*. Princeton University Press.

Kahneman, D., Diener, E., & Schwarz, N. (Eds.). (2003). *Well-Being: The Foundations of Hedonic Psychology*. New York, NY: Russell Sage Foundation.

Kahneman, D., Knetsch, J., & Thaler, R. (1990). Experimental tests of the endowment effect and the Coase theorem. *The Journal of Political Economy, 98*, 1325–1348. doi:10.1086/261737

Kahneman, D., Knetsch, J., & Thaler, R. (1991). Anomalies: The endowment effect, loss aversion, and status quo bias. *The Journal of Economic Perspectives, 5*(1), 193–206.

Kahneman, D., Ritov, I., & Schkade, D. (1999). Economic preferences or attitude expressions? An analysis of dollar responses to public issues. *Journal of Risk and Uncertainty, 19*, 203–235. doi:10.1023/A:1007835629236

Kahneman, D. (2003). Maps of Bounded Rationality: Psychology for Behavioral Economics. *The American Economic Review, 93*(5), 1449–1475. doi:10.1257/000282803322655392

Kahneman, D., & Tversky, A. (1979). Prospect Theory of Decisions under Risk. *Econometrica, 47*, 263–291. doi:10.2307/1914185

Kahneman, D., & Tversky, A. (1992). Advances in. prospect Theory: Cumulative representation of Uncertainty. *Journal of Risk and Uncertainty, 5*.

Kaizoji. T, Bornholdt, S. & Fujiwara.Y. (2002). Dynamics of price and trading volume in a spin model of stock markets with heterogeneous agent. *Physica A*.

Kambayashi, Y., & Takimoto, M. (2005). Higher-Order Mobile Agents for Controlling Intelligent Robots. *International Journal of Intelligent Information Technologies, 1*(2), 28–42.

Kambayashi, Y., Sato, O., Harada, Y., & Takimoto, M. (2009). Design of an Intelligent Cart System for Common Airports. In *Proceedings of 13th International Symposium on Consumer Electronics*. CD-ROM.

Kambayashi, Y., Tsujimura, Y., Yamachi, H., Takimoto, M., & Yamamoto, H. (2009). Design of a Multi-Robot System Using Mobile Agents with Ant Colony Clustering. In *Proceedings of Hawaii International Conference on System Sciences*. IEEE Computer Society. CD-ROM

Kawamura, H., Yamamoto, M., & Ohuchi, A. (2001). (in Japanese). Investigation of Evolutionary Pheromone Communication Based on External Measurement and Emergence of Swarm Intelligence. *Japanese Journal of the Society of Instrument and Control Engineers, 37*(5), 455–464.

Kawamura, H., & Yamamoto, M. Suzuki & Ohuchi, A. (1999). Ants War with Evolutive Pheromone Style Communication. In *Advances in Artificial Life, ECAL'99* (LNAI 1674, pp. 639-643).

Kennedy, J., & Eberhert, R. C. (2001). *Swarm Intelligence*. Morgan Kaufmann.

Kepecs, A., Uchida, N., & Mainen, Z. (2008, September). *How uncertainty boosts learning: Dynamic updating of decision strategies*. Paper presented at Annual Conference on Neuroeconomics, Park City, Utah.

Kimura, H., Yamamura, M., & Kobayashi, S. (1995). Reinforcement Learning by Stochastic Hill Climbing on Discounted Reward. In *Proceedings of the Twelfth International Conference on Machine Learning* (pp. 295-303).

Klucharev, V., Hytonen, K., Rijpkema, M., Smidts, A., & Fernandez, G. (2008, September). *Neural mechanisms of social decisions*. Paper presented at Annual Conference on Neuroeconomics, Park City, Utah.

Konda, V. R., & Tsitsiklis, J. N. (2000). Actor-Critic Algorithms. *Advances in Neural Information Processing Systems, 12*, 1008–1014.

Koritarov, V. S. (2004). *Real-World Market Representation with Agents* (pp. 39–46). IEEE Power and Energy Magazine.

Kovalerchuk, B. (1996). Context spaces as necessary frames for correct approximate reasoning. *International Journal of General Systems, 25*(1), 61–80. doi:10.1080/03081079608945135

Kovalerchuk, B., & Vityaev, E. (2000). *Data mining in finance: advances in relational and hybrid methods*. Kluwer.

Kovalerchuk, B. (1990). Analysis of Gaines' logic of uncertainty, In I.B. Turksen (Ed.), *Proceedings of NAFIPS '90* (Vol. 2, pp. 293-295).

Koza, J. (1992). *Genetic programming*. Cambridge, MA: The MIT Press.

Koza, J. R. (1992). *Genetic Programming: On the Programming of Computers by Means of Natural Selection*. MIT Press.

Krishna, V., & Ramesh, V. C. (1998). Intelligent agents for negotiations in market games. Part I. Model. *IEEE Transactions on Power Systems*, 1103–1108. doi:10.1109/59.709106

Krishna, V., & Ramesh, V. C. (1998a). Intelligent agents for negotiations in market games. Part II. Application. *IEEE Transactions on Power Systems*, 1109–1114. doi:10.1109/59.709107

Kuhlmann, G., & Stone, P. (2003). Progress in learning 3 vs. 2 keepaway. In *Proceedings of the RoboCup-2003 Symposium*.

Kuhnen, C., & Chiao, J. (2008, September). *Genetic determinants of financial risk taking*. Paper presented at Annual Conference on Neuroeconomics, Park City, Utah.

Kwak, K. J., Baryshnikov, Y. M., & Coffman, E. G. (2008). Self-Organizing Sleep-Wake Sensor Systems. In *Proc. the 2nd IEEE International Conference on Self-Adaptive and Self-Organizing Systems* (SASO2008) (pp. 393-402).

Kyle, A. S., & Wang, A. (1997). Speculation Duopoly with Agreement to Disagree: Can Overconfidence Survive the Market Test? *The Journal of Finance, 52*, 2073–2090. doi:10.2307/2329474

Laibson, D. (1997). Golden eggs and hyperbolic discounting. *The Quarterly Journal of Economics, 12*(2), 443–477. doi:10.1162/003355397555253

Lam, K., & Leung, H. (2004). An Adaptive Strategy for Trust/ Honesty Model in Multi-agent Semi- competitive Environments. In *Proceedings of the 16th IEEE International Conference on Tools with Artificial Intelligence*(ICTAI 2004)

Larsen, C. (1999). Cranbrook Institute of Science . *Bulletin, 64*, 1–30.

Lasseter, R., Akhil, A., Marnay, C., Stephens, J., Dagle, J., Guttromson, R., et al. (2002, April). White paper on Integration of consortium Energy Resources. The CERTS MicroGrid Concept. CERTS, CA, Rep.LBNL-50829.

Le Baron, B. (2001). A builder's guide to agent-based financial markets. *Quantitative Finance, 1*(2), 254–261. doi:10.1088/1469-7688/1/2/307

LeBaron, B. (2000). Agent-based Computational Finance: Suggested Readings and Early Research. *Journal of Economics & Control, 24*, 679–702. doi:10.1016/S0165-1889(99)00022-6

LeBaron, B., Arthur, W. B., & Palmer, R. (1999). Time Series Properties of an Artificial Stock Market. *Journal of Economics & Control, 23*, 1487–1516. doi:10.1016/S0165-1889(98)00081-5

Lerner, J., Small, D., & Loewenstein, G. (2004). Heart strings and purse strings: Carry-over effects of emotions on economic transactions. *Psychological Science, 15*, 337–341. doi:10.1111/j.0956-7976.2004.00679.x

Levy, M., Levy, H., & Solomon, S. (2000). *Microscopic Simulation of Financial Markets*. Academic Press.

Levy, M. Levy, H., & Solomon, S. (2000). *Microscopic Simulation of Financial Markets: From Investor Behavior to Market Phenomena*. San Diego: Academic Press.

Lewandowsky, S., Oberauer, K., Yang, L.-X., & Ecker, U. (2009). A working memory test battery for Matlab. under prepartion for being submitted to the *Journal of Behavioral Research Method*.

Lewis, C. (1994).. . *Quaternary Science Reviews, 13*, 891–922. doi:10.1016/0277-3791(94)90008-6

Lewis, C. (2007).. . *Journal of Paleolimnology, 37*, 435–452. doi:10.1007/s10933-006-9049-y

Lichtenstein, S., & Slovic, P. (Eds.). (2006). *The Construction of Preference*. Cambridge, UK: Cambridge University Press. doi:10.1017/CBO9780511618031

Lieberman, M. (2003). Reflective and reflexive judgment processes: A social cognitive neuroscience approach . In Forgas, J., Williams, K., & von Hippel, W. (Eds.), *Social Judgments: Explicit and Implicit Processes* (pp. 44–67). New York, NY: Cambridge University Press.

Liepins, G. E., Hilliard, M. R., Palmer, M., & Rangarajan, G. (1989). Alternatives for Classifier System Credit Assignment. In *Proceedings of the Eleventh International Joint Conference on Artificial Intelligent* (pp. 756-761).

Lin, C. C., & Liu, Y. T. (2008). Genetic Algorithms for Portfolio Selection Problems with Minimum Transaction Lots. *European Journal of Operational Research, 185*(1), 393–404. doi:10.1016/j.ejor.2006.12.024

Lin, C.-H., Chiu, Y.-C., Lin, Y.-K., & Hsieh, J.-C. (2008, September). *Brain maps of Soochow Gambling Task*. Paper presented at Annual Conference on Neuroeconomics, Park City, Utah.

Liu, Y., & Yao, X. (1996). A Population-Based Learning Algorithms Which Learns Both Architectures and Weights of Neural Networks. *Chinese Journal of Advanced Software Research, 3*(1), 54–65.

Lo, A. (2005). Reconciling efficient markets with behavioral finance: The adaptive market hypothesis. *The Journal of Investment Consulting, 7*(2), 21–44.

Loewenstein, G. (1988). Frames of mind in intertemporal choice. *Management Science, 34*, 200–214. doi:10.1287/mnsc.34.2.200

Loewenstein, G. (2005). Hot-cold empathy gaps and medical decision making. *Health Psychology, 24*(4), S49–S56. doi:10.1037/0278-6133.24.4.S49

Loewenstein, G., & Schkade, D. (2003). Wouldn't it be nice?: Predicting future feelings . In Kahneman, D., Diener, E., & Schwartz, N. (Eds.), *Hedonic Psychology: The Foundations of Hedonic Psychology* (pp. 85–105). New York, NY: Russell Sage Foundation.

Loewenstein, G., & O'Donoghue, T. (2005). Animal spirits: Affective and deliberative processes in economic behavior. Working Paper. Carnegie Mellon University, Pittsburgh.

Logenthiran, T., Srinivasan, D., & Wong, D. (2008). Multi-agent coordination for DER in MicroGrid. *IEEE International Conference on Sustainable Energy Technologies* (pp. 77-82).

Louie, K., Grattan, L., & Glimcher, P. (2008). Value-based gain control: Relative reward normalization in parietal cortex. Paper presented at Annual Conference on Neuroeconomics, Park City, Utah.

Lovis, W. (1989). *Michigan Cultural Resource Investigations Series 1*, East Lansing.

Ludwig, A., & Torsten, S. (2001). The impact of stock prices and house prices on consumption in OECD countries. Retrieved December 2008, from http://www.vwl.uni-mannheim.de/brownbag/ludwig.pdf.

Lumer, E. D., & Faieta, B. (1994). Diversity and Adaptation in Populations of Clustering Ants. In *From Animals to Animats 3: Proceedings of the 3rd International Conference on the Simulation of Adaptive Behavior* (pp. 501-508). Cambridge: MIT Press.

Lux, T., & Marchesi, M. (1999). Scaling and criticality in a stochastic multi-agent model of a financial market. *Nature, 397*, 498–500. doi:10.1038/17290

MacLean, P. (1990). *The Triune Brain in Evolution: Role in Paleocerebral Function*. New York, NY: Plenum Press.

Malkiel, B. (1995). Returns from Investing in Equity Mutual Funds 1971 to 1991. *The Journal of Finance, 50*, 549–572. doi:10.2307/2329419

Mamei, M., Roli, A., & Zambonelli, F. (2005). Emergence and Control of Macro-Spatial Structures in Perturbed Cellular Automata, and Implications for Pervasive Computing Systems. *IEEE Trans. Systems, Man, and Cybernetics . Part A: Systems and Humans, 35*(3), 337–348. doi:10.1109/TSMCA.2005.846379

Mamei, M. & Zambonelli, F. (2007). Pervasive pheromone-based interaction with RFID tags. *ACM Transactions on Autonomous and Adaptive Systems (TAAS) archive, 2*(2).

Manson, S. M. (2006). Bounded rationality in agent-based models: experiments with evolutionary programs. *International Journal of Geographical Information Science, 20*(9), 991–1012. doi:10.1080/13658810600830566

Markowitz, H. (1952). Portfolio Selection. *The Journal of Finance, 7*, 77–91. doi:10.2307/2975974

Markowitz, H. (1987). *Mean-Variance Analysis in Portfolio Choice and Capital Market*. New York: Basil Blackwell.

Marr, C., & Hütt, M. T. (2006). Similar Impact of Topological and Dynamic Noise on Complex Patterns. *Physics Letters. [Part A], 349*, 302–305. doi:10.1016/j.physleta.2005.08.096

McCallum, R. A. (1995). Instance-Based Utile Distinctions for Reinforcement Learning with Hidden State. In *Proceedings of the Twelfth International Conference on Machine Learning* (pp. 387-395).

McClure, S., Laibson, D., Loewenstein, G., & Cohen, J. (2004). Separate neural systems value immediate and delayed monetary rewards. *Science, 306*, 503–507. doi:10.1126/science.1100907

Merrick, K., & Maher, M. L. (2007). Motivated Reinforcement Learning for Adaptive Characters in Open-Ended Simulation Games. In *Proceedings of the International Conference on Advanced in Computer Entertainment Technology* (pp. 127-134).

Mitchell, M., Crutchfield, J. P., & Hraber, P. T. (1994). Evolving Cellular Automata to Perform Computations: Mechanisms and Impediments. *Physica D. Nonlinear Phenomena, 75*, 361–391. doi:10.1016/0167-2789(94)90293-3

Miyazaki, K., & Kobayashi, S. (2001). Rationality of Reward Sharing in Multi-agent Reinforcement Learning. *New Generation Computing, 91*, 157–172. doi:10.1007/BF03037252

Miyazaki, K., & Kobayashi, S. (2003). An Extension of Profit Sharing to Partially Observable Markov Decision Processes: Proposition of PS-r* and its Evaluation. [in Japanese]. *Journal of the Japanese Society for Artificial Intelligence, 18*(5), 286–296. doi:10.1527/tjsai.18.286

Miyazaki, K., & Kobayashi, S. (1998). Learning Deterministic Policies in Partially Observable Markov Decision Processes. In *Proceedings of the Fifth International Conference on Intelligent Autonomous System* (pp. 250-257).

Miyazaki, K., & Kobayashi, S. (2000). Reinforcement Learning for Penalty Avoiding Policy Making. In *Proceedings of the 2000 IEEE International Conference on Systems, Man and Cybernetics* (pp. 206-211).

Miyazaki, K., Yamaumra, M., & Kobayashi, S. (1994). On the Rationality of Profit Sharing in Reinforcement Learning. In *Proceedings of the Third International Conference on Fuzzy Logic, Neural Nets and Soft Computing* (pp. 285-288).

Modigliani, F., & Miller, M. H. (1958). The Cost of Capital, Corporation Finance and the Theory of Investment. *The American Economic Review, 48*(3), 261–297.

Mohr, P., Biele, G., & Heekeren, H. (2008, September). *Distinct neural representations of behavioral risk and reward risk.* Paper presented at Annual Conference on Neuroeconomics, Park City, Utah.

Monaghan, G., & Lovis, W. (2005). *Modeling Archaeological Site Burial in Southern Michigan.* East Lansing, MI: Michigan State Univ. Press.

Mondada, F., & Floreano, D. (1995). Evolution of neural control structures: Some experiments on mobile robots. *Robotics and Autonomous Systems, 16*(2-4), 183–195. doi:10.1016/0921-8890(96)81008-6

Money, C. N. N. com (2008). World economy on thin ice - U.N.: The United Nations blames dire situation on the decline of the U.S. housing and financial sectors. Retrieved December 2008, from http://money.cnn.com/2008/05/15/news/ international/global_economy.ap/.

Montero, J., Gomez, D., & Bustine, H. (2007). On the relevance of some families of fuzzy sets. *Fuzzy Sets and Systems, 16*, 2429–2442. doi:10.1016/j.fss.2007.04.021

Moody's. Economy.com (2008). Case-Shiller® Home Price Index forecasts. Moody's Analytics, Inc. Retrieved December 2008, from http://www.economy.com/home/products/case_shiller_indexes.asp.

Moore, T., Rea, D., Mayer, L., Lewis, C., & Dobson, D. (1994).. . *Canadian Journal of Earth Sciences, 31*, 1606–1617. doi:10.1139/e94-142

Murphy, R. R. (2000). *Introduction to AI robotics.* Cambridge: MIT Press.

Nagata, H., Morita, S., Yoshimura, J., Nitta, T., & Tainaka, K. (2008). Perturbation Experiments and Fluctuation Enhancement in Finite Size of Lattice Ecosystems: Uncertainty in Top-Predator Conservation. *Ecological Informatics, 3*(2), 191–201. doi:10.1016/j.ecoinf.2008.01.005

Nagata, T., Takimoto, M., & Kambayashi, Y. (2009). Suppressing the Total Costs of Executing Tasks Using Mobile Agents. In *Proceedings of the 42nd Hawaii International Conference on System Sciences*, IEEE Computer Society. CD-ROM.

Nakamura, M., & Kurumatani, K. (1997). Formation Mechanism of Pheromone Pattern and Control of Foraging Behavior in an Ant Colony Model. In *Proceedings of the Fifth International Workshop on the Synthesis and Simulation of Living Systems* (pp. 67 -74).

National Association of Home Builders, The Housing Policy Department. (2005). The local impact of home building in a typical metropolitan area: Income, jobs, and taxes generated. Retrieved December 2008, from http://www.nahb.org/fileUpload_details.aspx?contentTypeID=3&contentID= 35601& subContentID=28002.

NeuroSolutions™ (2002). The Neural Network Simulation Environment. Version 3, NeuroDimensions, Inc., Gainesville, FL.

Ng, A. Y. Ng & Russell, S. (2000). Algorithms for Inverse Reinforcement Learning. In *Proceedings of 17th International Conference on Machine Learning* (pp. 663-670). Morgan Kaufmann, San Francisco, CA.

Ng, A. Y., Harada, D., & Russell, S. J. (1999). Policy Invariance Under Reward Transformations: Theory and Application to Reward Shaping. In *Proceedings of the Sixteenth International Conference on Machine Learning* (pp. 278-287).

Nielsen, M., & Chuang, I. (2000). *Quantum Computation and Quantum Information*. Cambridge: Cambridge University Press.

Ninagawa, S., Yoneda, M., & Hirose, S. (1997). Cellular Automata in Dissipative Boundary Conditions [in Japanese]. *Transactions of Information Processing Society of Japan*, *38*(4), 927–930.

Ninagawa, S. (2005). Evolving Cellular Automata by 1/f Noise. In *Proc. the 8th European Conference on Artificial Life* (ECAL2005) (pp. 453-460).

Oh, K. J., Kim, T. Y., & Min, S. (2005). Using Genetic Algorithm to Support Portfolio Optimization for Index Fund Management . *Expert Systems with Applications*, *28*, 371–379. doi:10.1016/j.eswa.2004.10.014

Ohkura, K., Yasuda, T., Kawamatsu, Y., Matsumura, Y., & Ueda, K. (2007). MBEANN: Mutation-Based Evolving Artificial Neural Networks. In *Proceedings of the 9th European Conference in Artificial Life* (pp. 936-945).

Orito, Y., Takeda, M., & Yamamoto, H. (2009). Index Fund Optimization Using Genetic Algorithm and Scatter Diagram Based on Coefficients of Determination. *Studies in Computational Intelligence: Intelligent and Evolutionary Systems*, *187*, 1–11.

Orito, Y., & Yamamoto, H. (2007). Index Fund Optimization Using a Genetic Algorithm and a Heuristic Local Search Algorithm on Scatter Diagrams. In *Proceedings of 2007 IEEE Congress on Evolutionary Computation* (pp. 2562-2568).

Palmer, R. G., Arthur, W. B., Holland, J. H., LeBaron, B., & Tayler, P. (1994). Artificial economic life: a simple model of a stock market. *Physica D. Nonlinear Phenomena*, *75*(1-3), 264–274. doi:10.1016/0167-2789(94)90287-9

Palmer, R. G., Arthur, W. B., Holland, J. H., LeBaron, B., & Tayler, P. (1994). Artificial economic life: A simple model of a stock market. *Physica D. Nonlinear Phenomena*, *75*, 264–274. doi:10.1016/0167-2789(94)90287-9

Patel, J. (2007). *A Trust and Reputation Model For Agent-Based Virtual Organizations*. Phd thesis in the faculty of Engineering and Applied Science School of Electronics and Computer Sciences University of South Hampton January 2007.

Paulsen, D., Huettel, S., Platt, M., & Brannon, E. (2008, September). *Heterogeneity in risky decision making in 6-to-7-year-old children*. Paper presented at Annual Conference on Neuroeconomics, Park City, Utah.

Payne, J., Bettman, J., & Johnson, E. (1993). *The Adaptive Decision Maker*. Cambridge University Press.

Pearson, J., Hayden, B., Raghavachari, S., & Platt, M. (2008) *Firing rates of neurons in posterior cingulate cortex predict strategy-switching in a k-armed bandit task*. Paper presented at Annual Conference on Neuroeconomics, Park City, Utah.

Perkins, T. J. (2002). Reinforcement Learning for POMDPs based on Action Values and Stochastic Optimization. In *Proceedings of the Eighteenth National Conference on Artificial Intelligence* (pp. 199-204).

Praça, I., Ramos, C., Vale, Z., & Cordeiro, M. (2003). MASCEM: A Multi agent System That Simulates Competitive Electricity Markets. *IEEE International conference on Intelligent Systems* (pp. 54-60).

Preuschoff, K., Bossaerts, P., & Quartz, S. (2006). Neural Differentiation of Expected Reward and Risk in Human Subcortical Structures. *Neuron, 51*(3), 381–390. doi:10.1016/j.neuron.2006.06.024

Priest, G., & Tanaka, K. Paraconsistent Logic. (2004). Stanford Encyclopedia of Philosophy. http://plato.stanford.edu/entries/logic-paraconsistent.

Principe, J., Euliano, N., & Lefebvre, C. (2000). *Neural and Adaptive Systems: Fundamentals through Simulations*. New York: John Wiley & Sons, Inc.

Pushkarskaya, H., Liu, X., Smithson, M., & Joseph, J. (2008, September). *Neurobiological responses in individuals making choices in uncertain environments*: Ambiguity and conflict. Paper presented at Annual Conference on Neuroeconomics, Park City, Utah.

Quinn, M., & Noble, J. (2001). Modelling Animal Behaviour in Contests: Tactics, Information and Communication. In *Advances in Artificial Life: Sixth European Conference on Artificial Life (ECAL 01)*, (LNAI).

Raberto, M., Cincotti, S., Focardi, M., & Marchesi, M. (2001). Agent-based simulation of a financial market. *Physica A, 299*(1-2), 320–328. doi:10.1016/S0378-4371(01)00312-0

Rahman, S., Pipattanasomporn, M., & Teklu, Y. (2007). Intelligent Distributed Autonomous Power System (IDAPS). *IEEE Power Engineering Society General Meeting*.

Ramchurn, S. D., Huynh, D., & Jennings, N. R. (2004). Trust in multiagent Systems. *The Knowledge Engineering Review, 19*(1), 1–25. doi:10.1017/S0269888904000116

Rangel, A., Camerer, C., & Montague, R. (2008). A framework for studying the neurobiology of value-based decision making. *Nature Reviews. Neuroscience, 9*, 545–556. doi:10.1038/nrn2357

Resconi, G., & Jain, L. (2004). *Intelligent agents*. Springer Verlag.

Resconi, G., Klir, G. J., Harmanec, D., & St. Clair, U. (1996). Interpretation of various uncertainty theories using models of modal logic: a summary. *Fuzzy Sets and Systems, 80*, 7–14. doi:10.1016/0165-0114(95)00262-6

Resconi, G., Klir, G. J., & St. Clair, U. (1992). Hierarchical uncertainty metatheory based upon modal logic. *International Journal of General Systems, 21*, 23–50. doi:10.1080/03081079208945051

Resconi, G., Murai, T., & Shimbo, M. (2000). Field Theory and Modal Logic by Semantic field to make Uncertainty Emerge from Information. *International Journal of General Systems, 29*(5), 737–782. doi:10.1080/03081070008960971

Resconi, G., & Turksen, I. B. (2001). Canonical Forms of Fuzzy Truthoods by Meta-Theory Based Upon Modal Logic. *Information Sciences, 131*, 157–194. doi:10.1016/S0020-0255(00)00095-5

Resconi, G., & Kovalerchuk, B. (2006). The Logic of Uncertainty with Irrational Agents In *Proc. of JCIS-2006 Advances in Intelligent Systems Research, Taiwan*. Atlantis Press

Resconi, G., Klir, G.J., St. Clair, U., & Harmanec, D. (1993). The integration of uncertainty theories. *Intern. J. Uncertainty Fuzziness knowledge-Based Systems, 1*, 1-18.

Reynolds, R. G., & Ali, M. (2008). Computing with the Social Fabric: The Evolution of Social Intelligence within a Cultural Framework. *IEEE Computational Intelligence Magazine, 3*(1), 18–30. doi:10.1109/MCI.2007.913388

Reynolds, R. G., Ali, M., & Jayyousi, T. (2008). Mining the Social Fabric of Archaic Urban Centers with Cultural Algorithms. *IEEE Computer, 41*(1), 64–72.

Rocha, L. M. (2004). Evolving Memory: Logical Tasks for Cellular Automata. In *Proc. the 9th International Conference on the Simulation and Synthesis of Living Systems* (ALIFE9) (pp. 256-261).

Roli, A., & Zambonelli, F. (2002). Emergence of Macro Spatial Structures in Dissipative Cellular Automata. In *Proc. the 5th International Conference on Cellular Automata for Research and Industry* (ACRI2002) (pp. 144-155).

Roth, A. E., & Ockenfels, A. (2002). Last-minute bidding and the rules for ending second-price auction: evidence from Ebay and Amazon auctions on the Internet. *The American Economic Review, 92*, 1093–1103. doi:10.1257/00028280260344632

Ruspini, E. H. (1999). A new approach to clustering. *Information and Control, 15*, 22–32. doi:10.1016/S0019-9958(69)90591-9

Russell, S., & Norvig, P. (1995). *Artificial Intelligence*. Prentice-Hall.

Rust, J., Miller, J., & Palmer, R. (1994). Characterizing effective trading strategies: Insights from a computerized double auction tournament. *Journal of Economic Dynamics & Control, 18*, 61–96. doi:10.1016/0165-1889(94)90069-8

Rust, J., Miller, J., & Palmer, R. (1993). Behavior of trading automata in a computerized double auction market. In Friedman, D., & Rust, J. (Eds.), *Double Auction Markets: Theory, Institutions, and Laboratory Evidence*. Redwood City, CA: Addison Wesley.

Rutledge, R., Dean, M., Caplin, A., & Glimcher, P. (2008, September). *A neural representation of reward prediction error identified using an axiomatic model*. Paper presented at Annual Conference on Neuroeconomics, Park City, Utah.

Sakai, S., Nishinari, K., & Iida, S. (2006). A New Stochastic Cellular Automaton Model on Traffic Flow and Its Jamming Phase Transition. *Journal of Physics. A, Mathematical and General, 39*(50), 15327–15339. doi:10.1088/0305-4470/39/50/002

Samanez Larkin, G., Kuhnen, C., & Knutson, B. (2008). *Financial decision making across the adult life span*. Paper presented at Annual Conference on Neuroeconomics, Park City, Utah.

Sato, O., Ugajin, M., Tsujimura, Y., Yamamoto, H., & Kambayashi, Y. (2007). Analysis of the Behaviors of Multi-Robots that Implement Ant Colony Clustering Using Mobile Agents. In *Proceedings of the Eighth Asia Pacific Industrial Engineering and Management System*. CD-ROM.

Satoh, I. (1999). A Mobile Agent-Based Framework for Active Networks. In *Proceedings of IEEE Systems, Man, and Cybernetics Conference* (pp. 161-168).

Sauter, J., Matthews, R., Parunak, H., & Brueckner, S. (2005). Performance of digital pheromones for swarming vehicle control. In *Proceedings of the fourth international joint conference on Autonomous agents and multiagent systems* (pp. 903-910).

Schlosser, A., Voss, M., & Bruckner, L. (2004). *Comparing and evaluating metrics for reputation systems by simulation*. Paper presented at RAS-2004, A Workshop on Reputation in Agent Societies as part of 2004 IEEE/WIC/ACM International Joint Conference on Intelligent Agent Technology (IAT'04) and Web Intelligence (WI'04), Beijing China, September 2004.

Schwartz, B. (2003). *The Paradox of Choice: Why More Is Less*. New York, NY: Harper Perennial.

Shahidehpour, M., & Alomoush, M. (2001). *Restructured Electrical Power Systems: Operation, Trading, and Volatility*. Marcel Dekker Inc.

Shahidehpour, M., Yamin, H., & LI Z. (2002). *Market Operations in Electric Power Systems: Forecasting, Scheduling, and Risk Management*. Wiley-IEEE Press.

Shannon, C., & Weaver, W. (1964). *The Mathematical Theory of Communication*. The University of Illinois Press.

Sharot, T., De Martino, B., & Dolan, R. (2008, September) *Choice shapes, and reflects, expected hedonic outcome*. Paper presented at Annual Conference on Neuroeconomics, Park City, Utah.

Sharpe, W. F. (1964). Capital Asset Prices: A Theory of Market Equilibrium under condition of Risk. *The Journal of Finance, 19*, 425–442. doi:10.2307/2977928

Sheely, T. (1995). *The Wisdom of the Hive: The Social Physiology of Honey Bee Colonies*. Harvard University Press.

Shiller, R. J. (2000). *Irrational Exuberance*. Princeton University Press.

Shleifer, A. (2000). *Inefficient Markets*. Oxford University Press. doi:10.1093/0198292279.001.0001

Shott, M. (1999). Cranbrook Institute of Science . *Bulletin*, *64*, 71–82.

Shrestha, G. B., Song, K., & Goel, L. K. (2000). An Efficient Power Pool Simulator for the Study of Competitive Power Market. *Power Engineering Society Winter Meeting*.

Simon, H. (1955). A behavioral model of rational choice. *The Quarterly Journal of Economics*, *69*, 99–118. doi:10.2307/1884852

Simon, H. (1956). Rational choice and the structure of the environment. *Psychological Review*, *63*, 129–138. doi:10.1037/h0042769

Simon, H. (1965). The architecture of complexity. *General Systems*, *10*, 63–76.

Simon, H. (1981). Studying human intelligence by creating artificial intelligence. *American Scientist*, *69*, 300–309.

Simon, H. (1996). *The Sciences of the Artificial*. Cambridge, MA: MIT Press.

Simon, H. A. (1997). Behavioral economics and bounded rationality . In Simon, H. A. (Ed.), *Models of Bounded Rationality* (pp. 267–298). MIT Press.

Simon, H. (2005). Darwinism, altruism and economics. In: K. Dopfer (Ed.), *The Evolutionary Foundations of Economics* (89-104), Cambridge, UK: Cambridge University Press.

Singh, S. P., & Sutton, R. S. (1996). Reinforcement Learning with Replacing Eligibility Traces. *Machine Learning*, *22*(1-3), 123–158. doi:10.1007/BF00114726

Singh, S. P., Jaakkola, T., & Jordan, M. I. (1994). Learning Without State-Estimation in Partially Observable Markovian Decision Processes. In *Proceedings of the Eleventh International Conference on Machine Learning* (pp. 284-292).

Slovic, P. (1995). The construction of preference. *The American Psychologist*, *50*, 364–371. doi:10.1037/0003-066X.50.5.364

Smith, V. (1976). Experimental economics: induced value theory. *The American Economic Review*, *66*(2), 274–279.

Sole, R., Bonabeau, E., Delgado, J., Fernandez, P., & Marin, J. (2000). Pattern Formation and Optimization in Army Raids. [The MIT Press.]. *Artificial Life*, *6*(3), 219–226. doi:10.1162/106454600568843

Sornette, D. (2003). *Why stock markets crash*. Princeton University Press.

Standard & Poor's. (2008a). S&P/Case-Shiller® Home Price Indices Methodology. Standard & Poor's. Retrieved December 2008, from http://www2.standardandpoors.com/spf/pdf/index/SP_CS_Home_Price_Indices_Methodology_Web.pdf.

Standard & Poor's. (2008b). S&P/Case-Shiller Home Price Indices. Retrieved December 2008, from http://www2.standardandpoors.com/ portal/site/sp/en/us/page.topic/indices_csmahp/ 2,3,4,0,0,0,0,0,0,0,1,1,0,0,0,0,0.html.

Stanley, K., & Miikkulainen, R. (2002). Evolving neural networks through augmenting topologies . *Evolutionary Computation*, *10*(2), 99–127. doi:10.1162/106365602320169811

Stark, T. (2008). Survey of professional forecasters: May 13, 2008. Federal Reserve Bank of Philadelphia. Retrieved December 2008, from http://www.philadelphiafed.org/files/spf/survq208.html

Stolze, J., & Suter, D. (2004). *Quantum Computing*. Wiley-VCH. doi:10.1002/9783527617760

Stone, P., Sutton, R. S., & Kuhlmann, G. (2005). Reinforcement Learning for RoboCup Soccer Keepaway. *Adaptive Behavior*, 13(3), 165–188. doi:10.1177/105971230501300301

Stone, P., & Sutton, R. S. (2002). Keepaway Soccer: a machine learning testbed . In Birk, A., Coradeschi, S., & Tadokoro, S. (Eds.), *RoboCup-2001: Robot Soccer World Cup V* (pp. 214–223). doi:10.1007/3-540-45603-1_22

Stone, P., Kuhlmann, G., Taylor, M. E., & Liu, Y. (2006). Keepaway Soccer: From Machine Learning Testbed to Benchmark . In Noda, I., Jacoff, A., Bredenfeld, A., & Takahashi, Y. (Eds.), *RoboCup-2005: Robot Soccer World Cup IX*. Berlin: Springer Verlag. doi:10.1007/11780519_9

Streichert, F., & Tanaka-Yamawaki, M. (2006). The Effect of Local Search on the Constrained Portfolio Selection Problem. In *Proceedings of 2006 IEEE Congress on Evolutionary Computation* (pp. 2368-2374).

Sueyoshi, T., & Tadiparthi, G. R. (2007). Agent-based approach to handle business complexity in U.S. wholesale power trading. *IEEE Transactions on Power Systems*, 532–543. doi:10.1109/TPWRS.2007.894856

Sun, R., & Qi, D. (2001). Rationality Assumptions and Optimality of Co-learning, In *Design and Applications of Intelligent Agents* (LNCS 1881, pp. 61-75). Berlin/Heidelberg: Springer.

Sutton, R., & Barto, A. G. (1998). *Reinforcement Learning: An Introduction*. Cambridge, MA: MIT Press.

Sutton, R. S. (1988). Learning to Predict by the Methods of Temporal Differences. *Machine Learning*, 3, 9–44. doi:10.1007/BF00115009

Sutton, R. S., & Barto, A. (1998). *Reinforcement Learning: An Introduction*. Cambridge, MA: MIT Press.

Sutton, R. S., McAllester, D., Singh, S. P., & Mansour, Y. (2000). Policy Gradient Methods for Reinforcement Learning with Function Approximation. *Advances in Neural Information Processing Systems*, 12, 1057–1063.

Suzuki, K., & Ohuchi, A. (1997). Reorganization of Agents with Pheromone Style Communication in Mulltiple Monkey Banana Problem. In . *Proceedings of Intelligent Autonomous Systems*, 5, 615–622.

Takahashi, H., & Terano, T. (2003). Agent-Based Approach to Investors' Behavior and Asset Price Fluctuation in Financial Markets. *Journal of Artificial Societies and Social Simulation*, 6(3).

Takahashi, H., & Terano, T. (2004). Analysis of Micro-Macro Structure of Financial Markets via Agent-Based Model: Risk Management and Dynamics of Asset Pricing. *Electronics and Communications in Japan*, 87(7), 38–48.

Takahashi, H., Takahashi, S., & Terano, T. (2007). Analyzing the Influences of Passive Investment Strategies on Financial Markets via Agent-Based Modeling . In Edmonds, B., Hernandez, C., & Troutzsch, K. G. (Eds.), *Social Simulation- Technologies, Advances, and New Discoveries* (pp. 224–238). Hershey, PA: Information Science Reference.

Takahashi, H. (2010), "An Analysis of the Influence of Fundamental Values' Estimation Accuracy on Financial Markets, " *Journal of Probability and Statistics*, 2010.

Takahashi, H., & Terano, T. (2006a). Emergence of Overconfidence Investor in Financial markets. *5th International Conference on Computational Intelligence in Economics and Finance*.

Takahashi, H., & Terano, T. (2006b). Exploring Risks of Financial Markets through Agent-Based Modeling. In *Proc. SICE/ICASS 2006* (pp. 939-942).

Takimoto, M., Mizuno, M., Kurio, M., & Kambayashi, Y. (2007). Saving Energy Consumption of Multi-Robots Using Higher-Order Mobile Agents. In *Proceedings of the First KES International Symposium on Agent and Multi-Agent Systems: Technologies and Applications* (LNAI 4496, pp. 549-558).

Taniguchi, K., Nakajima, Y., & Hashimoto, F. (2004). A report of U-Mart experiments by human agents . In Shiratori, R., Arai, K., & Kato, F. (Eds.), *Gaming, Simulations, and Society: Research Scope and Perspective* (pp. 49–57). Springer.

Terano, T., Nishida, T., Namatame, A., Tsumoto, S., Ohsawa, Y., & Washio, T. (Eds.). (2001). *New Frontiers in Artificial Intelligence*. Springer Verlag. doi:10.1007/3-540-45548-5

Terano, T. (2007a). Exploring the Vast Parameter Space of Multi-Agent Based Simulation. In L. Antunes & K. Takadama (Eds.), *Proc. MABS 2006* (LNAI 4442, pp. 1-14).

Terano, T. (2007b). KAIZEN for Agent-Based Modeling. In S. Takahashi, D. Sallach, & J. Rouchier (Eds.), *Advancing Social Simulation -The First Congress-* (pp. 1-6). Springer Verlag.

Terano, T., Deguchi, H., & Takadama, K. (Eds.). (2003), *Meeting the Challenge of Social Problems via Agent-Based Simulation: Post Proceedings of The Second International Workshop on Agent-Based Approaches in Economic and Social Complex Systems*. Springer Verlag.

Tesfatsion, L. (2002). Agent-based computational economics: Growing economies from the bottom up. *Artificial Life*, *8*, 55–82. doi:10.1162/106454602753694765

Thomas, R., Kemp, A., & Lewis, C. (1973).. . *Canadian Journal of Earth Sciences*, *10*, 226–271.

Toyoda, Y., & Yano, F. (2004). Optimizing Movement of a Multi-Joint Robot Arm with Existence of Obstacles Using Multi-Purpose Genetic Algorithm. *Industrial Engineering and Management Systems*, *3*(1), 78–84.

Triani, V., et al. (2007). From Solitary to Collective Behaviours: Decision Making and Cooperation, In *Proceedings of the 9th European Conference in Artificial Life* (pp. 575-584).

Tsang, E., Li, J., & Butler, J. (1998). EDDIE beats the bookies. *Int. J. Software. Practice and Experience*, *28*, 1033–1043. doi:10.1002/(SICI)1097-024X(199808)28:10<1033::AID-SPE198>3.0.CO;2-1

U.S. Bureau of Economic Analysis. (2008). Regional economic accounts: State personal income. Retrieved December 2008, from http://www.bea.gov/regional/sqpi/default.cfm?sqtable=SQ1.

U.S. Census Bureau. (2008a). Housing vacancies and home ownership. Retrieved December 2008, from http://www.census.gov/hhes/ www/histt10.html.

U.S. Census Bureau. (2008b). New residential construction. Retrieved December 2008, from http://www.census.gov/const/www/newresconstindex_excel.html.

Ugajin, M., Sato, O., Tsujimura, Y., Yamamoto, H., Takimoto, M., & Kambayashi, Y. (2007). Integrating Ant Colony Clustering Method to Multi-Robots Using Mobile Agents. In *Proceedings of the Eigth Asia Pacific Industrial Engineering and Management System*. CD-ROM.

van Dinther, C. (2007). *Adaptive Bidding in Single-Sided Auctions under Uncertainty: An Agent-based Approach in Market Engineering (Whitestein Series in Software Agent Technologies and Autonomic Computing)*. Basel: Birkhäuser.

Vandersypen, L.M.K., Yannoni, C.S., & Chuang, I.L. (2000). *Liquid state NMR Quantum Computing*.

Vanstone, B., & Finnie, G. (2007). An empirical methodology for developing stockmarket trading systems using artificial neural networks. Retrieved December 2008, from http://epublications.bond.edu.au/cgi/ viewcontent.cgi? article=1022&context=infotech_pubs.

Von-Wun Soo. (2000). Agent Negotiation under Uncertainty and Risk In *Design and Applications of Intelligent Agents* (LNCS 1881, pp. 31-45). Berlin/Heidelberg: Springer.

Wang, T., & Zhang, H. (2004). Collective Sorting with Multi-Robot. In *Proceedings of the First IEEE International Conference on Robotics and Biomimetics* (pp. 716-720).

Warner, G., Hebda, R., & Hahn, B. (1984).. . *Palaeogeography, Palaeoclimatology, Palaeoecology*, *45*, 301–345. doi:10.1016/0031-0182(84)90010-5

Warren, M. (1994). Stock price prediction using genetic programming . In Koza, J. (Ed.), *Genetic Algorithms at Stanford 1994*. Stanford, CA: Stanford Bookstore.

Watkins, C. J. H., & Dayan, P. (1992). Technical note: Q-learning . *Machine Learning, 8*, 55–68. doi:10.1023/A:1022676722315

Weber, E., Johnson, E., Milch, K., Chang, H., Brodscholl, J., & Goldstein, D. (2007). Asymmetric discounting in intertemporal choice: A query-theory account. *Psychological Science, 18*, 516–523. doi:10.1111/j.1467-9280.2007.01932.x

Weber, B., Schupp, J., Reuter, M., Montag, C., Siegel, N., Dohmen, T., et al. (2008). *Combining panel data and genetics: Proof of principle and first results*. Paper presented at Annual Conference on Neuroeconomics, Park City, Utah.

Williams, R. J. (1992). Simple Statistical Gradient Following Algorithms for Connectionist Reinforcement Learning. *Machine Learning, 8*, 229–256. doi:10.1007/BF00992696

Wolfram, S. (2002). *A New Kind of Science*. Wolfram Media Inc.

Wood, W., & Kleb, W. (2002). Extreme Programming in a research environment . In Wells, D., & Williams, L. (Eds.), *XP/Agile Universe 2002* (pp. 89–99). doi:10.1007/3-540-45672-4_9

Wooldridge, M. (2000). *Reasoning about Rational Agents*. Cambridge, MA: The MIT Press.

Wu, W., Ekaette, E., & Far, B. H. (2003). Uncertainty Management Framework for Multi-Agent System, Proceedings of ATS http://www.enel.ucalgary.ca/People/far/pub/papers/2003/ATS2003-06.pdf

Xia, Y., Liu, B., Wang, S., & Lai, K. K. (2000). A Model for Portfolio Selection with Order of Expected Returns. *Computers & Operations Research, 27*, 409–422. doi:10.1016/S0305-0548(99)00059-3

Yao, X. (1999). Evolving artificial networks. *Proceedings of the IEEE, 87*(9), 1423–1447. doi:10.1109/5.784219

Yeh, C.-H., & Chen, S.-H. (2001). Market diversity and market efficiency: The approach based on genetic programming. *Journal of Artificial Simulation of Adaptive Behavior, 1*(1), 147–165.

Zacharia, G., & Maes, P. (2000). Trust management through reputation mechanisms. *Applied Artificial Intelligence Journal, 14*(9), 881–908. doi:10.1080/08839510050144868

Zhan, W., & Friedman, D. (2007). Markups in double auction markets. *Journal of Economic Dynamics & Control, 31*, 2984–3005. doi:10.1016/j.jedc.2006.10.004

About the Contributors

Shu-Heng Chen is a professor in the Department of Economics and Director of Center of International Education and Exchange at the National Chengchi University. He also serves as the Director of the AI-ECON Research Center, National Chengchi University, the editor- in-chief of the Journal of New Mathematics and Natural Computation (World Scientific), the associate editor of the Journal of Economic Behavior and Organization, and the editor of the Journal of Economic Interaction and Coordination. Dr. Chen holds an M.A. degree in mathematics and a Ph. D. in Economics from the University of California at Los Angeles. He has more than 150 publications in international journals, edited volumes and conference proceedings. He has been invited to give keynote speeches and plenary talks on many international conferences. He is also the editor of the volume "Evolutionary Computation in Economics and Finance" (Plysica-Verlag, 2002), "Genetic Algorithms and Genetic Programming in Computational Finance" (Kluwer, 2002), and the co-editor of the Volume I & II of "Computational Intelligence in Economics and Finance" (Springer-Verlag, 2002 & 2007), "Multi-Agent for Mass User Support" (Springer-Verlag, 2004), "Computational Economics: A Perspective from Computational Intelligence" (IGI publisher, 2005), and "Simulated Evolution and Learning," Lecture Notes in Computer Science, (LNCS 4247) (Springer, 2006), as well as the guest editor of Special Issue on Genetic Programming, International Journal on Knowledge Based Intelligent Engineering Systems (2008). His research interests are mainly on the applications of computational intelligence to the agent-based computational economics and finance as well as experimental economics. Details of Shu-Heng Chen can be found at http://www.aiecon.org/ or http://www.aiecon.org/staff/shc/E_Vita.htm.

Yasushi Kambayashi is an associate professor in the Department of Computer and Information Engineering at the Nippon Institute of Technology. He worked at Mitsubishi Research Institute as a staff researcher before joining the Institute. His research interests include theory of computation, theory and practice of programming languages, and political science. He received his PhD in Engineering from the University of Toledo, his MS in Computer Science from the University of Washington, and his BA in Law from Keio University. He is a committee member of IARIA International Multi-Conference on Computing in the Global Information Technology and IARIA International Conference on Advances in P2P Systems, a review committee member of Peer-to-Peer Networking and Applications, and a member of Tau Beta Pi, ACM, IEEE Computer Society, IPSJ, JSSST, IEICE System Society, IADIS, and Japan Flutist Association.

Hiroshi Sato is Assistant Professor of Department of Computer Science at National Defense Academy in Japan. He was previously Research Associate at Department of Mathematics and Information Sciences at Osaka Prefecture University in Japan. He holds the degrees of Physics from Keio University in Japan, and Master and Doctor of Engineering from Tokyo Institute of Technology in Japan. His research interests include agent-based simulation, evolutionary computation, and artificial intelligence. He is a member of Japanese Society for Artificial Intelligence, and Society for Economic Science with Heterogeneous Interacting Agents.

* * *

Akira Namatame is a Professor of Dept. of Computer Science, National Defense Academy of Japan. He holds the degrees of Engineering in Applied Physics from National Defense Academy, Master of Science in Operations Research and Ph.D, in Engineering-Economic System from Stanford University. His research interests include Multi-agents, Game Theory, Evolution and Learning, Complex Networks, Economic Sciences with Interaction Agents and A Science of Collectives.

Boris Kovalerchuk is a Professor of Computer Science at Central Washington University, USA. He received his Ph.D in Applied Mathematics and Computer Science in 1977 from the Russian Academy of Sciences. His research interests are in the agent theory, uncertainty and logic, neural networks, many-valued logic, fuzzy sets, data mining, machine learning, visual and spatial analytics, and applications in a variety of fields. Dr. Kovalerchuk published two books (Springer 2000, 2005) and multiple papers in these fields and chaired two International Computational Intelligence Conferences. He is collaborating with researchers at US National Laboratories, Industry, and Universities in the US, Russia, Italy, and UK.

Chung-Ching Tai received his Ph.D. degree in Economics from National Chengchi University, Taiwan, R.O.C. in 2008. He conducted his post-doctoral studies in AI-ECON Research Center, National Chengchi University, under Dr. Shu-Heng Chen from 2008 to 2009. He is currently an assistant professor in the Department of Economics at Tunghai University, Taiwan, R.O.C.

Dipti Srinivasan obtained her M.Eng. and Ph.D. degrees in Electrical Engineering from the National University of Singapore (NUS) in1991and1994, respectively. She worked at the University of California at Berkeley's Computer Science Division as a post-doctoral researcher from 1994 to1995. In June 1995, she joined the faculty of the Electrical and Computer Engineering department at the National University of Singapore, where she is an associate professor. From 1998 to 1999, she was a visiting faculty in the Department of Electrical and Computer Engineering at the Indian Institute of Science, Bangalore, India. Her main areas of interest are neural networks, evolutionary computation, intelligent multi-agent systems and application of computational intelligence techniques to engineering optimization, planning and control problems in intelligent transportation systems and power systems. Dipti Srinivasan is a senior member of IEEE and a member of IES, Singapore. She has published over 160 technical papers in international refereed journals and conferences. She currently serves as an associate editor of IEEE Transactions on Neural Networks, a social editor of IEEE Transactions Intelligent Transportation Systems, area editor of International Journal of Uncertainty, Fuzziness and Knowledge-based Systems, and as a managing guest editor of Neurocomputing.

Farshad Fotouhi received his Ph.D. in computer science from Michigan State University in 1988. He joined the faculty of Computer Science at Wayne State University in August 1988 where he is currently Professor and Chair of the department. Dr. Fotouhi's major areas of research include xml databases, semantic web, multimedia systems, and biocomputing. He has published over 100 papers in refereed journals and conference proceedings, served as program committee member of various database related conferences. Dr. Fotouhi is on the Editorial Boards of the IEEE Multimedia Magazine and the International Journal on Semantic Web and Information Systems and he serves as a member of the Steering Committee of the IEEE Transactions on Multimedia.

Germano Resconi is a Professor of Artificial Intelligence at the Department of Mathematics and Physics at the Catholic University in Brecia, Italy. He received his Ph.D degree in physics from the University of Milan in 1968. His research interests have led him to participate in a variety of activities, both in the theoretical and applied studies in the agent theory, uncertainty and logic, neural networks, morphic computing, many-valued logic, robotics, fuzzy sets, modal logic, quantum mechanics, quantum computation and tensor calculus. He is collaborating with researchers at the University of Pisa, University of Berkeley, University of Beijing, and other institutions.

Guy Meadows has been a faculty member at the University of Michigan since 1977. His areas of research include; field and analytical studies of marine environmental hydrodynamics with emphasis on mathematical modeling of nearshore waves, currents and shoreline evolution, active microwave remote sensing of ocean dynamics including wave/wave, wave/current, wave/topographic interactions with recent work in signatures of surface ship wakes and naturally occurring ocean surface processes and the development of in situ and remote oceanographic instrumentation and data acquisition systems designed to measure the spatial and temporal structure of coastal boundary layer flows.

Hafiz Farooq Ahmad is an Associate Professor at School of Electrical Engineering and Computer Science (SEECS), NUST Islamabad Pakistan and also has joint appointment as Consultant Engineer in DTS Inc, Tokyo, Japan. He received PhD from Tokyo Institute of Technology in 2002 under the supervision of Prof. Kinji Mori. His main research topics are autonomous decentralized systems, multi-agent systems, autonomous semantic grid and semantic web. He is a member of IEEE.

Hidemi Yamachi is an assistant professor in the Department of Computer and Information Engineering from the Nippon Institute of Technology, Japan. His research interests include optimization methods based on evolutional computation and visualization. He received his PhD from Tokyo Metropolitan University.

Hidenori Kawamura is an associate professor in the graduate school of Information Science and Technology, Hokkaido University, Japan. His research interests include information science, complex systems and multi-agent systems. He recieved his PhD, MS and BA in Information Engineering from Hokkaido University. Contact him at Graduate school of Infoarmation Science and Technology, Hokkaido University, Noth14 West9, Sapporo, Hokkaido, 060-0814, Japan; kawamura@complex.eng.hokudai.ac.jp

Hiroki Suguri is Professor of Information Systems at School of Project Design, Miyagi University, where he teaches systems design, object-oriented modeling, Java programming and information literacy. Professor Suguri received his Ph.D. in software information systems from Iwate Prefectural University in 2004. His research interest includes multi-agent systems, semantic grid/cloud infrastructure, management information systems, and computer-aided education of information literacy.

Hiroshi Takahashi is a associate professor at Graduate School of Business Administration, Keio University. He received his BA in 1994 from the University of Tokyo,Japan, MS and PhD degrees in 2002 and 2004 from the University of Tsukuba,Japan. He worked as a research scientist at Miyanodai technology development center of Fuji Photofilm co.,ltd., and Mitsui Asset Trust and Banking co.,ltd. His research interests include Finance, Financial Engineering, Time Series Analysis, Decision Theory, and Genetic Algorithm-based machine learning.

Hisashi Yamamoto is an associate professor in the Department of System Design at the Tokyo Metropolitan University. He received a BS degree, a MS degree, and a Dr. Eng. in Industrial Engineering from Tokyo Institute of Technology, Japan. His research interests include reliability engineering, operations research, and applied statistics. He received best paper awards from REAJ and IEEE Reliability Society Japan Chapter.

John O'Shea is a Professor of Anthropology and Curator of Great Lakes Archaeology in the Museum of Anthropology. He earned his Ph.D. in Prehistoric Archaeology from Cambridge University in 1978. His research focused on the ways in which the archaeological study of funerary customs could be used to recover information on the social organization of past cultures. O'Shea maintains active research interests in Eastern Europe and North America. His topical interests include: tribal societies, prehistoric ecology and economy, spatial analysis, ethnohistory, Native North America and later European Prehistory. His research in Native North America focuses on the late pre-contact and contact periods in the Upper Great Lakes and the Great Plains. In Europe, his research centers on the eastern Carpathian Basin region of Hungary, Romania and northern Yugoslavia during the later Neolithic through Bronze Age. Most recently, he has begun a program of research focused on the study of Nineteenth Century shipwrecks in the Great Lakes. In addition, O'Shea directs a series of programs in local archaeology, including the Archaeology in an Urban Setting project within the City of Ann Arbor, and the Vanishing Farmlands Survey in Washtenaw County. Within the profession, O'Shea is the editor-in-chief of the Journal of Anthropological Archaeology (Academic Press). He has also been active in the implementation of the Native American Grave Protection and Repatriation Act (NAGPRA) and was appointed in 1998 to a six-year term on the NAGPRA Review Committee by the Secretary of the Interior.

Kazuhiro Ohkura received a PhD degree in engineering from Hokkaido University in 1997. He is currently a professor in the graduate school of Mechanical Systems Engineering at Hiroshima University, Japan, and the leader of Manufacturing Systems Laboratory. His research interests include evolutionary algorithms, reinforcement learning and multiagent systems.

Keiji Suzuki is a professor in the graduate school of Information Science and Technology, Hokkaido University, Japan. His research interests include information science, complex systems and multi-agent systems. He recieved his PhD, MS and BA in Precision Engineering from Hokkaido University. Contact him at Graduate school of Infoarmation Science and Technology, Hokkaido University, Noth14 West9, Sapporo, Hokkaido, 060-0814, Japan; suzuki@complex.eng.hokudai.ac.jp

Mak Kaboudan is full professor of statistics in the School of Business, University of Redlands. Mak has an MS (1978) and a Ph. D. (1980) in Economics from West Virginia University. Before joining Redlands in 2001, he was tenured associate professor with Penn State, Smeal College of Business. Prior to joining Penn State, he worked as a management consultant for five years. His consulting work is mostly in economic and business planning as well as energy and macro-economic modeling. Mak's current research interests are focused on forecasting business, financial, and economic conditions using statistical and artificial intelligence modeling techniques. His work is published in many academic journals such as the Journal of Forecasting, Journal of Real Estate Literature, Journal of Applied Statistics, Computational Economics, Journal of Economic Dynamics and Control, Computers and Operations Research, and Journal of Geographical Systems.

Masanori Goka received a PhD degree in engineering from Kobe University in 2007. He is currently a researcher in the Hyogo Prefectural Institute of Technology, Japan. His research includes multiagent systems, emergence systems and embodied cognition.

Masao Kubo was graduated from precision engineering department, Hokkaido University, in 1991. He received his Ph.D. degree in computer Science from the Hokkaido University in 1996. He had been the research assistant of chaotic engineering Lab, Hokkaido university (1996-1999). He was the lecturer of Robotics lab, dep. of computer science, National Defense Academy, Japan. He was the visiting research fellow of Intelligent Autonomous Lab, university of the west of England (2003-2005). Now, he is the associate professor of Information system lab, dep. of computer science, National Defense Academy, Japan.

Munehiro Takimoto is an assistant professor in the Department of Information Sciences from Tokyo University of Science, Japan. His research interests include design and implementation of programming languages. He received his BS, MS and PhD in Engineering from Keio University.

Azzam ul Asar completed his BSc in Electrical Engineering and MSc in Electrical Power Engineering from NWFP UET Peshawar in 1979 and 1987 respectively. He completed his PhD in Artificial Neural Network from University of Strathclyde, Glasgow, UK in 1994 followed by post doctorate in Intelligent Systems in 2005 from New Jersey Institute of Technology, Newark, USA. He also served as a visiting Professor at New Jersey Institute of Technology, USA from June 2004 to June 2005. Currently he is acting as the chair of IEEE Peshawar Subsection and IEEE Power Engineering joint chapter. He is Dean faculty of Engineering and Technology Peshawar NWFP since November 2008.

R. Suzuki received his Ph.D. degree from Nagoya University in 2003. He is now an associate professor in the graduate school of information science, Nagoya University. His main research fields are artificial life and evolutionary computation. Especially he is investigating how evolutionary processes can be affected by ecological factors such as lifetime learning (phenotypic plasticity), niche construction, and network structures of interactions.

Ren-Jie Zeng is an assistant research fellow at Taiwan Institute of Economic Research starting from 2008 to the present. His current research is the macroeconomic and industrial studies of the Chinese Economy. He holds an M.A. degree in Economics from National Chengchi University in Taiwan.

Robert G. Reynolds received his Ph.D. degree in Computer Science, specializing in Artificial Intelligence, in 1979 from the University of Michigan, Ann Arbor. He is currently a professor of Computer Science and director of the Artificial Intelligence Laboratory at Wayne State University. He is an Adjunct Associate Research Scientist with the Museum of Anthropology at the University of Michigan-Ann Arbor. He is also affiliated with the Complex Systems Group at the University of Michigan-Ann Arbor and is a participant in the UM-WSU IGERT program on Incentive-Based Design. His interests are in the development of computational models of cultural evolution for use in the simulation of complex organizations and in computer gaming applications. Dr. Reynolds produced a framework, Cultural Algorithms, in which to express and computationally test various theories of social evolution using multi-agent simulation models. He has applied these techniques to problems concerning the origins of the state in the Valley of Oaxaca, Mexico, the emergence of prehistoric urban centers, the origins of language and culture, and the disappearance of the Ancient Anazazi in Southwestern Colorado using game programming techniques. He has co-authored three books; *Flocks of the Wamani* (1989, Academic Press), with Joyce Marcus and Kent V. Flannery; *The Acquisition of Software Engineering Knowledge* (2003, Academic Press), with George Cowan; and *Excavations at San Jose Mogote 1: The Household Archaeology* with Kent Flannery and Joyce Marcus (2005, Museum of Anthropology-University of Michigan Press). Dr. Reynolds has received funding from both government and industry to support his work. He has published over 250 papers on the evolution of social intelligence in journals, book chapters, and conference proceedings. He is currently an associate editor for the IEEE Transactions on Computational Intelligence in Games, IEEE Transactions on Evolutionary Computation, International Journal of Swarm Intelligence Research, International Journal of Artificial Intelligence Tools, International Journal of Computational and Mathematical Organization Theory, International Journal of Software Engineering and Knowledge Engineering, and the Journal of Semantic Computing.

Saba Mahmood is a PhD student at School of Electrical Engineering and Computer Science, NUST Islamabad Pakistan. Her MS research area was Reputation Systems for Open Multiagent Systems. Her PhD research is about Formalism of Trust in Dynamic Architectures. She served as the lecturer in the department of computer science at the Institute of Management Sciences Peshawar from 2004-2007. She is an active member of IEEE and remained academic chair of the IEEE Peshawar subsection.

Sachiyo Arai received the B.S degree in electrical engineering, the M.S. degree in control engineering and cognitive science, and Ph.D degree in artificial intelligence, from Tokyo Institute of Technology in 1998. She worked at Sony Corporation for 2 years after receiving the B.S degree. After receiving the Ph.D degree, she spent a year as a research associate in Tokyo Institute of Technology, and worked as a Postdoctoral Fellows at the Robotics Institute in Carnegie Mellon University 1999-2001, a visiting Associate Professor at the department of Social Informatics in Kyoto University 2001-2003. Currently, an Associate Professor of Urban Environment Systems, Faculty of Engineering, Chiba University.

Shu G. Wang is an associate professor in the Department of Economics and also serves as the Associate Director of the AI-ECON Research Center, National Chengchi University. Dr. Wang holds a Ph. D. in Economics from Purdue University. His research interests are mainly on microeconomics, institutional economics, law and economics and recently in agent-based computational economics and experimental economics.

T. Arita received his B.S. and Ph.D. degrees from the University of Tokyo in 1983 and 1988. He is now a professor in the graduate school of information science at Nagoya University. His research interest is in artificial life, in particular in the following areas: evolution of language, evolution of cooperation, interaction between evolution and learning, and swarm intelligence.

T. Logenthiran obtained his B.Eng. degree in the department of Electrical and Electronic Engineering, University of Peradeniya, Sri Lanka. He is currently pursuing Ph.D. degree in the department of Electrical and Computer Engineering, National University of Singapore. His main areas of interest are distributed power system and, application of intelligent multi-agent systems and computational intelligence techniques to power engineering optimization.

Takao Terano is a professor at Department of Computational Intelligence and Systems Science, Interdisciplinary Graduate School of Science and Engineering, Tokyo Institute of Technology. He received BA degree in Mathematical Engineering in 1976 and M. A. degree in Information Engineering in 1978 both from University of Tokyo, and Doctor of Engineering Degree in 1991 from Tokyo Institute of Technology. His research interests include Agent-based Modeling, Knowledge Systems, Evolutionary Computation, and Service Science. He is a member of the editorial board of major Artificial Intelligence- and System science- related academic societies in Japan and a member of IEEE, AAAI, and ACM. He is also the president of PAAA.

Tina Yu received the M.Sc. Degree in Computer Science from Northeastern University, Boston, MA in 1989. Between 1990 and 1995, she was a Member of Technical Staff at Bell-Atlantic (NYNEX) Science and Technology, White Plains, NY. She went to University College London in 1996 and completed her PhD in 1999. Between August 1999 and September 2005, she was with Math Modeling team at Chevron Information Technology Company, San Ramon, CA. She joined the Department of Computer Science, Memorial University of Newfoundland in October of 2005.

Tzai-Der Wang is a researcher who majors in Artificial Life, Evolutionary Computation, Genetic Algorithms, Artificial Immune Systems, and Estimation of Distribution Algorithms. He was the program chair in CIEF2007, a program co-chair in SEAL2006. He worked in AIECON, National Chengchi University, Taiwan ROC, as a post-doctor researcher in 2008 and is an assistant professor in department of industrial engineering and management, Cheng Shiu University, Taiwan ROC currently. He also holds international research relationship with Applied Computational Intelligence Research Unit of University of the West of Scotland, Scotland and works with other researchers together. He was supervised by Professor Colin Fyfe and graduated from University of Paisley in 2002 with the PhD in the evolution of cooperations in artificial communities.

Xiangdong Che received his Bachelosr degree in Electrical Engineering and an M.S. in Computer Science from Zhejiang University, China. He received his PhD in Computer Science in Wayne State University in 2009. He has been working as a computer engineer for 16 years. He is currently working for Wayne State University Computing and Information Technology Division. His research interests primarily focus on Cultural Algorithms, Socially Motivated Learning, Evolutionary Computation, Complex Systems, optimization, intelligent agents, and multi-agent simulation systems.

Y. Iwase received a B.S. degree from Toyama University in 2005 and a M.S. degree from Nagoya University in 2007. Now, he is a Ph.D. student in the Graduate School of Information Science at Nagoya University. His research interests include cellular automata, evolutionary computation and artificial life. Especially, he investigates cellular automata interacting with the external environment.

Yasuhiro Tsujimura is an associate professor in the Department of Computer and Information Engineering from the Nippon Institute of Technology, Japan. His research interests include evolutionary computations and their applications for operations research, reliability engineering and economics. Recently, he is interested in swarm intelligence, artificial life. He received his B.E., M.E. and Dr. Eng. in system safety engineering from Kogakuin University, Japan.

Yousof Gawasmeh is an Ph.D. student in computer science at the Wayne State University. He received his M.S. degree in computer science from New York Institute of Technology. He also holds B.S. degrees from the Yarmouk University. He is currently a Gradute Teacher Assistant in computer science at the Wayne State University. He was working as a lecturer in Phildelphia university-Jordan in Software Engineering Department. Yousof is interested in artificial intelligent systems that have to operate in games domains. Most of his research centers around techniques for learning and planning of teams of agents to act intelligently in their environments. He is concentrating on implementing the Cultural Algorithm in multi-agent syatems games to organise the agents and direct them toward the optimal solution. One of his applications is the landbridge game.

Yukiko Orito is a lecturer in the Graduate School of Social Sciences (Economics) at the Hiroshima University. Her research interests are the analysis of combinatorial optimization problems in financial research. She received Dr. Eng., MS and BA in Production and Information Engineering from the Tokyo Metropolitan Institute of Technology.

Kazuteru Miyazaki is an Associate professor at the Department of Assessment and Research for Degree Awarding, National Institution for Academic Degrees and University Evaluation. His other accomplishments include:1996- Assistant Professor, Tokyo Institute of Technology, 1998- Research Associate, Tokyo Institute of Technology, 1999- Associate Professor, National Institution for Academic Degrees and 2000- Associate Professor, National Institution for Academic Degrees and University Evaluation. Miyazaki's main works include : A Reinforcement Learning System for Penalty Avoiding in Continuous State Spaces, Journal of Advanced Computational Intelligence and Intelligent Informatics, Vol.11, No.6, pp.668-676, 2007 with S. Kobayashi and Development of a reinforcement learning system to play Othello, Artificial Life and Robotics, Vol.7, No.4, pp.177-181, 2004 with S. Tsuboi, and S. Kobayashi. Miyazaki is also a member of: The Japanese Society for Artificial Intelligence (JSAI), The Society of Instrument and Control Engineers (SICE), Information Processing Society of Japan (IPSJ), The Japan Society of Mechanical Engineers (JSME) The Robot Society of Japan (RSJ), and Japanese Association of Higher Education Research.

Index